AMBIENT IONIZATION MASS SPECTROMETRY IN LIFE SCIENCES

AMBIENT IONIZATION MASS SPECTROMETRY IN LIFE SCIENCES

Principles and Applications

Edited by

KEI ZAITSU

In Vivo Real-time Omics Laboratory,
Institute for Advanced Research, Nagoya University,
Nagoya, Japan

ELSEVIER

Elsevier
Radarweg 29, PO Box 211, 1000 AE Amsterdam, Netherlands
The Boulevard, Langford Lane, Kidlington, Oxford OX5 1GB, United Kingdom
50 Hampshire Street, 5th Floor, Cambridge, MA 02139, United States

Library of Congress Cataloging-in-Publication Data
A catalog record for this book is available from the Library of Congress

British Library Cataloguing-in-Publication Data
A catalogue record for this book is available from the British Library

ISBN: 978-0-12-817220-9

For information on all Elsevier publications visit our website at
https://www.elsevier.com/books-and-journals

Publisher: Charlotte Cockle
Acquisition Editor: Kathryn Eryilmaz
Editorial Project Manager: Susan Ikeda
Production Project Manager: James Selvam
Cover Designer: Alan Studholme

Typeset by TNQ Technologies

Contents

Contributors

Emma Bluemke
Techna Institute for the Advancement of Technology for Health, University Health Network, Toronto, ON, Canada; Department of Medical Biophysics, University of Toronto, Toronto, ON, Canada; Present address: Department of Engineering Science, Institute of Biomedical Engineering, University of Oxford, Old Road Campus Research Building, Oxford, United Kingdom

Cheng-Huang Lin
Department of Chemistry, National Taiwan Normal University, Taipei, Taiwan

Yea-Wenn Liou
Department of Chemistry, National Taiwan Normal University, Taipei, Taiwan

Tasuku Murata
MS Business Unit, Life Science Business Department, Analytical & Measuring Instruments Division, Shimadzu Corporation, Nakagyo-ku, Kyoto, Japan

Thanai Paxton
Nihon Waters K.K., Shinagawa-ku, Tokyo, Japan

Kanako Sekimoto
Yokohama City University, Graduate School of Nanobioscience, Yokohama, Japan

Makoto Suematsu
Department of Biochemistry, Keio University School of Medicine, Tokyo, Japan

Yuki Sugiura
Department of Biochemistry, Keio University School of Medicine, Tokyo, Japan

Eiji Sugiyama
Department of Biochemistry, Keio University School of Medicine, Tokyo, Japan

Wenjian Sun
Shimadzu Research Laboratory (Shanghai) Co., Ltd., Pudong New District, Shanghai, China

Alessandra Tata
Techna Institute for the Advancement of Technology for Health, University Health Network, Toronto, ON, Canada

Michael Woolman
Techna Institute for the Advancement of Technology for Health, University Health Network, Toronto, ON, Canada; Department of Medical Biophysics, University of Toronto, Toronto, ON, Canada

Kei Zaitsu
In Vivo Real-time Omics Laboratory, Institute for Advanced Research, Nagoya University, Chikusa-ku, Nagoya, Japan

Arash Zarrine-Afsar
Techna Institute for the Advancement of Technology for Health, University Health Network, Toronto, ON, Canada; Department of Medical Biophysics, University of Toronto, Toronto, ON, Canada; Department of Surgery, University of Toronto, Toronto, ON, Canada; Keenan Research Center for Biomedical Science & the Li Ka Shing Knowledge Institute, St. Michael's Hospital, Toronto, ON, Canada

Preface

Approximately 20 years ago, the two core ionization methods—desorption electrospray ionization (DESI) and direct analysis in real time (DART)—opened the door to a new mass spectrometric field: "ambient ionization." Since then, this field has grown rapidly. To date, numerous ambient ionization techniques have emerged, although this situation appears to be utterly chaotic, at least to me. Almost all analytical chemists and biologists who apply mass spectrometry in their research, including me, know the importance and convenience of ambient ionization techniques. However, most encounter difficulty understanding the techniques A to Z because of the vastness of the field.

In 2016, I received an email from Katy, who is an Elsevier acquisition editor. She provided me the opportunity to edit and write a book on ambient ionization. Approximately 3 years have passed since I met her at ASMS 2016. We have developed this book on ambient ionization mass spectrometry with great assistance from world-renowned experts in ambient ionization techniques. In this book, we have attempted to make it easy for readers to systematically grasp the point of ambient ionization techniques. In particular, it is necessary for us to understand the critical applications of ambient ionization techniques in the life sciences because many such techniques can potentially become essential tools in the life sciences. Fortunately, world-renowned experts in ambient ionization techniques provide the latest information and analytical/technical applications on selected topics, which makes the book highly valuable.

I hope this book will be helpful for readers to not only understand ambient ionization techniques and their applications in the life sciences but also to become deeply interested in ambient ionization techniques.

Kei Zaitsu

Acknowledgments

I would like to dedicate this book to my loving wife, Miwa, and our sons, Yu, Sho, and Kou. Moreover, I would like to express my gratitude to the continuous support from my loving parents, my father Nobuyuki and my mother Yayoi.

I would like to offer my special thanks to my mentor, Dr. Hitoshi Tsuchihashi, and my best comrades, Dr. Y. Hayashi and Mr. T. Murata.

Lastly, I am deeply grateful to all the people who have helped me with the preparation of this book.

CHAPTER 1

Introduction to ambient ionization mass spectrometry

Kei Zaitsu
In Vivo Real-time Omics Laboratory, Institute for Advanced Research, Nagoya University, Chikusa-ku, Nagoya, Japan

Contents

1.1 Definition of ambient ionization and classification

In a broad sense, ambient ionization is understood as *the ionization method under atmospheric pressure, which does not require tedious sample pretreatment,* though the definition of ambient ionization differs slightly among researchers. Indeed, Ifa et al. proposed a working definition of ambient ionization as "*ionization occurs externally to the mass spectrometer and that ions are introduced into the mass spectrometer*" [1]. Here, electrospray ionization (ESI), atmospheric-pressure chemical ionization (APCI), and atmospheric-pressure matrix-assisted laser desorption ionization (AP-MALDI) are excluded by this definition, although the scientific reasoning for excluding these methods is unclear. By contrast, Huang et al. defined ambient ionization mass spectrometry as *"a set of useful techniques for the analysis of samples under open-air conditions, and it allows direct, rapid, real-time, and high-throughput analyses with little or no sample pretreatment"* [2].

Ambient Ionization Mass Spectrometry in Life Sciences
ISBN 978-0-12-817220-9
https://doi.org/10.1016/B978-0-12-817220-9.00001-1

More recently, Javanshad and Venter have updated the definition of ambient ionization from "no sample preparation" to "sample preparation proximal and in real time with the ionization and analysis step"; that is, they have defined ambient ionization as *"a form of ionization where sample preparation takes place in real-time and proximal to the ionization and during the analysis of analytes"* [3].

Among the aforementioned definitions, that by Huang et al. may be more intuitive than the others because some ambient ionization techniques require little sample pretreatment (e.g., solid-phase microextraction (SPME), solvent addition, etc.); thus, in the present chapter, we define the ambient ionization technique as *an ionization technique for the analysis under open-air conditions allowing direct analysis of samples with little or no sample preparation*, consistent with the definition by Huang et al.

To date, more than 90 ambient ionization methods have been reported, although no exact method exists for classifying them because of their variation; thus, different researchers have proposed various categorization schemes [2—4].

On the basis of the principles of ionization, Huang et al. sorted ambient ionization methods into three main groups: (1) direct ionization methods, where analytes are directly ionized in a high electric field; (2) direct desorption/ionization methods, where analytes are desorbed/ionized by charged species such as electrosprayed droplets; and (3) two-step ionization methods, where analytes are desorbed/ablated using techniques such as laser irradiation, followed by ionization via postionization techniques [2].

Venter et al. demonstrated flowergrams [4], which are visually easy for readers to understand, to classify ambient ionization methods on the basis of the ionization mechanism and summary of the major classes of ambient ionization methods, where the methods are categorized on the basis of spray desorption, laser ablation, thermal desorption, liquid microjunction and substrate spray [3].

Harris et al. surveyed previous reports on ambient ionization and then adopted "two-step grouping" to categorize the reports [5]; they first categorized their selected references on the basis of technique principles and then assigned specific applications to each reference. Ambient ionization techniques were consequently grouped into the following categories: (1) spray- and solid-liquid extraction-based techniques that involve ESI or similar methods; (2) direct- and alternating-current plasma-based techniques involving chemical ionization (CI) mechanisms; (3) plasma-based techniques involving chemical sputtering-like desorption steps followed by CI; (4) multimode techniques; (5) laser desorption/ablation methods; (6) acoustic desorption methods; and (7) other techniques that do not fit into one of the aforementioned categories. Following "technique-centric" categorization, each reference is tagged with a specific application name as follows: (1) environmental samples; (2) food flavor and fragrances; (3) forensics; (4) homeland security; (5) molecular imaging; (6) pharmaceuticals; (7) oil, polymers, and additives; or (8) bioanalysis (e.g., clinical, metabolomics, or proteomics).

Van Berkel et al. sorted the vast array of ambient techniques into a few categories based on the approaches for surface sampling and ionization [6]: (1) thermal desorption/ionization; (2) laser desorption (ablation)/ionization; (3) atmospheric-pressure laser desorption (ablation) with secondary ionization (AP-LD/SI); (4) atmospheric-pressure matrix-assisted laser desorption ionization (AP-MALDI); (5) liquid and gas jet desorption/ionization; and (6) liquid extraction surface-sampling probe/ionization.

These reports are useful for understanding and overviewing ambient ionization methods; however, variants of ambient ionization methods such as hybrid-type and/or methods based on new concepts have been increasing. Thus, in this chapter, we more simply categorize the already-reported ambient ionization methods into the following six groups on the basis of their basic principles: (1) spray desorption/ionization-based methods; (2) laser ablation/desorption-based methods; (3) thermal desorption-based methods; (4) plasma-based methods; (5) substrate-based methods; and (6) hybrid/other methods.

Fig. 1.1 shows a conceptual schematic of ambient ionization; analytes are directly desorbed/ionized by a high-speed jet of ESI spray and a

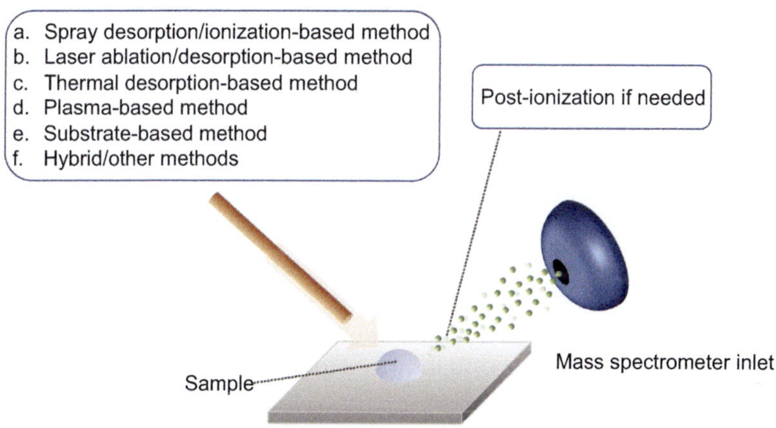

Figure 1.1 Conceptual schematic of ambient ionization techniques.

high-voltage impression via a substrate, whereas analytes are desorbed/ablated by laser irradiation, thermal energy, plasma, and other techniques such as acoustic irradiation, followed by ionization via a postionization method.

On the basis of this categorization concept, we classify the previously reported 93 ambient ionization methods into their respective groups, as shown in Table 1.1, where the classification name, acronym for each ionization method, technical name, and references regarding each ionization method are presented. Following such classification, we will outline the characteristics of each class in the next section.

1.2 Overview of ambient ionization methods

1.2.1 Spray desorption/ionization-based method

We classify desorption electrospray ionization (DESI) [7], DESI-related methods such as EADESI (EA-DESI) [10], LADESI (LA-DESI) [11], and extractive electrospray ionization (EESI) [12], air flow-assisted ionization (AFAI) [15], easy ambient sonic-spray ionization (EASI) [19], desorption electrospray/metastable-induced ionization (DEMI) [18], desorption ionization by charge exchange (DICE) [17], electrostatic spray ionization (ESTASI) [23], and similar methods into the group of spray desorption/ionization-based methods. The classification of ambient ionization methods is listed in Table 1.1.

DESI is one of the best-known ambient ionization techniques, as is direct analysis in real time (DART) [65]. DESI was developed by Cooks and coworkers in 2004 and it uses an electrosprayed jet of charged solvent droplets; analytes are ionized through interactions of the charged droplets with the desorbed analytes, as shown in Fig. 1.2. Notably, nanospray desorption electrospray ionization (nano-DESI) is classified into a different category in this chapter because the sampling and ionization principles of nano-DESI differ from those of DESI, as discussed later [111]. In DESI, ion suppression is reduced because analytes are ionized by such interactions, and the mass spectra obtained from DESI are similar to normal ESI mass spectra, demonstrating that the ionization principle of DESI is based on ESI. DESI has had a high impact not only in mass spectrometry but also in medical/pharmaceutical fields because of its strong applicability to imaging mass spectrometry; thus, applications and modified methods related to DESI have been increasing. In particular, matrix-free imaging mass spectrometry

Table 1.1 Classification of ambient ionization techniques.

Classification	Acronym	Technical name	Reference
(a) Spray desorption/ionization-based methods	DESI	Desorption electrospray ionization	[7]
	TM-DESI	Transmission-mode desorption electrospray ionization	[8]
	L-DESI	Liquid desorption electrospray ionization	[9]
	EADESI (EA-DESI)	Electrode-assisted desorption electrospray ionization	[10]
	LADESI (LA-DESI)	Laser-assisted desorption electrospray ionization	[11]
	EESI	Extractive electrospray ionization	[12]
	ND-EESI	Neutral desorption extractive electrospray ionization	[13,14]
	AFAI	Air-flow-assisted ionization	[15]
	AFA-DESI (AFADESI)	Air-flow-assisted desorption electrospray ionization	[16]
	DICE	Desorption ionization by charge exchange	[17]
	DEMI	Desorption electrospray/metastable-induced ionization	[18]
	EASI (DeSSI)	Easy ambient sonic-spray ionization (Desorption sonic spray ionization)	[19]
	SDC	Spray desorption collection	[20,21]
	DEFFI	Desorption electrospray-flow focusing ionization	[22]
	ESTASI	Electrostatic spray ionization	[23–25]
	RASTIR	Remote analyte sampling transport and ionization relay	[26]
	SESI	Secondary electrospray ionization	[27]
(b) Laser ablation/desorption-based methods			
(b)-1. Laser ablation–based methods	LAESI (LA-ESI)	Laser-ablation electrospray ionization	[28,29]
	PIR-LAESI	Picosecond infrared laser-ablation electrospray ionization	[30–32]
	LAAPCI (LA-APCI)	Laser-ablation atmospheric pressure chemical ionization	[33,34]
	LA-LMJ-SSP	Laser-ablation liquid microjunction surface-sampling probe	[35]
	LA-FAPA	Laser ablation flowing atmospheric-pressure afterglow	[36]

Continued

Table 1.1 Classification of ambient ionization techniques.—cont'd

Classification	Acronym	Technical name	Reference
	LA-ICP	Laser ablation inductively coupled plasma	[37]
	LSI	Laserspray ionization	[38]
	IR-LAMICI	Infrared laser-ablation metastable-induced chemical ionization	[39]
	LEMS	Laser electrospray (mass spectrometry)	[40,41]
(b)-2. Laser desorption–based methods	ELDI	Electrospray-assisted laser desorption/ionization	[42,43]
	MALDESI	Matrix-assisted laser desorption electrospray ionization	[44]
	IR-MALDESI	Infrared matrix-assisted laser desorption electrospray ionization	[45,46]
	LDSPI	Laser desorption spray postionization	[47]
	LD-APCI	Laser desorption atmospheric pressure chemical ionization	[48]
	LDESI (LD-ESI)	Laser desorption electrospray ionization	[49]
	CALDI	Charge-assisted laser desorption/ionization	[50]
	LIAD-ESI	Laser-induced acoustic desorption/electrospray ionization	[51]
	LIAD-APCI	Laser-induced acoustic desorption/atmospheric pressure chemical ionization	[52,53]
	LIAD-APPI	Laser-induced acoustic desorption/atmospheric pressure photoionization	[54]
	HALDI	High-voltage-assisted laser desorption ionization	[55]
	SpiderMass (remote IR-MALDI)	Remote infrared matrix-assisted laser desorption ionization	[56,57]
	AP fs-LDI (fs-LDI)	Atmospheric pressure femtosecond laser desorption ionization	[58]
(c) Thermal desorption–based methods	ASAP	Atmospheric pressure solids analysis probe	[59]
	LDTD-APCI	Laser diode thermal desorption atmospheric pressure chemical ionization	[60]
	LPTD	Leidenfrost phenomenon–assisted thermal desorption	[61]
	DIP-APCI	Direct inlet probe-atmospheric pressure chemical ionization	[62,63]
	TDAMS	Thermal desorption–based ambient mass spectrometry	[64]

(d) Plasma-based methods	DART	Direct analysis in real time	[65]
	LTP	Low-temperature plasma	[66,67]
	DBDI	Dielectric barrier discharge ionization	[68]
	DCBI	Desorption corona beam ionization	[69]
	DAPCI	Desorption atmospheric pressure chemical ionization	[70,71]
	GDI	Glow discharge ionization	[72,73]
	MFGDP	Microfabricated glow discharge plasma (desorption/ionization)	[74]
	APGD	Atmospheric pressure glow discharge desorption	[75]
	LS-APGD	Liquid sampling-atmospheric pressure glow discharge	[76]
	NAIMS	Nanotip ambient ionization mass spectrometry	[77]
	MIPDI	Microwave-induced plasma desorption/ionization	[78,79]
	PASIT	Plasma-based ambient sampling/ionization/transmission	[80]
	SwiFerr	Switched ferroelectric plasma ionizer	[81]
	FAPA	Flowing atmospheric pressure afterglow	[82]
	MPT	Microwave plasma torch	[83]
	MHCD	Microhollow cathode discharge	[84,85]
(e) Substrate-based methods	DEP	Direct electrospray probe (ionization)	[86]
	PESI	Probe electrospray ionization	[87]
	PSI	Paper spray ionization	[88–90]
	SPA-nanoESI	Solid probe-assisted nanoelectrospray ionization	[91]
	SSI	Sponge spray ionization	[92]
	LS	Leaf spray (ionization)	[93]
	WT-ESI	Wooden-tip electrospray ionization	[94–97]
	TS	Touch spray (ionization)	[98]
	BS	Brush-spray (ionization)	[99]
	FS	Fiber-spray (ionization)	[100]
	STS	Swab touch spray (ionization)	[101,102]
	CBS	Coated blade spray (ionization)	[103]

Continued

Table 1.1 Classification of ambient ionization techniques.—cont'd

Classification	Acronym	Technical name	Reference
(f) Hybrid/other methods	REIMS	Rapid evaporative ionization mass spectrometry	[104,105]
	LMJ-SSP	Liquid microjunction-surface sampling probe	[106–108]
	LESA	Liquid extraction surface analysis	[109,110]
	Nano-DESI	Nanospray desorption electrospray ionization	[111,112]
	MAII	Matrix-assisted inlet ionization	[113]
	DAPPI	Desorption atmospheric pressure photoionization	[114]
	FIDI	Field-induced droplet ionization	[115]
	PTC-ESI	Pipette tip column electrospray ionization	[116]
	SAWN	Surface acoustic wave nebulization	[117]
	PAUSI	Paper assisted ultrasonic spray ionization	[118]
	RADIO	Radio frequency acoustic desorption and ionization	[119]
	UASI	Ultrasonication-assisted spray ionization	[120]
	V-EASI	Venturi easy ambient sonic-spray ionization	[121]
	SPAMS	Single-particle aerosol mass spectrometry	[122]
	LA-DART	Laser-ablation direct analysis in real time	[123]
	LEMS	Laser electrospray mass spectrometry (laser vaporization and electrospray ionization)	[40,41]
	PADI	Plasma-assisted desorption ionization	[124]
	PALDI	Plasma-assisted laser desorption ionization	[125]
	PAMLDI	Plasma-assisted multiwavelength laser desorption ionization	[126]
	RoPPI-MS	Robotic plasma probe ionization mass spectrometry	[127]
	RoSA-MS	Robotic surface analysis mass spectrometry	[128]

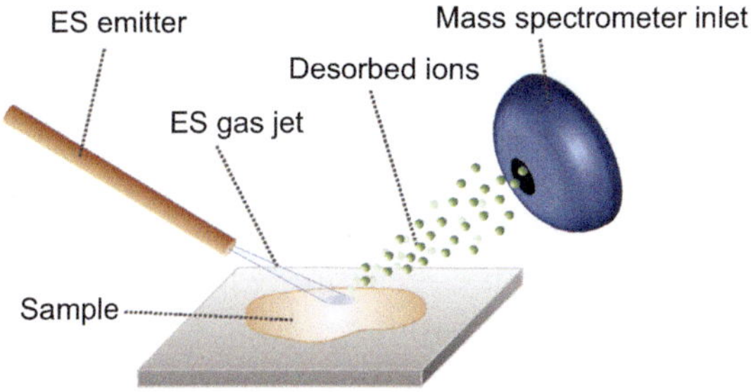

Figure 1.2 Schematic of the ionization mechanisms of DESI.

is achieved with DESI, which is advantageous in comparison with MALDI-based imaging mass spectrometry. More detailed information and applications of DESI are described in Chapter 4.

DESI and other desorption methods such as EASI, DEMI, and DICE are applicable only to solid samples because a liquid layer is necessary to ionize analytes with these methods. Dried spots on paper or other surfaces can also be used with these methods, although analyzing intact liquid samples with these methods is difficult.

By contrast, liquid-DESI and EESI have been improved to accommodate liquid samples [9,12]. EESI uses two separate sprayers, where one nebulizes the sample solution and the other produces an electrosprayed jet, and liquid—liquid interactions between the sample solution and the electrosprayed jet occur, followed by ionization of analytes in the liquid sample in the space in front of a mass spectrometer inlet. Chen in Zenobi group achieved sampling of living objects by combining EESI with a neutral gas desorption method [13,14], where the surface of human skin was directly analyzed in vivo.

1.2.2 Laser ablation/desorption-based methods

We classify electrospray-assisted laser desorption ionization (ELDI) [42,43], laser-ablation electrospray ionization (LAESI) [28,29], picosecond infrared laser-LAESI (PIRL-LAESI) [30—32], matrix-assisted laser desorption electrospray ionization (MALDESI) [44], charge-assisted laser desorption/ionization (CALDI) [50], laser-induced acoustic desorption electrospray ionization (LIAD-ESI) [51], and remote infrared matrix-assisted laser desorption ionization (Remote IR-MALDI, SpiderMass) [56,57] into the group of laser ablation/desorption-based methods. Because we provide more detailed information about laser-based ambient ionization methods in Chapter 5, an overview of the methods is presented in this section.

In general, the ionization process of the laser ablation/desorption-based ambient ionization methods is as follows: analytes are first ablated or desorbed under laser irradiation, followed by being ionized with postionization techniques such as electrospray plume. There are, however, several different technical terms (e.g., ELDI, LAESI, PIRL-LAESI, MALDESI, and IR-MALDESI) are found in the literature, which often confuses readers. Thus, Liu et al. have proposed that these techniques using various types of lasers (ultraviolet (UV) or infrared (IR) wavelength) for ablation/desorption and different postionization techniques (e.g., electrospray, sonic spray, electrosonic spray, desorption electrospray) can be merged under the term "laser desorption spray post-ionization (LDSPI)" [47].

Among the "LDSPI" methods, ELDI and LAESI are well known; a schematic of ELDI and LAESI is shown in Fig. 1.3. The ionization mechanism of ELDI and LAESI, which are two-step ionization techniques, differ from that of DESI. Unlike UV-MALDI and IR-MALDI, ELDI and LAESI do not require an external matrix because electrosprayed charged droplets ionize the ablated analytes, which are generated by laser irradiation.

ELDI was developed by Shiea and coworkers in 2005 [42]. A pulsed nitrogen laser of 337 nm, which is in the UV wavelength range, is used to ablate analytes, and the ablated particulates are ionized using electrospray plume. Shiea et al. succeeded in separating the desorption process from the ionization process.

Nemes and Vertes developed LAESI for in vivo and imaging mass spectrometry in 2007 [28]. This technique is now commercially available. As its name suggests, LAESI combines midinfrared laser ablation with ESI as a novel ionization source under atmospheric pressure. The use of a midinfrared laser enables the water in samples to act as a matrix because the

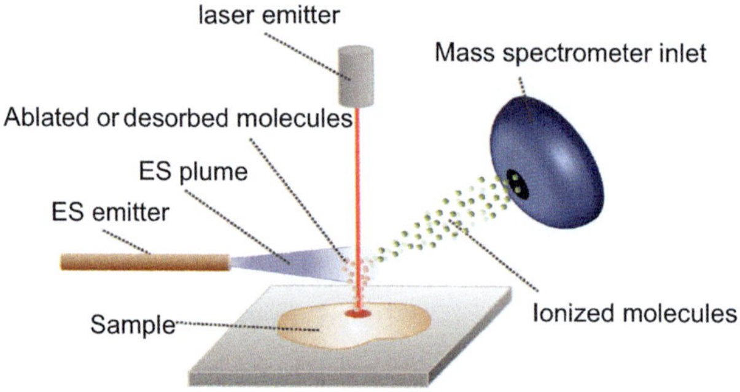

Figure 1.3 Schematic of the ionization mechanisms of ELDI/LAESI.

asymmetric O—H stretching vibration is excited by such a laser, facilitating desorption of analytes in the samples. Postionization using electrospray plume enhances ionization of the ablated molecules, many of which are neutral. Thus, because of its postionization technique, LAESI shows higher sensitivity than AP IR-MALDI.

More recently, Arash and coworkers coupled picosecond infrared laser (PIRL) with ESI to improve LAESI, which led to PIR-LAESI [30—32]. Although the details of PIR-LAESI are described in Chapter 5, PIR-LAESI demonstrated a threefold higher lateral resolution compared with that of conventional LAESI with a nanosecond optical parametric oscillator source at comparable laser fluence.

Sampson in the Muddiman group proposed MALDESI [44], in which a matrix method is required for enhancement of the ionization efficiency, and they examined whether a matrix is required for laser desorption-based ionization. Consequently, they demonstrated that MALDESI is a hybrid mechanism of MALDI, ESI, and ELDI, following which can be still disputable to research. They also demonstrated IR-MALDESI, where a mid-IR laser is used for ablation of analytes [45,46].

Jorabchi et al. used a pulsed corona discharge technique as a postionization technique in conjunction with UV-laser desorption; this technique is termed CALDI [50] because droplet charging is responsible for the generation of ions rather than ion-neutral reactions in the gas phase.

Cheng in Shiea group coupled laser-induced acoustic desorption (LIAD) with electrospray ionization (LIAD-ESI) [51]. In LIAD, a thin metal surface or a thermally insulating film is used and laser ablation of these films generates a large-amplitude acoustic wave (i.e., a shockwave). Thus, Shiea group used metal foils with low thermal conductivity coefficients (e.g., mercury and titanium) to create an acoustic wave; that is, LIAD-ESI is the combinational technique of laser-induced shockwave desorption of analytes with postionization using electrospray plume. The concept of LIAD ambient ionization was subsequently expanded, and the postionization method was modified to APCI or atmospheric-pressure photoionization (APPI), which correspond to LIAD-APCI and LIAD-APPI, respectively [52—54].

Finally, a new interesting instrument for in vivo real-time mass spectrometry using resonant infrared laser ablation, known as SpiderMass [56,57], is briefly introduced; the details of the instrument are also described in Chapter 5. The instrument comprises three parts: (1) a microsampling probe, (2) a transfer line, and (3) a mass spectrometer. The infrared laser ablation can be used for in vivo monitoring of human skin or a plant surface, with minimal damage. These studies demonstrate the strong potential of laser ablation/desorption-based ambient ionization mass spectrometry for medical applications.

1.2.3 Thermal desorption—based methods

In this class of technique, thermal desorption is used to produce analyte plume and the desorbed particulates are postionized with ESI or APCI.

Atmospheric-pressure solids analysis probe (ASAP) was developed by McEwen and coworkers in 2005 [59], where analytes are vaporized by a heated nitrogen gas stream generated from an ESI or APCI probe and then ionized by a corona discharge supplied with the APCI ion source. A schematic of an ASAP is shown in Fig. 1.4, and ASAP instrumentation is commercially available. Krieger et al. modified an ASAP to be applicable to solid samples using a direct inlet probe, resulting in direct inlet probe-APCI (DIP-APCI) [62,63].

Laser diode thermal desorption atmospheric-pressure chemical ionization (LDTD-APCI) [60], which was developed by Phytronix Technologies, Inc., is also a well-known technique with commercially available instrumentation. In LDTD, an infrared laser diode (980 nm) is focused onto the backside of a LazWell plate into which a small volume of sample (1—10 μL) is deposited in advance; analytes in the sample are then thermally desorbed by the infrared laser diode. The LazWell plate is made of polypropylene with proprietary stainless-steel inserts at the well bottom. The analytes are thermally desorbed, and gas-phase molecules are further postionized by corona discharge-induced CI. A schematic of the LDTD-APCI method is shown in Fig. 1.5.

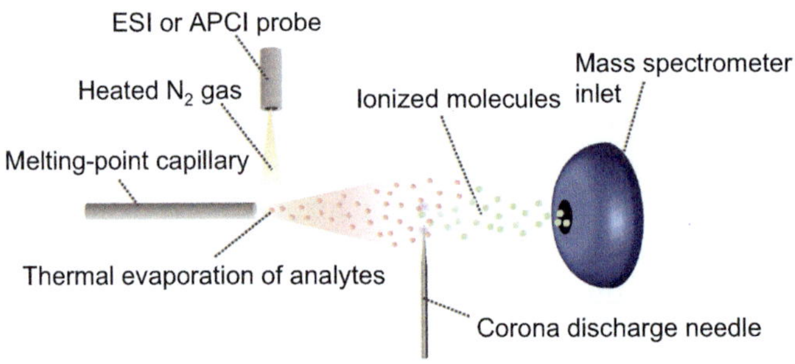

Figure 1.4 Schematic of the ionization mechanisms of ASAP.

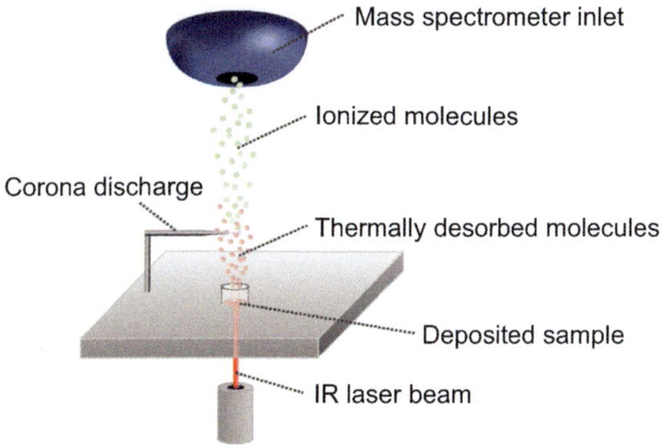

Figure 1.5 Schematic of ionization mechanisms of LDTD-APCI.

Lin et al. reported a new nanoparticle-assisted thermal desorption-based ambient mass spectrometry method, namely TDAMS [64], where a multilayer of a gold nanoparticle-based glass chip is used for the substrate and the analytes are thermally desorbed by a near-infrared laser diode (808 nm). Saha in Hiraoka group also demonstrated a new thermal desorption technique: the Leidenfrost phenomenon-assisted thermal desorption method, termed LPTD [61], where some open ion sources based on the Leidenfrost phenomenon are constructed with different ionization techniques such as low-temperature dielectric barrier discharge ionization, ESI, and APCI. During the analysis, the sample solution is spontaneously concentrated in the ion source, leading to the sensitive detection of analytes.

1.2.4 Plasma-based methods

We classify DART [65], low-temperature plasma (LTP) [66,67], dielectric barrier discharge ionization (DBDI) [68], desorption corona beam ionization (DCBI) [69,129,130], desorption atmospheric-pressure chemical ionization (DAPCI) [71,131], and plasma-assisted laser desorption ionization (PALDI) [125] into the group of plasma-based methods.

Cody and coworkers developed DART in 2003 [65]. Along with DESI, DART is one of the best known and the most influential ambient ionization techniques. The schematic of the ionization mechanism of DART is shown in Fig. 1.6. In DART, the gas (typically helium or nitrogen) is introduced into the DART source and then the gas is exposed to a high electric field with a glow discharge (or a DC point-to-plane glow discharge for the commercial DART source), generating electronic excited gas

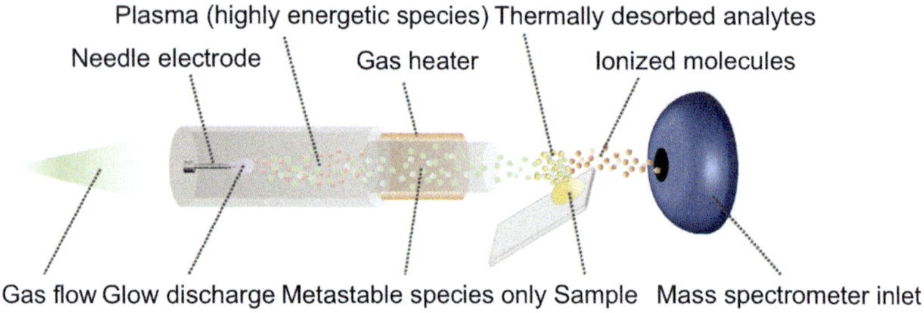

Figure 1.6 Schematic of the ionization mechanisms of DART.

species such as charged ions and metastable neutral species. Among them, some of the excited-state species can exit the DART source; these species induce ionization of analytes in samples. The mechanisms of ionization of DART and its technical/analytical applications are described in detail in Chapter 2.

Plasma generated from a corona discharge is nearly invisible; thus, it is somewhat difficult for users to determine the sampling point. By contrast, DBDI and LTP techniques use dielectric barrier discharge (DBD), which generates visible and low-temperature plasma from ambient air; the plasma generated by DBD secondarily causes desorption/ionization of the analytes. DBDI was developed by Na in Zhang group in 2007 [68]. LTP was developed by Harper et al. in 2008 [66]. Although the geometry of both techniques differs, the temperature of plasma generated by DBD is almost the same as the ambient temperature. A schematic of the LTP setup is shown in Fig. 1.7.

Wang et al. developed a novel new DCBI source in 2010 [69]. Unlike DART, DCBI uses a visible thin corona beam of helium formed with a hollow needle/ring electrode structure, which enables control of the temperature between room temperature and 450°C. A schematic of the

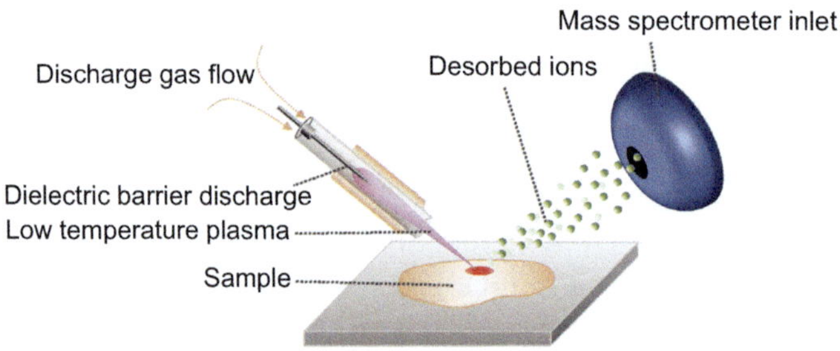

Figure 1.7 Schematic of the ionization mechanisms of LTP.

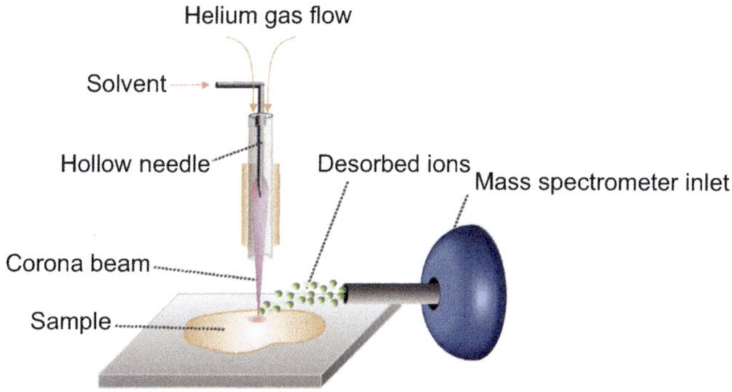

Figure 1.8 Schematic of the ionization mechanisms of DCBI.

DCBI method is shown Fig. 1.8; additional details of DCBI are presented in Chapter 3.

1.2.5 Substrate-based methods

We classify direct electrospray probe ionization (DEP) [86,132,133], probe electrospray ionization (PESI) [87], paper spray ionization (PSI) [88,89], wooden-tip electrospray ionization (WT-ESI) [94–97], and fiber-spray ionization (FS) [100] into the group of substrate-based methods.

The development of substrate-based methods has been recently increasing, and various "substrates" have been demonstrated for ambient ionization tools. As far as we know, Hong in Shiea group first demonstrated substrate-based ESI using a copper wire in 1999 [86], namely, DEP. They tested four different types of copper rings and demonstrated that the substrates with a high-voltage impression generated ESI.

In 2007, Hiraoka and coworkers developed PESI, where they employed an ultrafine needle probe used for acupuncture (needle therapy) as a probe [87]. A schematic of the PESI method is shown in Fig. 1.9. As previously mentioned, PESI uses a thin needle probe as a sampling and ionization unit and the needle moves vertically up and down. At the bottom position, the needle probe functions as a sampling unit without an applied voltage and the probe tip is held inside the sample for scores of milliseconds. The needle probe then rapidly moves to the upper position and the analytes are ionized from the probe tip via a high voltage applied to the needle; that is, PESI is a discontinuous ionization method, although PESI is applicable not only for wet samples but also for soft-solid samples such as tissues because the metal needle can be inserted into such samples. Moreover, PESI ordinarily uses an

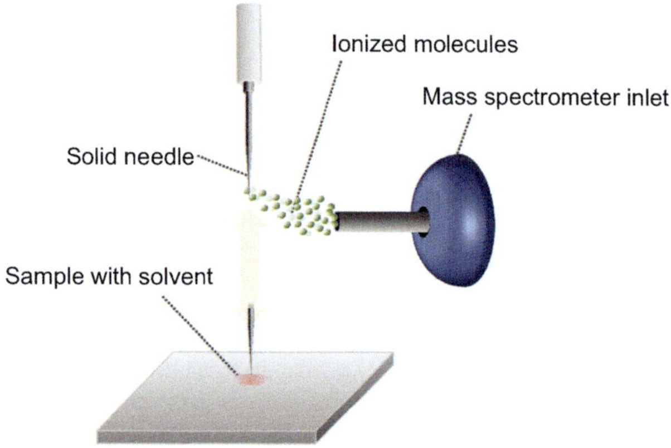

Figure 1.9 Schematic of the ionization mechanisms of PESI.

ultrathin needle (ca. 700 nm tip diameter), which enables low-invasive analyses. Thus, PESI achieves in situ analysis of a living cell or in vivo analysis of a living animal, resulting in various applications in the life sciences. The more detailed information about PESI is described in Chapter 6.

Gómez-Ríos et al. reported coated-blade spray (CBS) ionization, where a stainless-sheet is cut as a "gladius sword" and coated with biocompatible polymer (C_{18}-polyacrylonitrile) [103]. For sample preparation, the coated blade is immersed for SPME of analytes in the sample; the blade is then rinsed with water. The blade is placed on a holder, which is positioned in front of a mass spectrometer, and a high voltage is applied to the blade, thereby ionizing the analytes. Although this method requires a sample preparation step, the study of Gómez-Ríos et al. suggests that the adsorption of analytes onto the blade surface is an important factor for substrate-based ambient ionization methods.

As previously mentioned, metal substrates are used for ionization units in DEP, PESI, and CBS, although other materials are also available for substrate-based ESI sources. Liu et al. developed PSI [88]. As they noted, Nobel Laureate Fenn recognized that paper is a viable ESI source. A schematic of the PSI method is shown in Fig. 1.10. Porous filter paper was cut into a triangular shape, and a copper clip was set to hold the paper in front of the mass spectrometer inlet. A sample solution is pipetted onto the paper, to which a high voltage is then applied. The analytes are directly electrosprayed from the tip of the triangular paper; thus, PSI does not require a sheath gas. Because of its simplicity, numerous applications using PSI have been reported; more detailed information is therefore presented in Chapter 7.

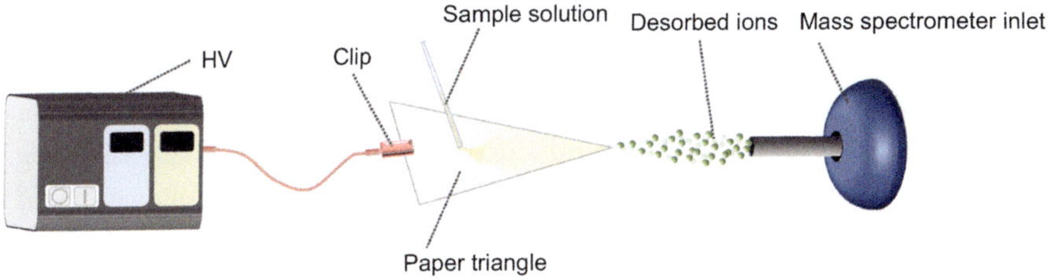

Figure 1.10 Schematic of the ionization mechanisms of PSI.

In 2011, Hu et al. reported an interesting substrate-based method: wooden tip-electrospray ionization (WT-ESI) [94]. As the name of the method suggests, the authors used a disposable wooden tip (wooden toothpick) as an ionization unit as well as the thin metal needle used in PESI. The authors formed a sharp tip-end by cutting the wooden toothpick and loaded a sample solution onto the tip-end. The wooden tip was then mounted onto the capillary holder for analysis. Notably, the wooden tip has a porous structure, which makes the entire wooden tip become conductive. Further details of WT-ESI are presented in Chapters 6 and 7.

In a sense, one of the ultimate substrate-based ambient ionization methods is leaf spray (LS) ionization [93], where a plant leaf itself serves as both the substrate and the sample. In brief, a plant leaf, which is obtained from a living plant in some cases, is cut and a copper clip is used to hold it. The researchers demonstrated that the addition of methanol onto the leaf led to good results; thus, applying methanol to the plant leaf is preferable.

As previously mentioned, substrate-based ambient ionization techniques have grown in diversity because anything being investigated can be a "substrate" for substrate-based ambient ionization. The applications of substrate-based ambient ionization methods in the life sciences are thus expected to expand dramatically.

1.2.6 Hybrid/other methods

We classify rapid evaporative ionization mass spectrometry (REIMS) [104,105], liquid microjunction surface-sampling probe (LMJ-SSP) [106–108], liquid extraction surface analysis (LESA) [109,110], Nano-DESI [111,112], robotic plasma probe ionization (RoPPI) [127], and robotic surface analysis mass spectrometry (RoSA-MS) [128] into the group of hybrid/other methods.

In 2010, Balog et al. developed REIMS, which enables in vivo, in situ mass spectrometric tissue analysis; the system of REIMS is highly ingenious [104]. They use an electrosurgical handpiece employing thermal ablation, where an actual cutting blade is embedded into the handpiece as a sampling unit. PTFE tubing is connected to the handpiece, and the aerosol evaporated from the surgical site is evacuated into the distant mass spectrometer through the PTFE tubing. A schematic of the REIMS method is shown in Fig. 1.11. The more detailed information of REIMS is described in Chapter 8.

Eikel et al. developed automated liquid extraction-based surface analysis platform, namely LESA, in 2010 [110]. The concept of LESA is based on their previous studies on applications of a liquid microjunction surface-sampling probe, termed LMJ-SSP [109]. In the LMJ-SSP, the hollow probe works for making a wall-less liquid microjunction for liquid—liquid extraction from the surface of samples such as tissues, and then aspirates the extracting liquid into the probe. Following sampling from the surface, the aspirated liquid is submitted to an APCI or ESI source via the probe acting as a liquid conduit. However, the formation of the probe-to-surface liquid microjunction plays a critical role in this strategy; thus, a mechanical system to form the liquid microjunction was mandatory for both generalization of the method and achievement of high-throughput analysis. Finally, they developed a fully automated system for LMJ-SSP by adapting the commercially available Advion NanoMate chip-based infusion nanoESI system, resulting in generation of LESA.

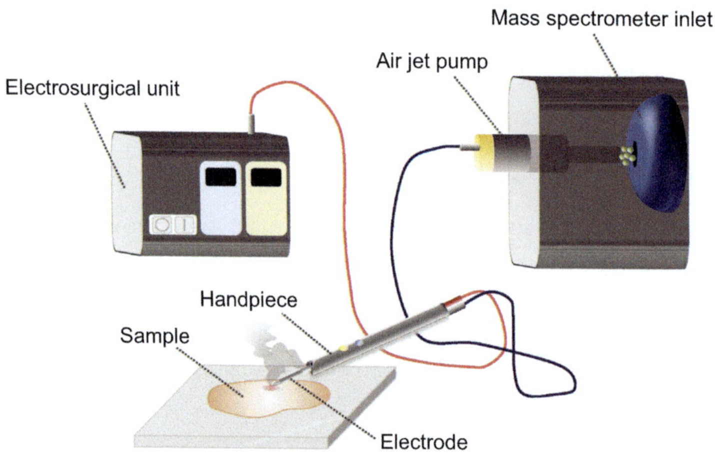

Figure 1.11 Schematic of the ionization mechanisms of REIMS.

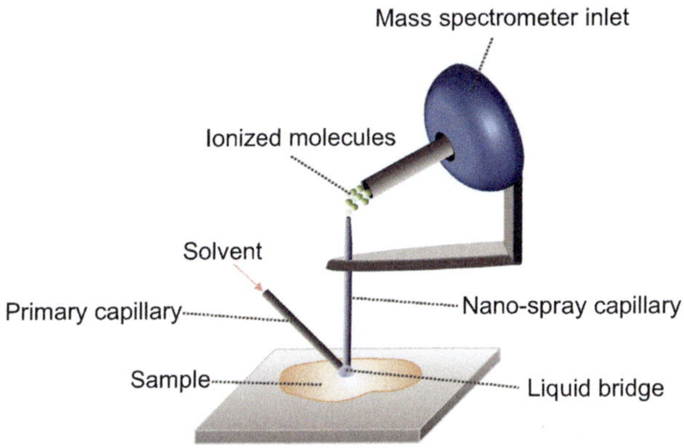

Figure 1.12 Schematic of the ionization mechanisms of Nano-DESI.

In addition to LESA/LMJ-SSP, Roach et al. also used a solvent bridge formed between primary and nanospray capillaries, where the bridge directly contacted the surface of sample. This technique was named nano-DESI [111]. A schematic of the nano-DESI method is shown in Fig. 1.12.

Nano-DESI is based on the ionization mechanisms of DESI. In DESI, the "droplet pick-up mechanism" has been suggested, where a charged droplet derived from an electrosprayed jet indirectly transfers analytes from the sample surface; thus, they developed an alternate approach using the solvent bridge to directly transport the analytes from the surface. This process is why the inventors named the process nanospray-DESI (nano-DESI). Notably, however, the concept of nano-DESI is quite different from that of DESI; thus, nano-DESI is not classified as a spray desorption/ionization-based method in the present chapter. This technique does not require momentum transfer from charged droplets, eliminating the electrosprayed jet and nebulizer gas used in DESI; the sampling area is easily controlled by changing the solvent flow or the capillary diameter.

In addition, extensive hybrid ambient ionization methods have also been reported, where the main ionization methods such as DESI and DART are combined with some assistive techniques, especially plasma-assisted- or laser ablation/desorption-assisted methods.

Here, PALDI [125] and plasma-assisted multiwavelength laser desorption ionization (PAMLDI) [126] are briefly introduced. These techniques use heated metastable plasma produced by DART, and laser desorption using one wavelength of 532 nm for PALDI or a Nd:YAG laser with a three-wavelength laser (1064, 532, and 355 nm) for PAMLDI. Analytes are desorbed by the pulse laser, followed by ionization of the heated metastable plasma, which induces a series of reactions such as Penning ionization or charge exchange in addition to LTP ionizations. Feng et al. reported imaging mass spectrometry using PALDI with automatic 3D-moving stage [125], achieving enhancement of spatial resolution.

As another direction of ambient ionization methods, we introduce the combinational use of an ambient ionization source and a robotic system. Fernández and coworkers demonstrated an interesting robotic sample introduction/ionization system for mass spectrometry and robotic surface analysis mass spectrometry, namely RoPPI-MS and RoSA-MS, respectively [127,128]. They coupled a robotic arm with the sampling probe attached, with DART and insisted that the concept of RoPPI-MS is based on the previous works on ambient ionization (e.g., DESI, LMJ-SSP, LAESI, IR-LAMICI, and PESI). Their work is a new direction of ambient ionization techniques to expand their applications more than ever because the combinational use of ambient ionization techniques with such a robotic system has strong potential to create a new analytical methodology. Indeed, RoPPI-MS uses the probe, such as PESI for the sampling unit, though solvent-free ionization is achieved because of plasma-based ionization by DART, which creates a new-generation 3D imaging mass spectrometry. More detailed information of RoPPI-MS is presented in Chapter 2. Fernández et al. have been improving the system and successfully developed a new system, namely RoSA-MS, in 2018. This system included a custom-built laser-scanner mounted for digitizing the surface topography of a 3D-sample. The number of new-generation mass spectrometric 3D imaging systems is expected to increase. The more detailed information of RoSA-MS is described in Chapter 6.

1.3 Objectives of this book and brief explanation of each chapter

The first objective of this book is to overview the previously reported ambient ionization techniques and to support readers in systematically understanding the techniques by classifying the methods into groups. In this chapter, 93 ambient ionization techniques are classified into one of six groups on the basis of their principles; the feature of each group were overviewed, which is helpful for readers to grasp the gist of ambient ionization.

The second objective is to provide readers with more detailed information about the selected ambient ionization techniques, especially their applications in the *life sciences*, in addition to the principle and/or ionization mechanisms. We use this approach because most of the latest ambient ionization techniques are applicable in the life sciences; examples include whole-body molecular imaging of a mouse by air flow-assisted desorption electrospray ionization (AFA-DESI) [16], SpiderMass [56], in vivo real-time monitoring of metabolites in a living mouse tissue by probe electrospray ionization/tandem mass spectrometry (PESI/MS/MS) [134,135], and intraoperative analysis by evaporative ionization mass spectrometry (REIMS) [105]. In particular, we selected the following seven interesting techniques: (1) DART (Chapter 2); (2) DCBI (Chapter 3); (3) DESI (Chapter 4); (4) laser-based ambient ionization (Chapter 5); (5) PESI (Chapter 6); (6) paper spray and fiber-spray ionization (Chapter 7); and (7) REIMS (Chapter 8).

These techniques were selected on the basis of both importance at present and their potential to be expanded in the future. In addition, these techniques have been receiving increasing attention in the life sciences. DART and DESI are undoubtedly some of the most important techniques in this field, and there are plentiful research results regarding DART and DESI.

Numerous studies on laser ablation/desorption-based techniques, which have become a substantial field in ambient ionization, have been reported. Thus, the laser-based methods are assigned to a chapter, where some laser-assisted hybrid methods are also included. Although plasma-based methods are important in the field, the most important plasma-based method, i.e., DART, is discussed in a separate chapter, and DCBI is selected as a representative plasma-based method, which is based on a corona discharge as well as LTP ambient ionization.

As previously mentioned, substrate-based methods have recently became diverse because they are well suited to analyses in the life sciences; for instance, non- or low-invasive analysis for both human and animals have been achieved via some substrate-based analyses. Here we present PESI and paper spray/fiber-spray ionization in separate chapters. PESI is well known to exhibit high applicability for any form of samples from liquids to solids such as tissues. In addition, applications in which paper spray/fiber-spray ionization techniques are used have been increasing in forensic or plant metabolome analysis.

Finally, we assign REIMS, which is directly connected to medical applications, to an independent chapter. As far as we know, this work is the first description of the details and applications of REIMS in book format.

Almost all of the ambient ionization techniques involve *no chromatographic processes* and *no or minimal sample preparation*, which would be generally disadvantageous in the analytical strategies for biological samples because the analytes of both endogenous and exogenous compounds coexist in complicated matrices [136,137]. Sample preparation and chromatographic separation prior to instrumental analysis are thus generally required for biological analyses. Beyond expectation, however, applications that involve ambient ionization techniques for biological analysis have been increasing; that is, numerous strong advantages are accessible only through ambient ionization techniques. For instance, metabolome analyses based on ambient ionization techniques can not only create a new high-throughput analytical platform but also enable direct analysis of intact analytes in biological samples [138]. Furthermore, in vivo real-time monitoring of biogenic compounds is one of the most difficult but exciting challenges, not only in analytical chemistry but also in biochemistry [139–141]; this "dream" is being realized through ambient ionization techniques.

In each chapter, the world-known experts for ambient ionization techniques provide the principle of the technique and introduce the latest applications in the life sciences on the basis of the professional view of each author.

Lastly, many beneficial review articles have been published; thus, we introduce some of them in brief.

1. Cooks et al. (2006) reported a technical review based on DESI techniques, where future possible uses of DESI for in vivo analysis or its adaptation to a portable mass spectrometer were discussed [142]. As mentioned in the first section, Ifa et al. (2010) also explained a detailed mechanism of DESI and related techniques [1].
2. Chen et al. (2009) reviewed application studies reported from 2005 to 2009, where they explained two case studies of using ambient ionization and proposed some topics on (1) estimation of matrix effects, (2) the possibility of in vivo analysis, and (3) the introduction of mechanistic studies of each ionization, all of which were topics to be further investigated at that time [143].
3. Weston (2010) reviewed key techniques developed between 2004 and 2010 and classified the ambient ionization methods to three groups as

follows: (1) ESI or spray-related techniques; (2) spray-based photon/energy techniques; or (3) APCI-related techniques [144].

4. Wu et al. (2013) provided detailed information about mass spectrometry imaging using DESI, LAESI, fs-LDI and LEMS, IR-LAMICI, LTP and LA-FAPA, DAPPI, PESI, LMJ-SSP, and LESA [145].
5. Smoluch et al. (2014) reported a review article on plasma-based ambient ionization in the bioanalysis sciences, where they provide a detailed mechanism for DART, FAPA, DBDI, LTP, ASAP, DCBI, PADI, MHCD, and MIPDI [146].
6. Kauppila et al. (2017) reviewed ambient ionization, which can be utilized for low-polarity compounds; their experiments provided detailed information about photoionization- or plasma-based ambient ionization techniques [147].
7. Cheng et al. (2017) reviewed laser-based ambient ionization techniques and classified the techniques to seven categories as follows: (1) laser desorption/ablation followed by ESI postionization; (2) laser desorption/ablation followed by APCI postionization; (3) laser desorption/ablation followed by APPI postionization; (4) laser desorption/ablation followed by ESI/APCI postionization; (5) LIAD followed by ESI postionization; (6) LIAD followed by APCI postionization; (7) LIAD followed by APPI postionization. Because this review provides details of the laser-based ambient ionizations and their classification, it is quite useful to outline these techniques for readers [148].
8. Klampfl (2018) reviewed the applications of ambient ionization using EASI, DART, LTP, DESI, LESA, PS, GDI, and DBDI in foodomics studies. Representative applications described in this review are as follows: metabolomics and lipidomics of food (e.g., wine, chicken meat, and fish), forensic discrimination of beers, and characterization of olive oil [149].

References

[1] D.R. Ifa, C. Wu, Z. Ouyang, R.G. Cooks, Desorption electrospray ionization and other ambient ionization methods: current progress and preview, Analyst 135 (2010) 669−681.

[2] M.-Z. Huang, S.-C. Cheng, Y.-T. Cho, J. Shiea, Ambient ionization mass spectrometry: a tutorial, Anal. Chim. Acta 702 (2011) 1−15.

[3] R. Javanshad, A.R. Venter, Ambient ionization mass spectrometry: real-time, proximal sample processing and ionization, Anal. Methods 9 (2017) 4896−4907.

[4] A. Venter, M. Nefliu, R. Graham Cooks, Ambient desorption ionization mass spectrometry, Trac. Trends Anal. Chem. 27 (2008) 284−290.

[5] G.A. Harris, A.S. Galhena, F.M. Fernández, Ambient sampling/ionization mass spectrometry: applications and current trends, Anal. Chem. 83 (2011) 4508–4538.
[6] G.J. Van Berkel, S.P. Pasilis, O. Ovchinnikova, Established and emerging atmospheric pressure surface sampling/ionization techniques for mass spectrometry, J. Mass. Spectrom 43 (2008) 1161–1180.
[7] Z. Takáts, J.M. Wiseman, B. Gologan, R.G. Cooks, Mass spectrometry sampling under ambient conditions with desorption electrospray ionization, Science 306 (2004) 471–473.
[8] J.E. Chipuk, J.S. Brodbelt, Transmission mode desorption electrospray ionization, J. Am. Soc. Mass. Spectrom 19 (2008) 1612–1620.
[9] Z. Miao, H. Chen, Direct analysis of liquid samples by desorption electrospray ionization-mass spectrometry (DESI-MS), J. Am. Soc. Mass. Spectrom 20 (2009) 10–19.
[10] A. Özdemir, C.-H. Chen, Electrode-assisted desorption electrospray ionization mass spectrometry, J. Mass. Spectrom 45 (2010) 1203–1211.
[11] Y.H. Rezenom, J. Dong, K.K. Murray, Infrared laser-assisted desorption electrospray ionization mass spectrometry, Analyst 133 (2008) 226–232.
[12] H. Chen, A. Venter, R.G. Cooks, Extractive electrospray ionization for direct analysis of undiluted urine, milk and other complex mixtures without sample preparation, Chem. Commun. (2006) 2042–2044.
[13] H. Chen, A. Wortmann, R. Zenobi, Neutral desorption sampling coupled to extractive electrospray ionization mass spectrometry for rapid differentiation of biosamples by metabolomic fingerprinting, J. Mass. Spectrom 42 (2007) 1123–1135.
[14] H. Chen, R. Zenobi, Neutral desorption sampling of biological surfaces for rapid chemical characterization by extractive electrospray ionization mass spectrometry, Nat. Protoc. 3 (2008) 1467.
[15] J. He, F. Tang, Z. Luo, Y. Chen, J. Xu, R. Zhang, X. Wang, Z. Abliz, Air flow assisted ionization for remote sampling of ambient mass spectrometry and its application, Rapid. Commun. Mass. Spectrum 25 (2011) 843–850.
[16] Z. Luo, J. He, Y. Chen, J. He, T. Gong, F. Tang, X. Wang, R. Zhang, L. Huang, L. Zhang, H. Lv, S. Ma, Z. Fu, X. Chen, S. Yu, Z. Abliz, Air flow-assisted ionization imaging mass spectrometry method for easy whole-body molecular imaging under ambient conditions, Anal. Chem. 85 (2013) 2977–2982.
[17] C.-C. Chan, M.S. Bolgar, S.A. Miller, A.B. Attygalle, Desorption ionization by charge exchange (DICE) for sample analysis under ambient conditions by mass spectrometry, J. Am. Soc. Mass. Spectrom 21 (2010) 1554–1560.
[18] L. Nyadong, A.S. Galhena, F.M. Fernández, Desorption electrospray/metastable-induced ionization: a flexible multimode ambient ion generation technique, Anal. Chem. 81 (2009) 7788–7794.
[19] R. Haddad, H.M.S. Milagre, R.R. Catharino, M.N. Eberlin, Easy ambient sonic-spray ionization mass spectrometry combined with thin-layer chromatography, Anal. Chem. 80 (2008) 2744–2750.
[20] S. Jain, A. Heiser, A.R. Venter, Spray desorption collection: an alternative to swabbing for pharmaceutical cleaning validation, Analyst 136 (2011) 1298–1301.
[21] K.A. Douglass, S. Jain, W.R. Brandt, A.R. Venter, Deconstructing desorption electrospray ionization: independent optimization of desorption and ionization by spray desorption collection, J. Am. Soc. Mass. Spectrom 23 (2012) 1896–1902.
[22] T.P. Forbes, T.M. Brewer, G. Gillen, Desorption electro-flow focusing ionization of explosives and narcotics for ambient pressure mass spectrometry, Analyst 138 (2013) 5665–5673.
[23] L. Qiao, R. Sartor, N. Gasilova, Y. Lu, E. Tobolkina, B. Liu, H.H. Girault, Electrostatic-spray ionization mass spectrometry, Anal. Chem. 84 (2012) 7422–7430.

[24] L. Qiao, E. Tobolkina, A. Lesch, A. Bondarenko, X. Zhong, B. Liu, H. Pick, H. Vogel, H.H. Girault, Electrostatic spray ionization mass spectrometry imaging, Anal. Chem. 86 (2014) 2033–2041.

[25] L. Qiao, X. Zhong, E. Belghith, Y. Deng, T.-E. Lin, E. Tobolkina, B. Liu, H.H. Girault, Electrostatic spray ionization from 384-well microtiter plates for mass spectrometry analysis-based enzyme assay and drug metabolism screening, Anal. Chem. 89 (2017) 5983–5990.

[26] R.B. Dixon, J.S. Sampson, A.M. Hawkridge, D.C. Muddiman, Ambient aerodynamic ionization source for remote analyte sampling and mass spectrometric analysis, Anal. Chem. 80 (2008) 5266–5271.

[27] C. Wu, W.F. Siems, H.H. Hill, Secondary electrospray ionization ion mobility spectrometry/mass spectrometry of illicit drugs, Anal. Chem. 72 (2000) 396–403.

[28] P. Nemes, A. Vertes, Laser ablation electrospray ionization for atmospheric pressure, in vivo, and imaging mass spectrometry, Anal. Chem. 79 (2007) 8098–8106.

[29] B. Shrestha, A. Vertes, In situ metabolic profiling of single cells by laser ablation electrospray ionization mass spectrometry, Anal. Chem. 81 (2009) 8265–8271.

[30] J. Zou, F. Talbot, A. Tata, L. Ermini, K. Franjic, M. Ventura, J. Zheng, H. Ginsberg, M. Post, D.R. Ifa, D. Jaffray, R.J.D. Miller, A. Zarrine-Afsar, Ambient mass spectrometry imaging with picosecond infrared laser ablation electrospray ionization (PIR-LAESI), Anal. Chem. 87 (2015) 12071–12079.

[31] M. Woolman, A. Gribble, E. Bluemke, J. Zou, M. Ventura, N. Bernards, M. Wu, H.J. Ginsberg, S. Das, A. Vitkin, A. Zarrine-Afsar, Optimized mass spectrometry analysis workflow with polarimetric guidance for ex vivo and in situ sampling of biological tissues, Sci. Rep. 7 (2017) 468.

[32] M. Woolman, I. Ferry, C.M. Kuzan-Fischer, M. Wu, J. Zou, T. Kiyota, S. Isik, D. Dara, A. Aman, S. Das, M.D. Taylor, J.T. Rutka, H.J. Ginsberg, A. Zarrine-Afsar, Rapid determination of medulloblastoma subgroup affiliation with mass spectrometry using a handheld picosecond infrared laser desorption probe, Chem. Sci. 8 (2017) 6508–6519.

[33] C. Herdering, O. Reifschneider, C.A. Wehe, M. Sperling, U. Karst, Ambient molecular imaging by laser ablation atmospheric pressure chemical ionization mass spectrometry, Rapid. Commun. Mass. Spectrom 27 (2013) 2595–2600.

[34] C. Herdering, C.A. Wehe, O. Reifschneider, I. Raj, G. Ciarimboli, K. Diebold, C. Becker, M. Sperling, U. Karst, Laser ablation based bioimaging with simultaneous elemental and molecular mass spectrometry: towards spatially resolved speciation analysis, Rapid. Commun. Mass. Spectrom 27 (2013) 2588–2594.

[35] O.S. Ovchinnikova, V. Kertesz, G.J. Van Berkel, Combining laser ablation/liquid phase collection surface sampling and high-performance liquid Chromatography–Electrospray ionization-mass spectrometry, Anal. Chem. 83 (2011) 1874–1878.

[36] J.T. Shelley, S.J. Ray, G.M. Hieftje, Laser ablation coupled to a flowing atmospheric pressure afterglow for ambient mass spectral imaging, Anal. Chem. 80 (2008) 8308–8313.

[37] S.E. Jackson, N.J. Pearson, W.L. Griffin, E.A. Belousova, The application of laser ablation-inductively coupled plasma-mass spectrometry to in situ U–Pb zircon geochronology, Chem. Geol. 211 (2004) 47–69.

[38] E.D. Inutan, S. Trimpin, Laserspray ionization-ion mobility Spectrometry–Mass spectrometry: baseline separation of isomeric amyloids without the use of solvents desorbed and ionized directly from a surface, J. Proteome Res. 9 (2010) 6077–6081.

[39] A.S. Galhena, G.A. Harris, L. Nyadong, K.K. Murray, F.M. Fernández, Small molecule ambient mass spectrometry imaging by infrared laser ablation metastable-induced chemical ionization, Anal. Chem. 82 (2010) 2178–2181.

[40] E.J. Judge, J.J. Brady, R.J. Levis, Mass analysis of biological macromolecules at atmospheric pressure using nonresonant femtosecond laser vaporization and electrospray ionization, Anal. Chem. 82 (2010) 10203–10207.
[41] E.J. Judge, J.J. Brady, D. Dalton, R.J. Levis, Analysis of pharmaceutical compounds from glass, fabric, steel, and wood surfaces at atmospheric pressure using spatially resolved, nonresonant femtosecond laser vaporization electrospray mass spectrometry, Anal. Chem. 82 (2010) 3231–3238.
[42] J. Shiea, M.-Z. Huang, H.-J. HSu, C.-Y. Lee, C.-H. Yuan, I. Beech, J. Sunner, Electrospray-assisted laser desorption/ionization mass spectrometry for direct ambient analysis of solids, Rapid. Commun. Mass. Spectrom 19 (2005) 3701–3704.
[43] M.-Z. Huang, H.-J. Hsu, J.-Y. Lee, J. Jeng, J. Shiea, Direct protein detection from biological media through electrospray-assisted laser desorption ionization/mass spectrometry, J. Proteome Res. 5 (2006) 1107–1116.
[44] J.S. Sampson, A.M. Hawkridge, D.C. Muddiman, Generation and detection of multiply-charged peptides and proteins by matrix-assisted laser desorption electrospray ionization (MALDESI) fourier transform ion cyclotron resonance mass spectrometry, J. Am. Soc. Mass. Spectrom 17 (2006) 1712–1716.
[45] J.A. Barry, D.C. Muddiman, Global optimization of the infrared matrix-assisted laser desorption electrospray ionization (IR MALDESI) source for mass spectrometry using statistical design of experiments, Rapid. Commun. Mass. Spectrom 25 (2011) 3527–3536.
[46] G. Robichaud, J.A. Barry, K.P. Garrard, D.C. Muddiman, Infrared matrix-assisted laser desorption electrospray ionization (IR-MALDESI) imaging source coupled to a FT-ICR mass spectrometer, J. Am. Soc. Mass. Spectrom 24 (2013) 92–100.
[47] J. Liu, B. Qiu, H. Luo, Fingerprinting of yogurt products by laser desorption spray post-ionization mass spectrometry, Rapid. Commun. Mass. Spectrom 24 (2010) 1365–1370.
[48] J.J. Coon, W.W. Harrison, Laser desorption-atmospheric pressure chemical ionization mass spectrometry for the analysis of peptides from aqueous solutions, Anal. Chem. 74 (2002) 5600–5605.
[49] J.S. Sampson, D.C. Muddiman, Atmospheric pressure infrared (10.6 μm) laser desorption electrospray ionization (IR-LDESI) coupled to a LTQ Fourier transform ion cyclotron resonance mass spectrometer, Rapid. Commun. Mass. Spectrom 23 (2009) 1989–1992.
[50] K. Jorabchi, M.S. Westphall, L.M. Smith, Charge assisted laser desorption/ionization mass spectrometry of droplets, J. Am. Soc. Mass. Spectrom 19 (2008) 833–840.
[51] S.-C. Cheng, T.-L. Cheng, H.-C. Chang, J. Shiea, Using laser-induced acoustic desorption/electrospray ionization mass spectrometry to characterize small organic and large biological compounds in the solid state and in solution under ambient conditions, Anal. Chem. 81 (2009) 868–874.
[52] J. Gao, D.J. Borton, B.C. Owen, Z. Jin, M. Hurt, L.M. Amundson, J.T. Madden, K. Qian, H.I. Kenttämaa, Laser-induced acoustic desorption/atmospheric pressure chemical ionization mass spectrometry, J. Am. Soc. Mass. Spectrom 22 (2011) 531–538.
[53] D.J. Borton, L.M. Amundson, M.R. Hurt, A. Dow, J.T. Madden, G.J. Simpson, H.I. Kenttämaa, Development of a high-throughput laser-induced acoustic desorption probe and raster sampling for laser-induced acoustic desorption/atmospheric pressure chemical ionization, Anal. Chem. 85 (2013) 5720–5726.
[54] K. Benham, R. Hodyss, F.M. Fernández, T.M. Orlando, Laser-induced acoustic desorption atmospheric pressure photoionization via VUV-generating microplasmas, J. Am. Soc. Mass. Spectrom 27 (2016) 1805–1812.

[55] X. Ren, J. Liu, C. Zhang, H. Luo, Direct analysis of samples under ambient condition by high-voltage-assisted laser desorption ionization mass spectrometry in both positive and negative ion mode, Rapid. Commun. Mass. Spectrom 27 (2013) 613–620.
[56] B. Fatou, P. Saudemont, E. Leblanc, D. Vinatier, V. Mesdag, M. Wisztorski, C. Focsa, M. Salzet, M. Ziskind, I. Fournier, In vivo real-time mass spectrometry for guided surgery application, Sci. Rep. 6 (2016) 25919.
[57] B. Fatou, P. Saudemont, M. Duhamel, M. Ziskind, C. Focsa, M. Salzet, I. Fournier, Real time and in vivo pharmaceutical and environmental studies with SpiderMass instrument, J. Biotechnol. 281 (2018) 61–66.
[58] Y. Coello, A.D. Jones, T.C. Gunaratne, M. Dantus, Atmospheric pressure femtosecond laser imaging mass spectrometry, Anal. Chem. 82 (2010) 2753–2758.
[59] C.N. McEwen, R.G. McKay, B.S. Larsen, Analysis of solids, liquids, and biological tissues using solids probe introduction at atmospheric pressure on commercial LC/MS instruments, Anal. Chem. 77 (2005) 7826–7831.
[60] J. Wu, C.S. Hughes, P. Picard, S. Letarte, M. Gaudreault, J.-F. Lévesque, D.A. Nicoll-Griffith, K.P. Bateman, High-throughput cytochrome P450 inhibition assays using laser diode thermal desorption-atmospheric pressure chemical ionization-tandem mass spectrometry, Anal. Chem. 79 (2007) 4657–4665.
[61] S. Saha, L.C. Chen, M.K. Mandal, K.J. Hiraoka, Leidenfrost Phenomenon-assisted Thermal Desorption (LPTD) and its application to open ion sources at atmospheric pressure mass spectrometry, J. Am. Soc. Mass. Spectrom 24 (2013) 341–347.
[62] S. Krieger, H. Hayen, O.J.J.A. Schmitz, B. Chemistry, Quantification of coumarin in cinnamon and woodruff beverages using DIP-APCI-MS and LC-MS, Anal. Bioanal. Chem 405 (2013) 8337–8345.
[63] S. Krieger, A. von Trotha, K.S.-Y. Leung, O.J.J.A. Schmitz, B. Chemistry, Development, optimization, and use of an APCI source with temperature-controlled vaporization of solid and liquid samples, Anal. Bioanal. Chem 405 (2013) 1373–1381.
[64] J.-Y. Lin, T.-Y. Chen, J.-Y. Chen, Y.-C. Chen, Multilayer gold nanoparticle-assisted thermal desorption ambient mass spectrometry for the analysis of small organics, Analyst 135 (2010) 2668–2675.
[65] R.B. Cody, J.A. Laramée, H.D. Durst, Versatile new ion source for the analysis of materials in open air under ambient conditions, Anal. Chem. 77 (2005) 2297–2302.
[66] J.D. Harper, N.A. Charipar, C.C. Mulligan, X. Zhang, R.G. Cooks, Z. Ouyang, Low-temperature plasma probe for ambient desorption ionization, Anal. Chem. 80 (2008) 9097–9104.
[67] J.F. Garcia-Reyes, J.D. Harper, G.A. Salazar, N.A. Charipar, Z. Ouyang, R.G. Cooks, Detection of explosives and related compounds by low-temperature plasma ambient ionization mass spectrometry, Anal. Chem. 83 (2011) 1084–1092.
[68] N. Na, M. Zhao, S. Zhang, C. Yang, X.J. Zhang, Development of a dielectric barrier discharge ion source for ambient mass spectrometry, J. Am. Soc. Mass. Spectrom 18 (2007) 1859–1862.
[69] H. Wang, W. Sun, J. Zhang, X. Yang, T. Lin, L. Ding, Desorption corona beam ionization source for mass spectrometry, Analyst 135 (2010) 688–695.
[70] J. Jeng, J. Shiea, Electrospray ionization from a droplet deposited on a surface-modified glass rod, Rapid. Commun. Mass. Spectrom 17 (2003) 1709–1713.
[71] Y. Song, R.G. Cooks, Atmospheric pressure ion/molecule reactions for the selective detection of nitroaromatic explosives using acetonitrile and air as reagents, Rapid. Commun. Mass. Spectrom 20 (2006) 3130–3138.
[72] S.A. McLuckey, G.L. Glish, K.G. Asano, B.C. Grant, Atmospheric sampling glow discharge ionization source for the determination of trace organic compounds in ambient air, Anal. Chem. 60 (1988) 2220–2227.

[73] C. Zhen, Y. Zhou, N. Zhang, J. Wang, C. Xiong, S. Chen, Z. Nie, Differentiation of Chinese liquors by using ambient glow discharge ionization mass spectrometry, Analyst 138 (2013) 3830–3835.
[74] B. Wang, X. Ding, Z. Zhao, Y. Duan, Method development for directly screening pesticide residues in foodstuffs using ambient microfabricated glow discharge plasma (MFGDP) desorption/ionization mass spectrometry, Int. J. Mass Spectrom. 377 (2015) 507–514.
[75] M.C. Jecklin, G. Gamez, D. Touboul, R. Zenobi, Atmospheric pressure glow discharge desorption mass spectrometry for rapid screening of pesticides in food, Rapid. Commun. Mass. Spectrom 22 (2008) 2791–2798.
[76] C.D. Quarles, A.J. Carado, C.J. Barinaga, D.W. Koppenaal, R.K.J.A. Marcus, B. Chemistry, Liquid sampling–atmospheric pressure glow discharge (LS-APGD) ionization source for elemental mass spectrometry: preliminary parametric evaluation and figures of merit, Anal. Bioanal. Chem. 402 (2012) 261–268.
[77] Z. Zhou, J.K. Lee, S.C. Kim, R.N. Zare, Nanotip ambient ionization mass spectrometry, Anal. Chem. 88 (2016) 5542–5548.
[78] X. Zhan, Z. Zhao, X. Yuan, Q. Wang, D. Li, H. Xie, X. Li, M. Zhou, Y. Duan, Microwave-induced plasma desorption/ionization source for ambient mass spectrometry, Anal. Chem. 85 (2013) 4512–4519.
[79] Z. Zhao, Y. Duan, Microwave induced plasma desorption ionization (MIPDI) mass spectrometry for qualitative and quantitative analysis of preservatives in cosmetics, RSC Adv. 5 (2015) 40636–40646.
[80] Y. Zhou, N. Zhang, Y. Li, C. Xiong, S. Chen, Y. Chen, Z. Nie, Plasma-based ambient sampling/ionization/transmission integrated source for mass spectrometry, Analyst 139 (2014) 5387–5392.
[81] E.L. Neidholdt, J.L. Beauchamp, Switched ferroelectric plasma ionizer (SwiFerr) for ambient mass spectrometry, Anal. Chem. 83 (2011) 38–43.
[82] F.J. Andrade, J.T. Shelley, W.C. Wetzel, M.R. Webb, G. Gamez, S.J. Ray, G.M. Hieftje, Atmospheric pressure chemical ionization source. 1. Ionization of compounds in the gas phase, Anal. Chem. 80 (2008) 2646–2653.
[83] T. Zhang, W. Zhou, W. Jin, J. Zhou, E. Handberg, Z. Zhu, H. Chen, Q. Jin, Direct desorption/ionization of analytes by microwave plasma torch for ambient mass spectrometric analysis, J. Mass. Spectrom. 48 (2013) 669–676.
[84] M. Miclea, K. Kunze, J. Franzke, K. Niemax, Microplasma jet mass spectrometry of halogenated organic compounds, J. Anal. At. Spectrom. 19 (2004) 990–994.
[85] J.M. Symonds, A.S. Galhena, F.M. Fernández, T.M. Orlando, Microplasma discharge ionization source for ambient mass spectrometry, Anal. Chem. 82 (2010) 621–627.
[86] C.-M. Hong, C.-T. Lee, Y.-M. Lee, C.-P. Kuo, C.-H. Yuan, J. Shiea, Generating electrospray from solutions predeposited on a copper wire, Rapid. Commun. Mass. Spectrom 13 (1999) 21–25.
[87] K. Hiraoka, K. Nishidate, K. Mori, D. Asakawa, S. Suzuki, Development of probe electrospray using a solid needle, Rapid. Commun. Mass. Spectrom 21 (2007) 3139–3144.
[88] J. Liu, H. Wang, N.E. Manicke, J.-M. Lin, R.G. Cooks, Z. Ouyang, Development, characterization, and application of paper spray ionization, Anal. Chem. 82 (2010) 2463–2471.
[89] H. Wang, J. Liu, R.G. Cooks, Z. Ouyang, Paper spray for direct analysis of complex mixtures using mass spectrometry, Angew. Chem. Int. Ed. Engl. 122 (2010) 889–892.
[90] P.-H. Lai, P.-C. Chen, Y.-W. Liao, J.-T. Liu, C.-C. Chen, C.-H. Lin, Comparison of gampi paper and nanofibers to chromatography paper used in paper spray-mass spectrometry, Int. J. Mass Spectrom. 375 (2015) 14–17.

[91] M.K. Mandal, K. Yoshimura, S. Saha, S. Ninomiya, M.O. Rahman, Z. Yu, L.C. Chen, Y. Shida, S. Takeda, H. Nonami, K. Hiraoka, Solid probe assisted nanoelectrospray ionization mass spectrometry for biological tissue diagnostics, Analyst 137 (2012) 4658–4661.
[92] M. Hecht, H. Evard, K. Takkis, R.t. Veigure, R. Aro, R. Lohmus, K. Herodes, I. Leito, K. Kipper, Sponge spray—reaching new dimensions of direct sampling and analysis by MS, Anal. Chem. 89 (2017) 11592–11597.
[93] J. Liu, H. Wang, R.G. Cooks, Z. Ouyang, Leaf spray: direct chemical analysis of plant material and living plants by mass spectrometry, Anal. Chem. 83 (2011) 7608–7613.
[94] B. Hu, P.-K. So, H. Chen, Z.-P. Yao, Electrospray ionization using wooden tips, Anal. Chem. 83 (2011) 8201–8207.
[95] B. Hu, P.-K. So, Z.-P. Yao, Analytical properties of solid-substrate electrospray ionization mass spectrometry, J. Am. Soc. Mass. Spectrom 24 (2013) 57–65.
[96] J. Deng, T. Yu, Y. Yao, Q. Peng, L. Luo, B. Chen, X. Wang, Y. Yang, T. Luan, Surface-coated wooden-tip electrospray ionization mass spectrometry for determination of trace fluoroquinolone and macrolide antibiotics in water, Anal. Chim. Acta 954 (2017) 52–59.
[97] B. Hu, P.-K. So, Y. Yang, J. Deng, Y.-C. Choi, T. Luan, Z.-P. Yao, Surface-modified wooden-tip electrospray ionization mass spectrometry for enhanced detection of analytes in complex samples, Anal. Chem. 90 (2018) 1759–1766.
[98] K.S. Kerian, A.K. Jarmusch, R.G. Cooks, Touch spray mass spectrometry for in situ analysis of complex samples, Analyst 139 (2014) 2714–2720.
[99] J.-Y. Liu, P.-C. Chen, Y.-W. Liou, K.-Y. Chang, C.-H. Lin, Development and application of a brush-spray derived from a calligraphy-brush-style synthetic hair pen for use in ESI/MS, Mass Spectrom. 6 (2017). S0058-S0058.
[100] Y.-W. Liou, J.-S. Wang, C.-C. Chen, C.-H. Lin, Development of an on-line microextraction method for use in fiber-spray/mass spectrometry, Int. J. Mass Spectrom. 421 (2017) 178–183.
[101] A.K. Jarmusch, V. Pirro, K.S. Kerian, R.G. Cooks, Detection of strep throat causing bacterium directly from medical swabs by touch spray-mass spectrometry, Analyst 139 (2014) 4785–4789.
[102] A.K. Jarmusch, V. Pirro, D.L. Logsdon, R.G. Cooks, Direct ion generation from swabs, Talanta 184 (2018) 356–363.
[103] G.A. Gómez-Ríos, J. Pawliszyn, Development of coated blade spray ionization mass spectrometry for the quantitation of target analytes present in complex matrices, Angew. Chem. Int. Ed. Engl. 53 (2014) 14503–14507.
[104] J. Balog, T. Szaniszlo, K.-C. Schaefer, J. Denes, A. Lopata, L. Godorhazy, D. Szalay, L. Balogh, L. Sasi-Szabo, M. Toth, Z. Takats, Identification of biological tissues by rapid evaporative ionization mass spectrometry, Anal. Chem. 82 (2010) 7343–7350.
[105] J. Balog, L. Sasi-Szabó, J. Kinross, M.R. Lewis, L.J. Muirhead, K. Veselkov, R. Mirnezami, B. Dezső, L. Damjanovich, A. Darzi, J.K. Nicholson, Z. Takáts, Intraoperative tissue identification using rapid evaporative ionization mass spectrometry, Sci. Transl. Med. 5 (2013), 194ra193-194ra193.
[106] T. Wachs, J. Henion, Electrospray device for coupling microscale separations and other miniaturized devices with electrospray mass spectrometry, Anal. Chem. 73 (2001) 632–638.
[107] G.J. Van Berkel, A.D. Sanchez, J.M.E. Quirke, Thin-layer chromatography and electrospray mass spectrometry coupled using a surface sampling probe, Anal. Chem. 74 (2002) 6216–6223.
[108] T. Wachs, J. Henion, A device for automated direct sampling and quantitation from solid-phase sorbent extraction cards by electrospray tandem mass spectrometry, Anal. Chem. 75 (2003) 1769–1775.

[109] V. Kertesz, G.J. Van Berkel, Fully automated liquid extraction-based surface sampling and ionization using a chip-based robotic nanoelectrospray platform, J. Mass. Spectrom. 45 (2010) 252–260.

[110] D. Eikel, M. Vavrek, S. Smith, C. Bason, S. Yeh, W.A. Korfmacher, J.D. Henion, Liquid extraction surface analysis mass spectrometry (LESA-MS) as a novel profiling tool for drug distribution and metabolism analysis: The terfenadine example, Rapid. Commun. Mass. Spectrom 25 (2011) 3587–3596.

[111] P.J. Roach, J. Laskin, A. Laskin, Nanospray desorption electrospray ionization: an ambient method for liquid-extraction surface sampling in mass spectrometry, Analyst 135 (2010) 2233–2236.

[112] L. Hartmanova, V. Ranc, B. Papouskova, P. Bednar, V. Havlicek, K. Lemr, Fast profiling of anthocyanins in wine by desorption nano-electrospray ionization mass spectrometry, J. Chromatogr. A 1217 (2010) 4223–4228.

[113] C.N. McEwen, V.S. Pagnotti, E.D. Inutan, S. Trimpin, New paradigm in ionization: multiply charged ion formation from a solid matrix without a laser or voltage, Anal. Chem. 82 (2010) 9164–9168.

[114] M. Haapala, J. Pól, V. Saarela, V. Arvola, T. Kotiaho, R.A. Ketola, S. Franssila, T.J. Kauppila, R. Kostiainen, Desorption atmospheric pressure photoionization, Anal. Chem. 79 (2007) 7867–7872.

[115] R.L. Grimm, J.L. Beauchamp, Field-induced droplet ionization mass spectrometry, J. Phys. Chem. B 107 (2003) 14161–14163.

[116] Y.-Q. Huang, J.-Q. You, B.-F. Yuan, Y.-Q. Feng, Sample preparation and direct electrospray ionization on a tip column for rapid mass spectrometry analysis of complex samples, Analyst 137 (2012) 4593–4597.

[117] S.R. Heron, R. Wilson, S.A. Shaffer, D.R. Goodlett, J.M. Cooper, Surface acoustic wave nebulization of peptides as a microfluidic interface for mass spectrometry, Anal. Chem. 82 (2010) 3985–3989.

[118] H. Zhu, G. Li, G.J. Huang, Screening of complicated matrixes with paper assisted ultrasonic spray ionization mass spectrometry, J. Am. Soc. Mass. Spectrom 25 (2014) 935–942.

[119] R.B. Dixon, J.S. Sampson, D.C. Muddiman, Generation of multiply charged peptides and proteins by radio frequency acoustic desorption and ionization for mass spectrometric detection, J. Am. Soc. Mass. Spectrom 20 (2009) 597–600.

[120] T.-Y. Chen, C.-S. Chao, K.-K.T. Mong, Y.-C. Chen, Ultrasonication-assisted spray ionization mass spectrometry for on-line monitoring of organic reactions, Chem. Commun. 46 (2010) 8347–8349.

[121] V.G. Santos, T. Regiani, F.F.G. Dias, W. Romão, J.L.P. Jara, C.F. Klitzke, F. Coelho, M.N. Eberlin, Venturi easy ambient sonic-spray ionization, Anal. Chem. 83 (2011) 1375–1380.

[122] P.T. Steele, G.R. Farquar, A.N. Martin, K.R. Coffee, V.J. Riot, S.I. Martin, D.P. Fergenson, E.E. Gard, M. Frank, Autonomous, broad-spectrum detection of hazardous aerosols in seconds, Anal. Chem. 80 (2008) 4583–4589.

[123] K.L. Fowble, K. Teramoto, R.B. Cody, D. Edwards, D. Guarrera, R.A. Musah, Development of "laser ablation direct analysis in real time imaging" mass spectrometry: application to spatial distribution mapping of metabolites along the biosynthetic cascade leading to synthesis of atropine and scopolamine in plant tissue, Anal. Chem. 89 (2017) 3421–3429.

[124] L.V. Ratcliffe, F.J.M. Rutten, D.A. Barrett, T. Whitmore, D. Seymour, C. Greenwood, Y. Aranda-Gonzalvo, S. Robinson, M. McCoustra, Surface analysis under ambient conditions using plasma-assisted desorption/ionization mass spectrometry, Anal. Chem. 79 (2007) 6094–6101.

[125] B. Feng, J. Zhang, C. Chang, L. Li, M. Li, X. Xiong, C. Guo, F. Tang, Y. Bai, H. Liu, Ambient mass spectrometry imaging: plasma assisted laser desorption ionization mass spectrometry imaging and its applications, Anal. Chem. 86 (2014) 4164–4169.
[126] J. Zhang, Z. Zhou, J. Yang, W. Zhang, Y. Bai, H. Liu, Thin layer chromatography/plasma assisted multiwavelength laser desorption ionization mass spectrometry for facile separation and selective identification of low molecular weight compounds, Anal. Chem. 84 (2012) 1496–1503.
[127] R.V. Bennett, E.M. Morzan, J.O. Huckaby, M.E. Monge, H.I. Christensen, F.M. Fernández, Robotic plasma probe ionization mass spectrometry (RoPPI-MS) of non-planar surfaces, Analyst 139 (2014) 2658–2662.
[128] A. Li, M.R.L. Paine, S. Zambrzycki, R.B. Stryffeler, J. Wu, M. Bouza, J. Huckaby, C.-Y. Chang, M. Kumar, P. Mukhija, F.M. Fernández, Robotic surface analysis mass spectrometry (RoSA-MS) of three-dimensional objects, Anal. Chem. 90 (2018) 3981–3986.
[129] D. Chen, Y.-N. Hu, D. Hussain, G.-T. Zhu, Y.-Q. Huang, Y.-Q. Feng, Electrospun fibrous thin film microextraction coupled with desorption corona beam ionization-mass spectrometry for rapid analysis of antidepressants in human plasma, Talanta 152 (2016) 188–195.
[130] D. Chen, H.-B. Zheng, Y.-Q. Huang, Y.-N. Hu, Q.-W. Yu, B.-F. Yuan, Y.-Q. Feng, Magnetic solid phase extraction coupled with desorption corona beam ionization-mass spectrometry for rapid analysis of antidepressants in human body fluids, Analyst 140 (2015) 5662–5670.
[131] Z. Takáts, I. Cotte-Rodriguez, N. Talaty, H. Chen, R.G. Cooks, Direct, trace level detection of explosives on ambient surfaces by desorption electrospray ionization mass spectrometry, Chem. Commun. (2005) 1950–1952.
[132] C.-P. Kuo, J. Shiea, Application of direct electrospray probe to analyze biological compounds and to couple to solid-phase microextraction to detect trace surfactants in aqueous solution, Anal. Chem. 71 (1999) 4413–4417.
[133] J. Jeng, C.-H. Lin, J. Shiea, Electrospray from nanostructured tungsten oxide surfaces with ultralow sample volume, Anal. Chem. 77 (2005) 8170–8173.
[134] K. Zaitsu, Y. Hayashi, T. Murata, T. Ohara, K. Nakagiri, M. Kusano, H. Nakajima, T. Nakajima, T. Ishikawa, H. Tsuchihashi, A. Ishii, Intact endogenous metabolite analysis of mice liver by probe electrospray ionization/triple quadrupole tandem mass spectrometry and its preliminary application to in vivo real-time analysis, Anal. Chem. 88 (2016) 3556–3561.
[135] K. Zaitsu, Y. Hayashi, T. Murata, K. Yokota, T. Ohara, M. Kusano, H. Tsuchihashi, T. Ishikawa, A. Ishii, K. Ogata, H. Tanihata, In vivo real-time monitoring system using probe electrospray ionization/tandem mass spectrometry for metabolites in mouse brain, Anal. Chem. 90 (2018) 4695–4701.
[136] W. Römisch-Margl, C. Prehn, R. Bogumil, C. Röhring, K. Suhre, J.J.M. Adamski, Procedure for tissue sample preparation and metabolite extraction for high-throughput targeted metabolomics, Metabolomics 8 (2012) 133–142.
[137] R. Dams, M.A. Huestis, W.E. Lambert, C.M. Murphy, Matrix effect in bio-analysis of illicit drugs with LC-MS/MS: influence of ionization type, sample preparation, and biofluid, J. Am. Soc. Mass. Spectrom 14 (2003) 1290–1294.
[138] M. Zampieri, K. Sekar, N. Zamboni, U. Sauer, Frontiers of high-throughput metabolomics, Curr. Opin. Chem. Biol. 36 (2017) 15–23.
[139] C.-C. Hsu, M.S. ElNaggar, Y. Peng, J. Fang, L.M. Sanchez, S.J. Mascuch, K.A. Møller, E.K. Alazzeh, J. Pikula, R.A. Quinn, Y. Zeng, B.E. Wolfe, R.J. Dutton, L. Gerwick, L. Zhang, X. Liu, M. Månsson, P.C. Dorrestein, Real-time

metabolomics on living microorganisms using ambient electrospray ionization flow-probe, Anal. Chem. 85 (2013) 7014–7018.
[140] H. Link, T. Fuhrer, L. Gerosa, N. Zamboni, U. Sauer, Real-time metabolome profiling of the metabolic switch between starvation and growth, Nat. Methods 12 (2015) 1091.
[141] P. Martinez-Lozano Sinues, L. Tarokh, X. Li, M. Kohler, S.A. Brown, R. Zenobi, R. Dallmann, Circadian variation of the human metabolome captured by real-time breath analysis, PLoS One 9 (2015) e114422.
[142] R.G. Cooks, Z. Ouyang, Z. Takats, J.M. Wiseman, Ambient Mass Spectrometry, vol. 311, 2006, pp. 1566–1570.
[143] H. Chen, G. Gamez, R. Zenobi, What can we learn from ambient ionization techniques? J. Am. Soc. Mass Spectrom. 20 (2009) 1947–1963.
[144] D.J. Weston, Ambient ionization mass spectrometry: current understanding of mechanistic theory; analytical performance and application areas, Analyst 135 (2010) 661–668.
[145] C. Wu, A.L. Dill, L.S. Eberlin, R.G. Cooks, D.R. Ifa, Mass spectrometry imaging under ambient conditions, Mass. Spectrom. Rev. 32 (2013) 218–243.
[146] M. Smoluch, P. Mielczarek, J. Silberring, Plasma-based ambient ionization mass spectrometry in bioanalytical sciences, Mass. Spectrom. Rev. 35 (2016) 22–34.
[147] T.J. Kauppila, R. Kostiainen, Ambient mass spectrometry in the analysis of compounds of low polarity, Anal. Methods 9 (2017) 4936–4953.
[148] S.-C. Cheng, C. Shiea, Y.-L. Huang, C.-H. Wang, Y.-T. Cho, J. Shiea, Laser-based ambient mass spectrometry, Anal. Methods 9 (2017) 4924–4935.
[149] C.W. Klampfl, Ambient mass spectrometry in foodomics studies, Curr. Opin. Food Sci. 22 (2018) 137–144.

CHAPTER 2

Direct analysis in real time

Kanako Sekimoto
Yokohama City University, Graduate School of Nanobioscience, Yokohama, Japan

Contents

2.1 Introduction

Direct analysis in real time (DART), first reported by Robert B. Cody et al. in 2005 [1], is the origin of ambient ionization technique, together with desorption electrospray ionization (DESI) [2]. This technique is classified with "plasma-based" or "atmospheric pressure chemical ionization (APCI)-like" methods because DART combines separate desorption and ionization processes into a single method. The desorption is carried out using hot helium gases, whereas the ionization goes through gas-phase reactions involving Penning ionization by excited helium atoms generated in glow discharge.

The chronology of events that led to development of the DART ion source started in 2001 with work that was intended to provide an

Ambient Ionization Mass Spectrometry in Life Sciences
ISBN 978-0-12-817220-9
https://doi.org/10.1016/B978-0-12-817220-9.00002-3

atmospheric pressure thermal electron source for electron-capture ionization to be used in chemical agent monitoring, industrial chemical sensors, and similar devices. When a prototype was being operated in front of the atmospheric pressure interface of a mass spectrometer, ion signals were observed that could only be explained in terms of traces of solvent vapor present in the laboratory atmosphere. Having recognized the potential of the ionization source under development, James A. Laramée and Robert B. Cody directed their work toward the development of what became known as the DART source since their seminal publication in 2005 [1].

Ever since, DART has become an established technique for rapid mass spectral analysis of numerous types of bulk materials, e.g., pesticide monitoring on vegetables and fruits [3,4], forensic and safety applications, for example screening for traces of explosives [1,5,6], warfare agents [7] or illicit drugs [8–16], on luggage, cloths or bank notes, and so on. In the life sciences, DART serves for the rapid analysis of fatty acid methyl esters from whole cells [17], for clinical studies of compounds from plasma and urine [18,19], for the surface analysis of living organisms [20], and for metabolomic fingerprinting [21,22]. Furthermore, this technique has been recently used for analysis of aerosols and micrometer-sized particles [23,24], as well as flavors and fragrances [25,26]. Table 2.1 summarizes representative DART applications published so far. Further information and details regarding the entire set of DART applications can be found in the earlier reviews [67–70] and books [71,72].

The aim of this chapter is to introduce (1) the principles of DART, including detailed ionization mechanism, (2) some technical applications for improving DART performance/sensitivity, and (3) a novel discharge ionization technique using argon gas, whose performance is comparable to DART.

2.2 DART ion source

2.2.1 Basic setup

A DART ion source consists of the following five segments: (1) a glow discharge chamber including a needle electrode and a first perforated disk electrode, (2) a heating chamber with a second perforated electrode and a heater, (3) DART source exit with a grid electrode, (4) sampling and ionization area, and (5) an ion transfer tube of Vapur Interface or an orifice of mass spectrometer (Fig. 2.1). Ground-state helium gases with flow rates of 1–3 L/min are guided into the glow discharge chamber through a tube

Table 2.1 Representative DART applications published so far.

Field of application		Methodology	Mass spectrometer	Samples/analytes	Reference
Forensics	Detection of explosives	He-DART	JEOL AccuTOF	Explosives (nitroglycerin, di- and trinitrotoluenes, trinitrobenzene etc.)	[1]
	or warfare agents	He-DART	JEOL AccuTOF	Explosives (di- and trinitrotoluenes, trinitrobenzene etc.)	[5]
		He-DART	JEOL AccuTOF	Chemical warfare agents (sulfur mustard, tabun, sarin, and VX)	[7]
		He-DART	JEOL AccuTOF	Explosives (trinitrotoluene, methyl tetranitroaniline, tetranitro tetraazacyclooctane etc.)	[6]
	Drugs	He-DART	JEOL AccuTOF	Drugs of abuse (alprazolam, heroin, hydromorphone, cocaine, etc.)	[8]
		He-DART	JEOL AccuTOF	Illicit drugs	[9]
		He-DART	JEOL AccuTOF	Synthetic cannabinoids (AM-2201, 251, JWH-122, 203, 210, 015, RCS-4)	[10–12]
		He-DART	JEOL AccuTOF	Synthetic cannabinoids (JWH-018) from herbal marijuana alternatives	[13]

Continued

Table 2.1 Representative DART applications published so far.—cont'd

Field of application		Methodology	Mass spectrometer	Samples/analytes	Reference
		He-DART	JEOL AccuTOF	"Bath salt" cathinone drug mixtures	[14]
		He-DART	Sciex API-4000 triple-quadrupole	Cocaine and methadone in urine	[15]
		He-DART	Thermo (Q-) Exactive orbitrap	Delta-9-tetrahydrocannabinol in hair	[16]
	Others	He-DART	JEOL AccuTOF	Ballpoint pen inks on paper	[27]
		He-DART	JEOL AccuTOF	Inkjet inks	[28]
		ionRocket + He-DART	JEOL AccuTOF	Automotive clear coats	[29]
Life sciences	Biological studies	He-DART	JEOL AccuTOF	Hydrocarbons on live flies before and after courtship	[20]
		He-DART	JEOL AccuTOF	Juvenile hormone III and its terpene precursors from insects (farnesol, farnesoic acid, and methyl farnesoate)	[30]
	Metabolomics	He-DART	JEOL AccuTOF	Metabolome of beer	[21]
		He-DART	Bruker micrOTOF-Q II	Untargeted blood serum metabolic fingerprinting	[22]
	Clinic, pharmaceutical, and cosmetic application	He-DART	JEOL AccuTOF	Fatty acid methyl esters from bacterial whole cells	[17]
		He-DART	Sciex API-4000 triple-quadrupole	Compounds from plasma	[18]

		He-DART	JEOL AccuTOF	Compounds from urine	[19]
		He-DART	JEOL AccuTOF	Organic UV filters and parabens in water and cosmetic products	[31,32]
		He-DART	Waters Xevo triple-quadrupole	Dried blood spot samples from phenylketonuria newborns	[33]
		He-DART	Sciex API-4000 triple-quadrupole	Traditional Chinese medicine	[34]
		He-DART	Thermo Exactive orbitrap	Toxic glycols (ethylene glycol and diethylene glycol) in glycerin-containing products	[35]
		Ultrasonic Mist + He-DART	Bruker micrOTOF-Q II	Nonvolatile and heat-unstable compounds in creams and lotions (e.g., glycyrrhizic acid)	[36]
Quality control		He-DART	Thermo TSQ Quantum triple-quadrupole	Perfumery raw materials (cashmeran, galaxolide, manzanate, phantolide, decal, damascone, etc.)	[25]
		He-DART	JEOL AccuTOF	Polypropylene in plastic samples	[37]
		He-DART	JEOL AccuTOF	Insecticide-treated bednets (deltamethrin, permethrin)	[38]
		He-DART	JEOL AccuTOF	Tackifier resins	[39]
		He-DART	Bruker MicroTOF-Q type Qq-TOF	Phthalic acid esters in poly(vinyl chloride) samples	[40]
		He-DART	Agilent MSD ToF	Triazine herbicides	[41,42]

Continued

Table 2.1 Representative DART applications published so far.—cont'd

Field of application	Methodology	Mass spectrometer	Samples/analytes	Reference
Food quality and authentication	Volatimeship + He-DART	Shimadzu LCMS-8040	Food flavors (e.g., raw and dry pepper)	[26]
	He-DART	JEOL AccuTOF	Olive oils	[43]
	He-DART	Thermo Exactive orbitrap	Mycotoxins in beer (*Fusarium* and *Alternaria* toxins, aflatoxins, ergot alkaloids, ochratoxins, and sterigmatocystin)	[44]
	He-DART	JEOL AccuTOF	Organosulfur species formed upon crushing garlic and onion	[45,46]
	He-DART	JEOL AccuTOF	Animal fats	[47]
	He-DART	Thermo Exactive orbitrap	Pesticides on vegetables and fruits	[3]
	He-DART	JEOL AccuTOF	Caffeine in coffee brews	[48]
	He-DART	Thermo Exactive orbitrap	Flavonoids and other phenolic compounds in propolis	[49]
	He-DART	Thermo Exactive orbitrap	Pesticides on vegetables and fruits	[4]
	He-DART	Thermo Exactive orbitrap	Thermally induced oxidation products in vegetable oils	[50]
	He-DART	Thermo Exactive orbitrap	Aflatoxin AFB1 from corn	[51]
	He-DART	Thermo Exactive orbitrap	Polysorbates in fruit juice	[52]

		He-DART	Waters qDa	Milk samples from assorted animals and vegetal sources	[53]
Environmental analysis		He-DART	Thermo LTQ XL ion-trap	Sugars in saccharification matrix	[54]
		He-DART	Sciex QSTAR XL triple-quadrupole TOF	Submicrometer organic particles including alkanes, alkenes, carboxylic acids, esters, alcohols, aldehydes, and amino acid	[23,24]
General chemical analysis		He-DART	JEOL AccuTOF	Contaminants in soils	[55]
		He-DART	JEOL AccuTOF	Nonpolar compounds	[56]
		He-DART, Ar-DART	Sciex API 150 EX single-quadrupole	Acetaminophen	[57]
		He-DART	JEOL AccuTOF	Hydrocarbons, alcohols, and other nonpolar compounds	[58]
		Dopant Ar-DART	Thermo LTQ ion-trap	Labile compounds (nucleosides, alkaloids, glucose, and other small molecules)	[59]
		RoPPI + He-DART	Bruker micrOTOF-Q II	Rhodamine 6G and Nile blue chloride	[60]
		Drift tube ion mobility spectrometer + He-DART	JEOL AccuTOF	2,6-Di-tert-butylpyridine, acetaminophen	[61]

Continued

Table 2.1 Representative DART applications published so far.—cont'd

Field of application		Methodology	Mass spectrometer	Samples/analytes	Reference
		He-DART	Thermo LCQ ion-trap	20 alpha-amino acids	[62]
		Corona^{++} + He-DART	Thermo LCQ ion-trap, Shimadzu LCMS-2020, Thermo Q-Exactive orbitrap	Alpha-amino acids, *n*-alkanes, and agricultural chemicals	[63,64]
		Dopant Ar-DART	JEOL AccuTOF	Polycyclic aromatic hydrocarbons, diesel fuel, delta-9-tetrahydrocannabinol, cannabidiol, trinitrotoluene	[65]
		Atmospheric pressure dark current argon discharge	Thermo LCQ ion-trap, Shimadzu LCMS-2020	Alpha-amino acids and *n*-alkanes	[66]

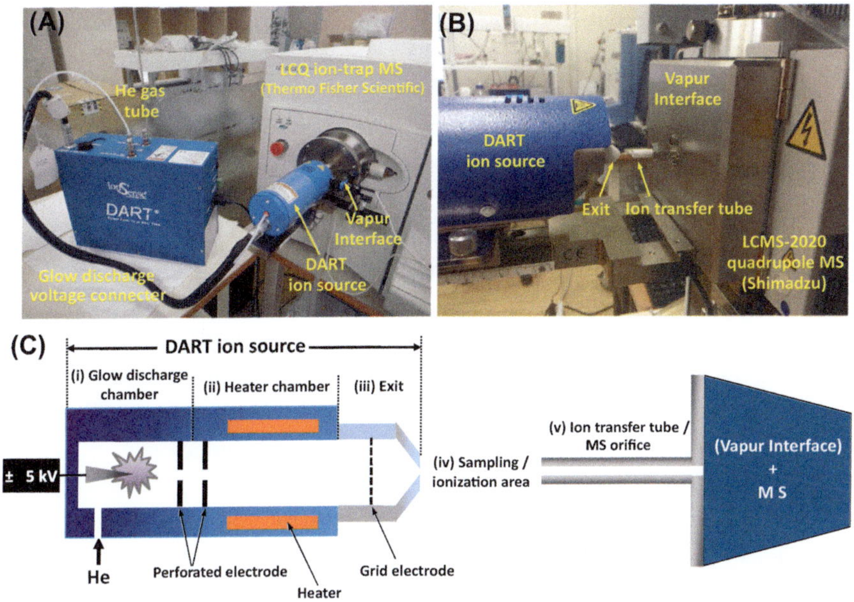

Figure 2.1 (A), (B) Photographs of Direct Analysis in Real Time (DART) system. (C) Schematic illustration of DART ion source. *(Photographs by courtesy of AMR Inc.)*

((i) in Fig. 2.1C). A glow discharge between the needle and perforated electrodes excites and/or ionizes a part of helium gases to form plasma gas stream including helium-related ions and excited atoms as well as electrons. Typical needle voltage and resulting electric current for the glow discharge are ± 5 kV and 3 mA, respectively. The perforated electrode is grounded. The plasma gas stream and residual ground-state helium atoms pass the second perforated electrode, the heater (50–500°C), and the grid electrode ((ii) and (iii) in Fig. 2.1C). The voltages of the second perforated electrode and the grid electrode are biased to positive voltages for positive-ion mode and to negative voltages for negative-ion mode in the order of a 100 volts. The adjustment of the polarity of these electrodes allows removing charged species from the plasma gas stream. The residual excited and ground-state helium atoms are flowed in open area where samples are inserted ((iv) in Fig. 2.1C). When they hit the sample surface or vaporized samples, analytes of interest can be desorbed and/or ionized. The resulting gas-phase analyte ions are introduced into the mass spectrometer through the ion transfer tube of Vapur Interface and/or the mass spectrometer orifice ((v) in Fig. 2.1C). A typical distance between the DART source exit and the entrance of the transfer tube/orifice is 5–25 mm.

Vapur Interface, an additional vacuum pumping system, should be installed in mass spectrometers other than the JEOL AccuTOF instrument. In contrast to the gases normally contained in air, helium presents a true challenge for all types of vacuum pumps. In a general mass spectrometer, multiple turbomolecular pumps create a vacuum by flicking gas molecules through a rotation of vanes. However, because helium is a light gas with a low atomic mass of 4, it can sneak through the vanes, resulting in a situation in which there is a decrease in the vacuum. Therefore, in order to eliminate the helium gas, Vapur Interface is attached in front of the atmospheric pressure interface of a mass spectrometer. The AccuTOF instrument has some zig-zag ion path in the ion source optics that guides ions to the orthogonal TOF analyzer while it deflects neutrals. Thus, the use of helium is easily accepted in this instrument without Vapur Interface.

2.2.2 Multiple configurations for sampling

DART can be used to analyze bulk materials, either solid or liquid, as well as volatile compounds that are spontaneously vaporized into the atmosphere. The classical sampling method is direct injection in the excited helium gas flow (Fig. 2.2A). More or less planar macroscopic objects have been analyzed by this method, for example tablets in counterfeit drug detection, leaves, pieces of paper or cloths, films deposited on to glass surfaces, and viscous liquids and powders applied to melting point tubes. Successfully doing it requires the sample to be carefully positioned, to enable the helium gas stream to flow tangentially to the surface while avoiding blockage of the flow of desorbed analyte ions into the mass spectrometer [38,73].

Based on the issue of the direct injection method, transmission mode has been introduced to rapidly and reproducibly position samples. In this case, a sample is applied to a surface of a fine metal mesh or glass tube placed into the path between the DART source exit and mass spectrometer (Fig. 2.2B). Liquid-type samples are suitable for this method: deposition of solution is more reproducible in that it enables replicate analysis from a stock solution of known concentration or, at least, repeated application of a constant volume of solution, e.g., to find the optimum desorption temperature, to check reproducibility (variation of 1%–2% can be achieved), etc [68]. The DART ion source can be adjusted with an angle (Fig. 2.2C), as necessary for the analysis of larger objects, or to scan surfaces along the x axis (thin-layer chromatography [TLC] plates or glass slides) or on the xy-plane

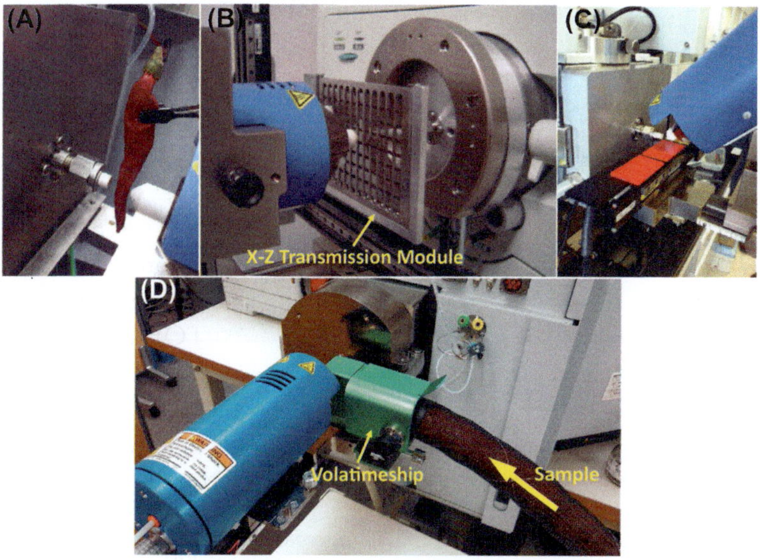

Figure 2.2 Multiple configurations for sampling in Direct Analysis in Real Time (DART). (A) Direct injection of samples (in this case, red pepper as a food sample), (B) Sampling by transmission mode. Here DART is combined with X-Z Transmission Module (IonSense Inc.). (C) DART mounted at 45 degrees. (D) Volatimeship (BioChromato Inc.) to introduce volatile samples. *(Photographs by courtesy of AMR Inc. and BioChromato Inc.)*

(extended objects, for example bank notes). Various sampling options for transmission mode and the angle option have been commercialized from IonSense Inc.

Volatile samples such as flavors and fragrances can be analyzed using a closed system, Volatimeship (BioChromato Inc.) (Fig. 2.2D). This system allows effective ionization of volatile compounds which are easily dispersed while adjusting the total gas flow rates involving helium gas flow from the DART ion source, sample flow from Volatimeship, and evacuated flow by pumps in mass spectrometer [26].

The 3D surface analysis of native, irregularly shaped, or curved samples has been allowed by coupling robotic probe surface sampling system (Fig. 2.3) [60]. The larger ionization region accommodates minor precision errors in the robotic placement of the sample prove while still allowing for efficient ionization.

DART is also compatible with separation techniques. High-performance liquid chromatography (HPLC) is coupled to a DART source; the effluent of the HPLC system is guided into a PEEK transfer

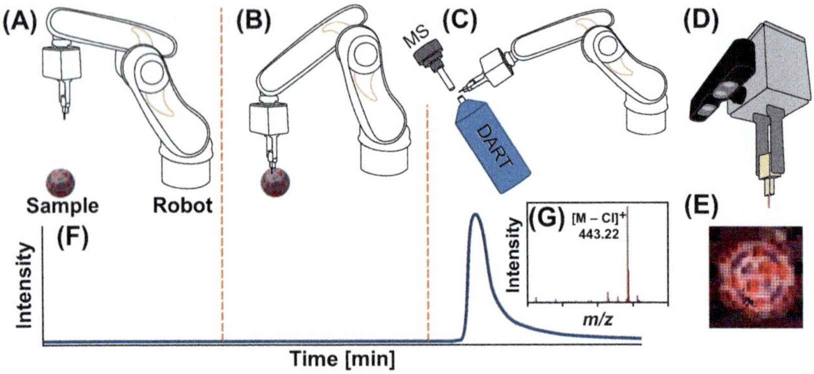

Figure 2.3 Schematic illustration of the process of the 3D surface analysis using robotic probe surface sampling system with Direct Analysis in Real Time. The sample used was hand-painted polystyrene hemisphere. Starting at a "home position" (A), a robotic arm maneuvers a probe to stab the surface of the sample (B). The probe is then reproducibly inserted into the ionization area (C). The position sampled by the robotic probe is selected by using a 3D camera (D) that creates an (*x*, *y*, *z*) point cloud (E) for accurate sample coordinate determination. A selected ion chronogram at *m/z* 443 ([M – Cl]$^+$ for rhodamine 6G) observed for each of the analysis steps described, and the subsequent mass spectrum are shown in (F) and (G). By probing multiple points across the sample surface, the acquired mass spectra (and corresponding peak areas from extracted ion chronograms) are correlated with the known *x*, *y*, and *z* position. *(Reproduced with permission from R.V. Bennett, E.M. Morzan, J.O. Huckaby, M.E. Monge, H.I. Christensen, F.M. Fernández, Robotic plasma probe ionization mass spectrometry (RoPPI-MS) of non-planar surfaces, Analyst 139 (2014) 2658–2662, https://doi.org/10.1039/c4an00277f. Copyright © 2014 Royal Society of Chemistry.)*

capillary to which either a stainless steel capillary or a piece of fused silica capillary is attached via a zero-dead-volume junction. This capillary is adjusted to expose the eluting liquid to the sampling area of the DART source. Online coupling of capillary electrophoresis is also coupled with DART. The analytes eluting from the electrophoresis capillary are directly ionized by elution into the DART source. The ability of DART to directly analyze gases has been exploited in gas chromatography (GC)-DART coupling. The gas chromatographic capillary column is extended outside the GC oven through copper tubing wrapped with heating tape and heated. The DART gas is set to the same temperature. The tubing is then positioned directly in front of the mass spectrometer orifice.

2.3 Ionization processes in DART

DART is an APCI-like ambient ionization technique. In common DART operated with helium gas (denoted as He-DART), gas-phase analyte ions detected by mass spectrometer are formed via the following four steps: (1) generation of excited helium atoms by glow discharges, (2) formation of reagents (including ions and neutrals) via Penning ionization of ambient air constituents with excited helium, (3) desorption of analytes by hot helium gas flows, and (4) ionization of desorbed analytes via reactions with reagents. Step (1) occurs inside the DART chamber ((i) in Fig. 2.1), whereas steps (2)—(4) take place in the sampling and ionization area ((iv) in Fig. 2.1). In this section, the formation mechanism of reagents, analyte ionization reactions, and ionization characteristics of He-DART are described.

2.3.1 Formation mechanism of reagents

The major reagents involved in analyte ionization in He-DART are hydronium ion H_3O^+, superoxide $O_2^{\cdot-}$, hydroxyl radical HO, ozone O_3, and slow electrons e^-_{slow}. The formation mechanism of these reagents is as follows. A glow discharge in helium produces a stream of gas containing electronically excited helium atoms (He*), ions (He_n^+), and electrons (e^-):

$$\text{Glow discharge} + \text{He} \rightarrow \text{He}^*/\text{He}_n^+/\text{e}^- \quad (n = 1,\ 2) \tag{2.1}$$

Charged species are removed from a stream of gas through the perforated disk or grid electrodes in the DART chamber. Thus, excited atoms, mostly metastable 2^3S states with internal energy of 19.8 eV, exit to the sampling area surrounded by ambient air. As ionization energies (IEs) of the major air constituents are lower than internal energy of He(2^3S) (e.g., IEs of N_2, O_2, and H_2O are 15.6, 12.1, and 12.6 eV, respectively), Penning ionization occurs to form molecular cations of air constituents:

$$\text{He}(2^3S) + N_2/O_2/H_2O \rightarrow N_2^{\cdot+}/O_2^{\cdot+}/H_2O^{\cdot+} + \text{He} + e^-_{slow} \tag{2.2}$$

This reaction is the trigger of the formation of reagents in He-DART. The major positive reagent ion is hydronium ion H_3O^+ and its water clusters $H_3O^+(H_2O)_n$ (Fig. 2.4A). The $H_3O^+(H_2O)_n$ ions can be formed via several pathways of successive reactions involving the molecular cations $N_2^{\cdot+}/O_2^{\cdot+}/H_2O^{\cdot+}$ as well as water molecules H_2O (i.e., reaction sequences f-i-c, g (or h)-b-c, g (or h)-n-k-l-m, and k-l-m as shown in Fig. 2.5 and Table 2.2). Of these, the dominant pathway originates from $H_2O^{\cdot+}$, i.e., hydrogen abstraction from H_2O by $H_2O^{\cdot+}$ and subsequent water cluster formation:

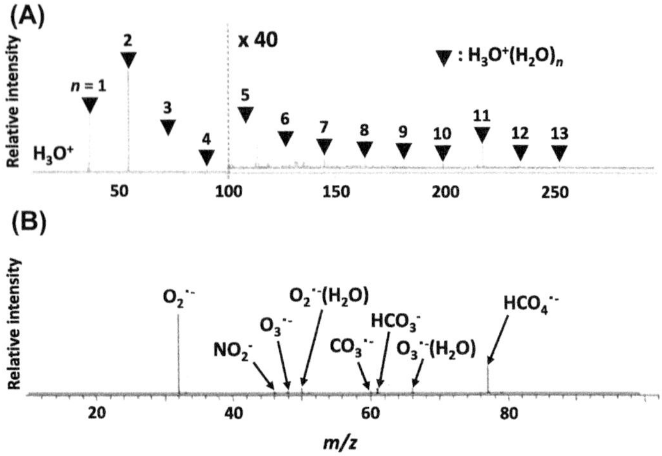

Figure 2.4 (A) Positive and (B) negative background ion mass spectra obtained by helium Direct Analysis in Real Time combined with a JEOL AccuTOF mass spectrometer. In (A), the relatively high intensity of $H_3O^+(H_2O)_{11}$ is due to a background interference and does not indicate a "magic-number" water cluster. *(Reproduced with permission from R.B. Cody, J.A. Laramée, H.D. Durst, Versatile new ion source for the analysis of materials in open air under ambient conditions, Anal. Chem. 77 (2005) 2297–2302, https://doi.org/10.1021/ac050162j. Copyright © 2005 American Chemical Society.)*

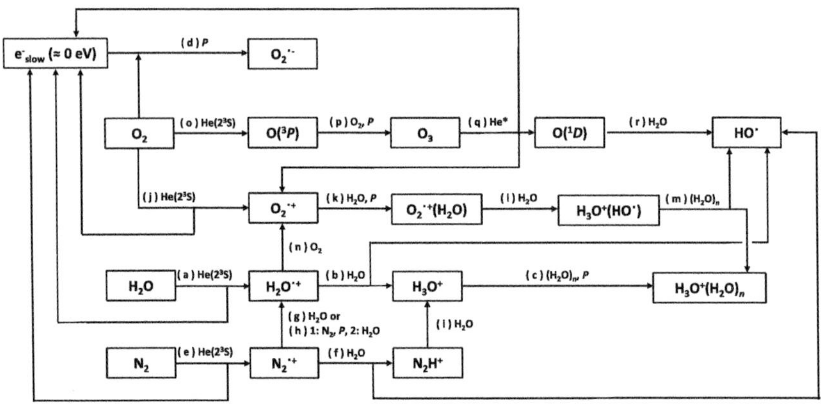

Figure 2.5 Sequential reactions for the formation of reagents in helium Direct Analysis in Real Time. *(Reproduced with permission from K. Sekimoto, M. Sakakura, T. Kawamukai, H. Hike, T. Shiota, F. Usui, Y. Bando, M. Takayama, Ionization characteristics of amino acids in direct analysis in real time mass spectrometry, Analyst 139 (2014) 2589–2599, https://doi.org/10.1039/c3an02193a. Copyright © 2014 Royal Society of Chemistry.)*

Table 2.2 Elementary processes for the formation of reagents in helium DART (Fig. 2.5).

	Reaction	Reaction rate constant	Reference
a	$H_2O + He(2^3S) \rightarrow H_2O^{\cdot +} + e^-_{slow} + He(1^1S)$	—	[1]
b	$H_2O^{\cdot +} + H_2O \rightarrow H_3O^+ + HO^{\cdot}$	1.9×10^{-9} cm^3/s	[1]
c	$H_3O^+ + (H_2O)_n + P \rightarrow H_3O^+(H_2O)_n + P$	$\approx 1.0 \times 10^{-28}$ cm^6/s ($n = 1$; $P = N_2, O_2$)	[1]
d	$O_2 + e^-_{slow} (\approx 0 \text{ eV}) + P \rightarrow O_2^{\cdot -} + P$	1.9×10^{-30} cm^6/s ($P = O_2$) 1.0×10^{-31} cm^6/s ($P = N_2$)	[74] [75]
e	$N_2 + He(2^3S) \rightarrow N_2^{\cdot +} + e^-_{slow} + He(1^1S)$	—	-
f	$N_2^{\cdot +} + H_2O \rightarrow N_2H^+ + HO^{\cdot}$	(2.4–2.8) $\times 10^{-9}$ cm^3/s	[76]
g	$N_2^{\cdot +} + H_2O \rightarrow H_2O^{\cdot +} + N_2$	(2.4–2.8) $\times 10^{-9}$ cm^3/s	[76]
h-1	$N_2^{\cdot +} + N_2 + P \rightarrow N_4^{\cdot +} + P$	$\approx 1.0 \times 10^{-28}$ cm^6/s ($P = N_2$)	[77]
h-2	$N_4^{\cdot +} + H_2O \rightarrow H_2O^{\cdot +} + 2N_2$	—	-
i	$N_2H^+ + H_2O \rightarrow H_3O^+ + N_2$	2.6×10^{-9} cm^3/s	[78]
j	$O_2 + He(2^3S) \rightarrow O_2^{\cdot +} + e^-_{slow} + He(1^1S)$	—	-
k	$O_2^{\cdot +} + H_2O + P \rightarrow O_2^{\cdot +}(H_2O) + P$	(2.3–2.8) $\times 10^{-28}$ cm^6/s ($P = N_2, O_2$)	[79]
l	$O_2^{\cdot +}(H_2O) + H_2O \rightarrow H_3O^+ + HO^{\cdot} + O_2 \rightarrow H_3O^+(HO^{\cdot}) + O_2$	2.2×10^{-9} cm^3/s	[79]
m	$H_3O^+(HO^{\cdot}) + (H_2O)_n \rightarrow H_3O^+(H_2O)_n + HO^{\cdot}$	3.2×10^{-9} cm^3/s ($n = 1$)	[79]
n	$H_2O^{\cdot +} + O_2 \rightarrow O_2^{\cdot +} + H_2O$	3.3×10^{-10} cm^3/s	[78]
o	$O_2 + He(2^3S) \rightarrow O(^3P) + O(^1D) + He(1^1S)$	—	[80]
p	$O(^3P) + O_2 + P \rightarrow O_3 + P$	—	[80]
q	$O_3 + He(2^3S) \rightarrow O(^1D) + O_2^{\cdot +} + e^-_{slow} + He(1^1S)$	—	—
r	$O(^1D) + H_2O \rightarrow 2HO^{\cdot}$	—	[80]

$$H_2O^{\cdot+} + H_2O \rightarrow H_3O^+ + HO^{\cdot} \tag{2.3}$$

$$H_3O^+ + (H_2O)_n + P \rightarrow H_3O^+(H_2O)_n + P \quad (P\text{: a third body such as } N_2 \text{and } O_2) \tag{2.4}$$

The oxygen molecular cation $O_2^{\cdot+}$ may also serve as reagent ion in specific analyte ionization (e.g., molecular ion formation discussed in the next section).

The dominant negative reagent ion is superoxide $O_2^{\cdot-}$ (Fig. 2.4B). The $O_2^{\cdot-}$ ions are formed via the attachment of thermal electrons that have kinetic energy at ambient temperature, i.e., e^-_{slow} (≈ 0 eV), to O_2:

$$O_2 + e^-_{slow}(\approx 0\text{ eV}) + P \rightarrow O_2^{\cdot-} + P \quad (P\text{: } N_2 \text{ or } O_2) \tag{2.5}$$

The production of thermal electrons is attributed to Penning ionization (Eq. 2.2).

Reagent neutrals mainly involved in DART ionization are ozone O_3 and hydroxyl radical $HO^{\cdot}$. O_3 can be generated via the collision of $He(2^3S)$ with O_2 to form an O atom in a ground-level triplet state, $O(^3P)$, and the subsequent combination of $O(^3P)$ with O_2 (Eqs. 2.6 and 2.7; reactions o and p in Fig. 2.5).

$$O_2 + H(2^3S) \rightarrow O(^3P) + O(^1D) + He \tag{2.6}$$

$$O(^3P) + O_2 + P \rightarrow O_3 + P \quad (P\text{: } N_2 \text{or } O_2) \tag{2.7}$$

The formation of O_3 is evidenced by the existence of the negative background ions $O_3^{\cdot-}$ and $CO_3^{\cdot-}$ (Fig. 2.4B). According to the difference in the electron affinities of O_2 (0.5 eV) and O_3 (2.1 eV), charge transfer from $O_2^{\cdot-}$ to O_3 easily occurs, and the resulting $O_3^{\cdot-}$ ions react with CO_2 in ambient air to generate $CO_3^{\cdot-}$. The radical species $HO^{\cdot}$ are formed via the formation pathways of $H_3O^+(H_2O)_n$ (e.g., Eq. (2.3); reactions m, b, and f in Fig. 2.5), as well as the dissociation of O_3 by $He(2^3S)$ and subsequent reaction of resulting $O(^1D)$ with H_2O (Eqs. 2.8 and 2.9; reactions q and r in Fig. 2.5)

$$O_3 + He(2^3S) \rightarrow O(^1D) + O_2^{\cdot+} + e^-_{slow} + He \tag{2.8}$$

$$O(^1D) + H_2O \rightarrow 2HO^{\cdot} \quad (2.9)$$

2.3.2 Analyte ionization reactions

APCI-type DART is a soft ionization method, so that analytes are dominantly ionized via protonation and deprotonation; however, other reactions (i.e., molecular ion formation, fragmentation, oxidation, and background ion attachment) simultaneously occur depending on properties of analytes (Fig. 2.6). Herein a description of analyte ionization reactions in He-DART is presented, based on systematic analyses of 20 α-amino acids [62] and other polar and nonpolar compounds [5,56,58].

(De)protonation: Protonation and deprotonation of analytes in H-DART involve the reagent ions H_3O^+ and $O_2^{\cdot -}$. If a given gas-phase analyte (M_{gas}) has proton affinity (PA) higher than H_2O (691.0 kJ/mol) or lower than $O_2^{\cdot -}$ (1477 ± 2.9 kJ/mol), it can be ionized as protonated or deprotonated molecule:

$$M_{gas} + H_3O^+ \rightarrow [M_{gas} + H]^+ + H_2O \quad (\text{if } PA(M_{gas}) > PA(H_2O)) \quad (2.10)$$

$$M_{gas} + O_2^{\cdot -} \rightarrow [M_{gas} - H]^- + HO_2^{\cdot} \quad (\text{if } PA([M_{gas} - H]^-) < PA(O_2^{\cdot -})) \quad (2.11)$$

Deprotonation also occurs via resonant capture of slow electrons with kinetic energies of 1–1.5 eV by M_{gas}:

$$M_{gas} + e^-_{slow}\ (1\text{–}1.5\ eV) \rightarrow M_{gas}^{\cdot -*} \rightarrow [M_{gas} - H]^- + H^{\cdot} \quad (2.12)$$

Molecular ion formation: Positive-ion He-DART brings about the formation of molecular cations $M_{gas}^{\cdot +}$ for almost all the amino acids [62], as well as other polar and nonpolar compounds (e.g., dibenzosuberone and *n*-hexadecane) [56]. The mechanism involved in forming $M_{gas}^{\cdot +}$ is not fully understood. But, according to IEs of analytes and correlation in ion intensity between $O_2^{\cdot +}$ and $M_{gas}^{\cdot +}$, $M_{gas}^{\cdot +}$ may be formed via Penning

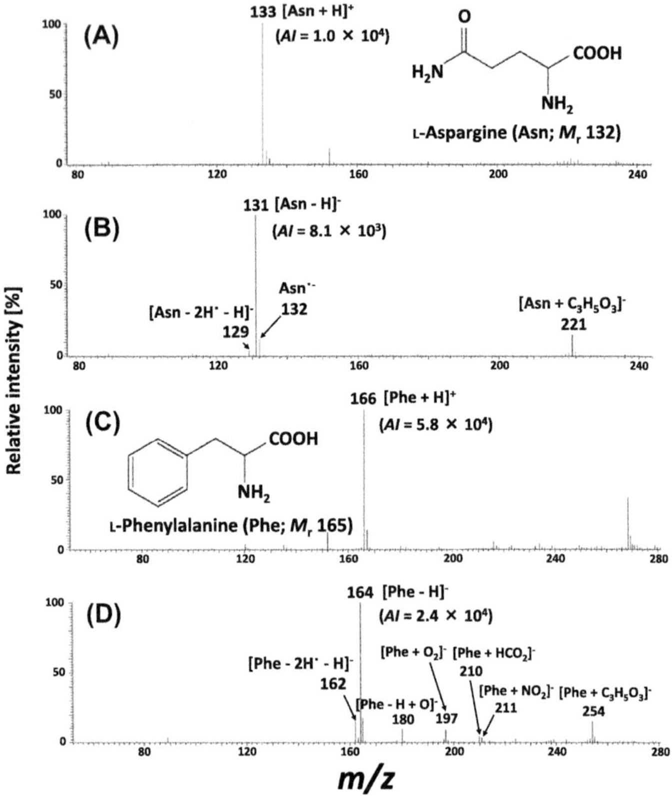

Figure 2.6 Helium Direct Analysis in Real Time mass spectra of (A) L-asparagine in positive-ion mode, (B) L-asparagine in negative-ion mode, (C) L-phenylalanine in positive-ion mode, and (D) L-phenylalanine in negative-ion mode, obtained with a Thermo Fisher Scientific LCQ ion-trap mass spectrometer. *(Reproduced with permission from K. Sekimoto, M. Sakakura, T. Kawamukai, H. Hike, T. Shiota, F. Usui, Y. Bando, M. Takayama, Ionization characteristics of amino acids in direct analysis in real time mass spectrometry, Analyst 139 (2014) 2589–2599, https://doi.org/10.1039/c3an02193a. Copyright © 2014 Royal Society of Chemistry.)*

ionization of $He(2^3S)$ (Eq. 2.13) and charge transfer reaction from $O_2^{\cdot+}$ to M_{gas} (Eq. 2.14) [56]:

$$M_{gas} + He(2^3S) \rightarrow M_{gas}^{\cdot+} + e^-_{slow} + He \quad (\text{if IE}(M_{gas}) < 19.8\text{ eV}) \tag{2.13}$$

$$M_{gas} + O_2^{\cdot+} \rightarrow M_{gas}^{\cdot+} + O_2 \quad (\text{if IE}(M_{gas}) < 12.1\text{ eV}) \tag{2.14}$$

In contrast, molecular anions form the resonant capture of thermal electrons by analytes:

$$M_{gas} + e^{-}_{slow}(\approx 0\ eV) + P \rightarrow M_{gas}^{\cdot -} + P\ (P\text{: } N_2 \text{ and } O_2) \tag{2.15}$$

Fragmentation: Fragmentation occurs due to (1) pyrolysis during thermal desorption and/or (2) excess energy obtained via ionization reactions. In the case of α-amino acids, glutamic acid (Glu) and glutamine (Gln) are ionized with fragmentation (i): Initially, H_2O and NH_3 are lost from Glu and Gln to form pyroglutamic acid (PGA) during the thermal desorption process (Eq. 2.16). Subsequently, the resulting PGA is (de)protonated via proton transfer reactions with H_3O^+ and $O_2^{\cdot -}$ and/or resonant electron capture (Eq. 2.17). Note that in case of fragmentation (i), the same neutral fragments are observed in both positive- and negative-ion modes (Fig. 2.7A and B).

Fragmentation (i):

$$M_{gas/liquid/solid} \xrightarrow{\text{(thermal desorption)}} (M\text{—Fragment})_{gas} + \text{Fragment} \tag{2.16}$$

$$\begin{aligned}&(M - \text{Fragment})_{gas} + H_3O^+/\ O_2^{\cdot -}/e^{-}_{slow}\ (1\text{–}1.5\ eV)\\ &\qquad [(M - \text{Fragment}\ )_{gas} \pm H]^{\pm} + H_2O/\ HO_2^{\cdot}/\ H^{\cdot}\end{aligned} \tag{2.17}$$

In contrast, ionization of tyrosine (Tyr), aspartic acid (Asp), and alcohols including cholesterol leads to several fragment ions, i.e., protonated Tyr with CO_2 loss, protonated Asp with H_2O loss, deprotonated Asp with NH_3 loss, and protonated alcohols with H_2O loss (Fig. 2.7C) [56,62]. Neutral fragments change by varying the polarity of ionization. This suggests that those fragmentations result from the excess energy obtained via (de)protonation (Eqs. 2.18 and 2.19).

Fragmentation (ii):

$$M_{gas} + H_3O^+ \rightarrow [M_{gas} + H]^{+*} + H_2O \tag{2.18a}$$

$$[M_{gas} + H]^{+*} \rightarrow [M_{gas} + H\text{—Fragment}]^{+} + \text{Fragment} \tag{2.18b}$$

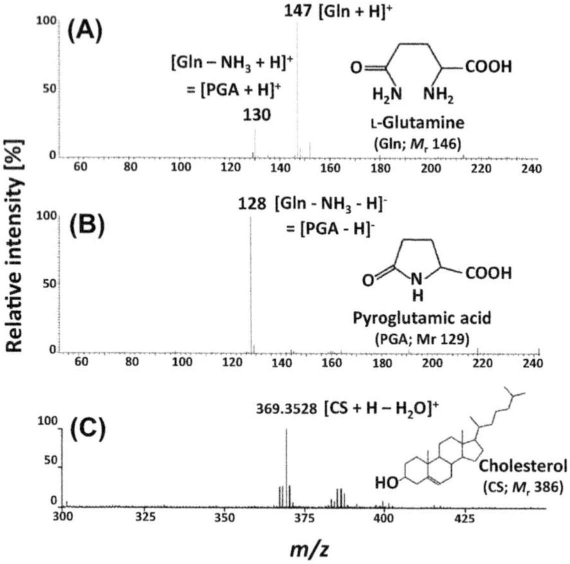

Figure 2.7 Helium Direct Analysis in Real Time mass spectra of (A) L-glutamine in positive-ion mode, (B) L-glutamine in negative-ion mode, and (C) cholesterol in positive-ion mode. (A) and (B) were obtained with a Thermo Fisher Scientific LCQ ion-trap mass spectrometer, whereas (C) was acquired with a JEOL AccuTOF instrument. *(Reproduced with permission from R.B. Cody, Observation of molecular ions and analysis of nonpolar compounds with the direct analysis in real time ion source, Anal. Chem. 81 (2009) 1101–1107, https://doi.org/10.1021/ac8022108; K. Sekimoto, M. Sakakura, T. Kawamukai, H. Hike, T. Shiota, F. Usui, Y. Bando, M. Takayama, Ionization characteristics of amino acids in direct analysis in real time mass spectrometry, Analyst 139 (2014) 2589–2599, https://doi.org/10.1039/c3an02193a. Copyright © 2014 Royal Society of Chemistry and 2009 American Chemistry Society.)*

$$M_{gas} + O_2^{\cdot -} \rightarrow [M_{gas} - H]^{-*} + HO_2^{\cdot} \tag{2.19a}$$

$$[M_{gas}{-}H]^{-*} \rightarrow [M_{gas}{-}H{-}\text{Fragment}]^{-} + \text{Fragment} \tag{2.19b}$$

Oxidation: He-DART ionization leads to two types of oxidation, (1) oxygen attachment and (2) hydrogen loss, which are most likely due to ozone O_3 and/or hydroxyl radical $HO^{\cdot}$. In case of oxidation (i), one and two oxygen atoms attach to specific amino acids, i.e., sulfur-containing, aromatic, and aromatic heterocyclic amino acids such as methionine (Met), phenylalanine (Phe; Fig. 2.6D), tyrosine (Tyr), tryptophan (Trp), and histidine (His) (Eq. 2.20). Then, the oxidized compounds ($M_{gas} + nO$) are (de)protonated (Eq. 2.21). The oxidation reactivity order observed in He-DART is Trp > Met > Phe > Tyr > His. This order is in agreement with that observed in ozonated solutions with pH 5.8 [81].

Oxidation (i):

$$M_{gas} + nO \rightarrow (M_{gas} + nO) \tag{2.20}$$

$$\begin{aligned}&(M_{gas} + nO) + H_3O^+/\ O_2^{\cdot -}/\ e^-_{slow}\ (1\text{–}1.5\ eV)\\&\rightarrow [(M_{gas} + nO) \pm H]^{\pm} + H_2O/\ HO_2^{\cdot}/\ H^{\cdot a}\end{aligned} \tag{2.21}$$

Oxidation (ii) is observed in negative-ion He-DART. Deprotonated analytes with loss of two hydrogens (2H$^{\cdot}$), $[M_{gas} - 2H^{\cdot} - H]^-$, are formed when ionizing amino acids with aliphatic or nonionic side chains (such as asparagine and phenylalanine in Fig. 2.6B and D), as well as *n*-alkanes [5,58]. The precise mechanism of the 2H$^{\cdot}$ loss reactions in He-DART has not yet been clarified. However, it is noted that hydrogen loss is observed in negative-ion mode only. Based on the studies regarding oxidation of amino acids in aqueous solutions with HO$^{\cdot}$ and O_2 [82], deprotonated amino acids with 2H$^{\cdot}$ loss observed here may correspond to deprotonated imino acids generated via the following steps: (1) deprotonation of amino acids by $O_2^{\cdot -}$ (Eq. 2.22; here amino acid is described as $RCH(NH_2)COOH$), (2) one hydrogen loss from alpha-carbon by HO$^{\cdot}$ (Eq. 2.23), and (3) one more hydrogen loss from amino group by O_2 to form deprotonated imino acids $RC(NH)COO^-$ (Eq. 2.24).

Possible processes of Oxidation (ii):

$$RCH(NH_2)COOH + O_2^{\cdot -} \rightarrow RCH(NH_2)COO^- + HO_2^{\cdot} \tag{2.22}$$

$$RCH(NH_2)(COO^-) + HO^{\cdot} \rightarrow RC^{\cdot}(NH_2)COO^- + H_2O \tag{2.23}$$

$$RC^{\cdot}(NH_2)(COO^-) + O_2 \rightarrow RC(NH)COO^- + HO_2^{\cdot} \tag{2.24}$$

Background ion attachment: Ionization due to background ion attachment is frequently observed in DART when using mass spectrometers whose ion optics allow forming clusters (Fig. 2.6B and D). Background ions attached ($R^{\pm}$) are, for example, ammonium ion NH_4^+, superoxide $O_2^{\cdot-}$, nitrogen dioxide anion NO_2^-, deprotonated formic acid $HCOO^-$, and deprotonated lactic acid $C_3H_5O_3^-$. These ions can be generated from major air constituents and ubiquitous impurities in the sample and/or laboratory atmosphere. The adduct ions $[M_{gas} + R]^{\pm}$ are formed via three-body reactions with a third body P such as N_2 and O_2:

$$M_{gas} + R^{\pm} + P \rightarrow [M_{gas} + R]^{\pm} + P \quad (P: N_2, O_2) \tag{2.25}$$

The $O_2^{\cdot-}$ attachment is useful for soft ionization of large polarizable *n*-alkanes (approximately C18 and larger), alcohols, and other nonpolar compounds (Fig. 2.8) [58].

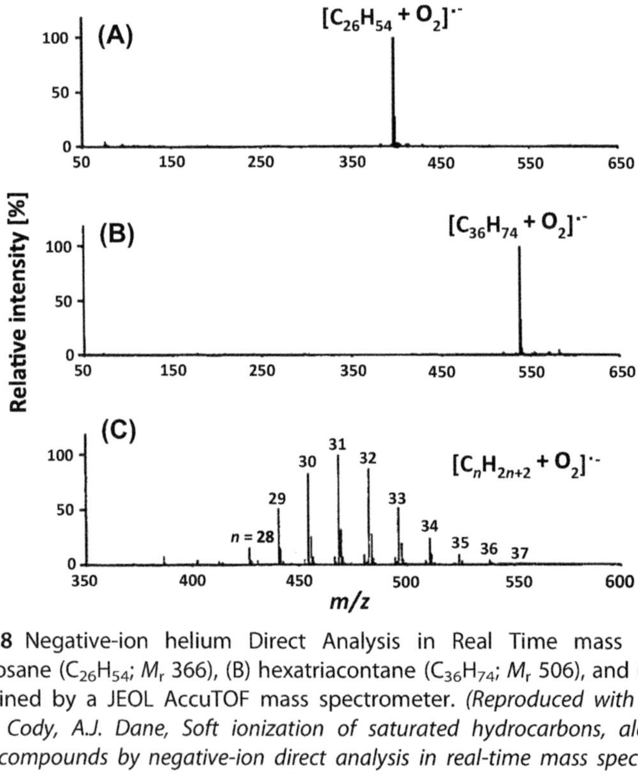

Figure 2.8 Negative-ion helium Direct Analysis in Real Time mass spectra of (A) hexacosane ($C_{26}H_{54}$; M_r 366), (B) hexatriacontane ($C_{36}H_{74}$; M_r 506), and (C) paraffin film, obtained by a JEOL AccuTOF mass spectrometer. *(Reproduced with permission from R.B. Cody, A.J. Dane, Soft ionization of saturated hydrocarbons, alcohols and nonpolar compounds by negative-ion direct analysis in real-time mass spectrometry, J. Am. Soc. Mass Spectrom. 24 (3) (2013) 329–334, https://doi.org/10.1007/s13361-012-0569-6. Copyright © 2013 American Society for Mass Spectrometry.)*

2.3.3 Ionization characteristics

The ionization processes and resulting product ions described above are summarized in Fig. 2.9. The analyte ionization area is at atmospheric pressure, and the electric field applied in that area is negligibly small. These conditions lead to a significantly long reaction time in which a number of ion–molecule reactions occur and ions can alter more stable ion species. This means that analytes with specific physicochemical properties (e.g., higher proton affinity and electron affinity) can be ionized preferentially. Also, a long reaction time results in thermalization of analyte ions. Thus, DART can make "soft" and highly sensitive analyte measurements feasible, if analytes of interest have specific properties suitable for the ionization processes shown in Fig. 2.9.

DART is capable of semiquantitative and quantitative analysis if internal standards are used. For example, calibration curves spanning 0–1500, 0–300, and 0–30 ng/mL for sarin in dichloromethane with correlation coefficients ranging from 0.992 to 0.999 have been reported in a study on chemical warfare agents [7]. Another example is quantification of six sugars in a solution used for the enzymatic hydrolysis of switchgrass (*Panicum virgatum*) with [D_2]glucose as the internal standard [54]. Here, linear calibration curves were reported in the range 10–3000 μmol/L with correlation coefficients better than 0.997 and recoveries of 95%–103.0%. Limits of detection and quantitation were 5.8×10^{-6} mol/L and 2.0×10^{-5} mol/L, respectively.

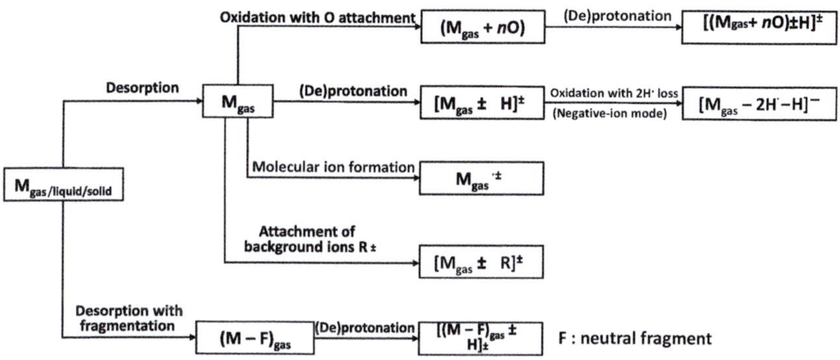

Figure 2.9 Summary of ionization processes occurring in helium Direct Analysis in Real Time. *(Reproduced with permission from K. Sekimoto, M. Sakakura, T. Kawamukai, H. Hlke, T. Shiota, F. Usui, Y. Bando, M. Takayama, Ionization characteristics of amino acids in direct analysis in real time mass spectrometry, Analyst 139 (2014) 2589–2599, https://doi.org/10.1039/c3an02193a. Copyright © 2014 Royal Society of Chemistry.)*

2.4 Technical applications for improving DART performance/sensitivity

The broad range of analytical applications using He-DART has necessitated a continual development and refinement of techniques to enhance its performance and sensitivity. In the case of DART, ion intensities detected by mass spectrometer is primarily determined by the following four factors: (1) the efficiency of desorption and (2) the efficiency of ionization of the analyte in the hot and excited helium gas flow, respectively, as well as (3) the efficiency of the resulting ion collection and transmission to the mass spectrometer, all of which are strongly dependent upon (4) the analyte position in the gas flow. Various techniques to improve these factors have been developed, e.g., the combination of an ultrasonic atomizer with nonthermal He-DART to effectively desorb nonvolatile and/or heat-unstable compounds without hot helium gas, and commercial systems to optimize the analyte position (e.g., the 12-Dip-it Holder and Tweezer Module, IonSense Inc.), to enhance the efficiency of the ion collection and transmission to a mass spectrometer (e.g., the SVA-45A, IonSense Inc.), to selectively and separately desorb analytes of interest by gradually increasing sample temperature (ionRocket, BioChromato Inc.), and to improve the ionization efficiency of He-DART by combining a corona discharge system ($Corona^{++}$, AMR Inc.). In this section, the three techniques for improvement of the desorption and ionization efficiencies are introduced.

2.4.1 ionRocket

In general, molecules have specific boiling points. Consider a situation where there are several analytes of interest in a given bulk material and analytes have different boiling points. If analytes can be desorbed separately, selective ionization of each analyte is achieved. This concept led to the development of ionRocket (BioChromato Inc.) (Fig. 2.10). It is a temperature gradient system for analyte desorption and makes selective ionization possible by controlling desorption rates of analytes. Samples are mounted onto a copper sampling "pot" (Fig. 2.10D), which is placed onto the heating block and positioned in the sample gap (Fig. 2.10C). Typically, the temperature program is 50°C with a given hold time, followed by a ramp rate of 100°C/min to a final temperature of 600°C.

Recently, ionRocket has been successfully used to analyze automotive clear coats [29]. The thermal desorption plots were highly reproducible. By examining the mass spectral information from different temperature regions

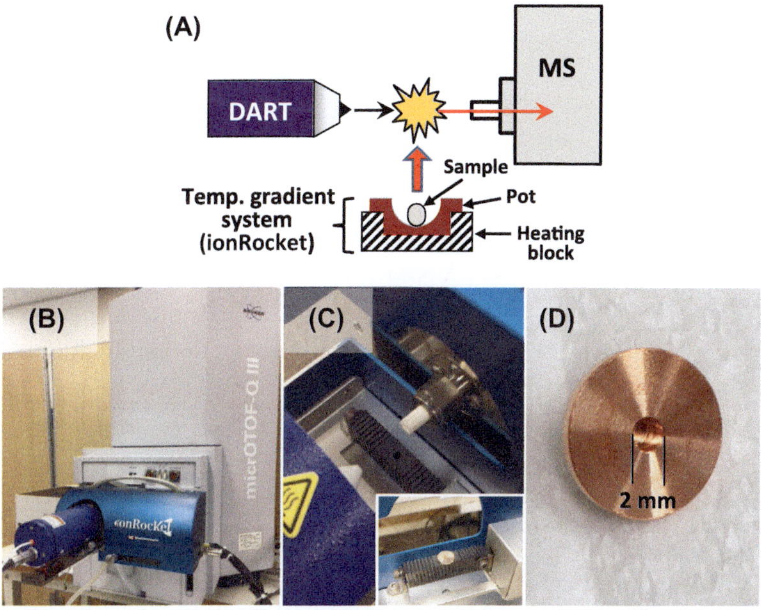

Figure 2.10 (A) Schematic illustration of Direct Analysis in Real Time mass spectrometry with ionRocket. (B)–(D) Photographs of the ionRocket system, heating block inside ionRocket, and copper sampling "pot". *(Photographs by courtesy of BioChromato Inc.)*

of the thermal desorption plot, the authors were able to find latent compounds that was not readily identifiable by traditional methodologies for automotive paint characterization (e.g., pyrolysis-GC) under a specific temperature range (Fig. 2.11).

2.4.2 Nonthermal desorption system using ultrasonic atomization [36].

As mentioned in Section 2.3.2, He-DART leads to pyrolytic fragmentation during desorption by hot helium gas. Thus, it is difficult to detect heat-unstable compounds, as well as nonvolatile compounds. To solve this issue, a nonthermal desorption system has been developed, using an ultrasonic atomizer (Fig. 2.12). The atomizer is placed on the bottom of the water vessel. A bottle containing sample solution is installed on the sample holder. The atomized liquid droplets are transported to the part for mixing through the hole in the tube by air supplied by a compressor. The liquid droplets are mixed with metastable helium gas (but not heated), which is released from the DART source exit. The mixed gas is transported to the mass spectrometer.

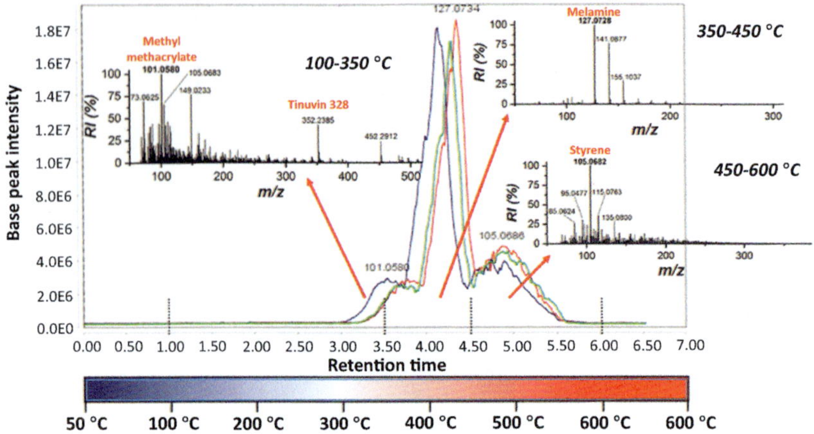

Figure 2.11 Thermal desorption plots for a sample automotive paint obtained by ionRocket and He-Direct Analysis in Real Time combined with a JEOL AccuTOF mass spectrometer. Mass spectral data were acquired across three temperature profiles (i.e., 100–350, 350–450, and 450–600°C). Three lines in the plots show base peak chronograms (methyl methacrylate in blue, melamine in green, and styrene in red). These reflect changes in the molecular base peak of the mass spectrum as a function of time, which in this instance is temperature dependent. A peak at *m/z* 352.2385 observed under 100–350°C can be identified as hydroxyphenylbenzotriazole ultraviolet absorber known as Tinuvin 328. This compound is not observed by pyrolysis GCMS and He-DART only. *(Reprinted with permission from M. Maric, J. Marano, R.B. Cody, C. Bridge, DART-MS: a new analytical technique for forensic paint analysis, Anal. Chem. 90 (2018) 6877–6884, https://doi.org/10.1012/acs.analchem.8b01067. Copyright © 2018 American Chemical Society.)*

The performance of this technique has been evaluated using glycyrrhizic acid, which is an anti-inflammatory agent widely used in pharmaceutical creams and is unstable at high temperature (>200°C). When glycyrrhizic acid was analyzed by conventional He-DART with hot helium gas, no molecular-related ions were detected (Fig. 2.13A). Instead, pyrolytic products such as the aglycone of glycyrrhizic acid were observed (Fig. 2.13A). In contrast, the combination of the ultrasonic atomization with nonthermal He-DART made it possible to ionize it as a deprotonated molecule (Fig. 2.13B). Further advantage of this technique is that samples do not need to be completely dissolved; the cream- and lotion-type samples can be analyzed as a dispersion in water (Fig. 2.13C and D). This technique enables us to obtain the composition of pharmaceutical products simply and rapidly.

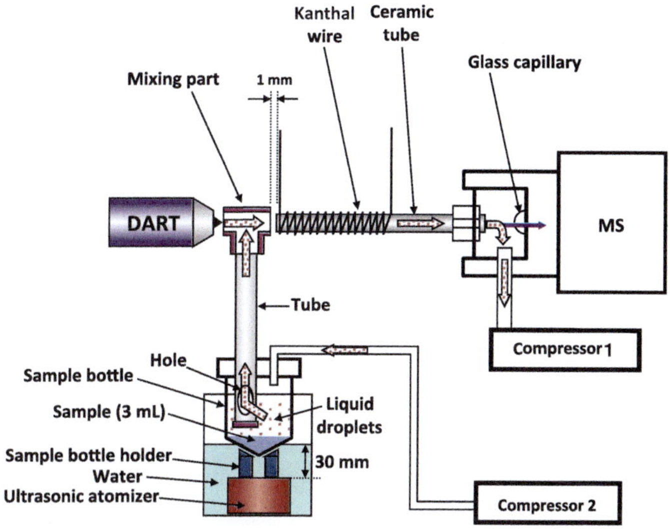

Figure 2.12 Schematic illustration of Direct Analysis in Real Time mass spectrometer combined with nonthermal desorption system using ultrasonic atomization. *(Reproduced with permission from H. Shimada, K. Maeno, K. Kinoshita, Y. Shida, Rapid analysis of ingredients in cream using ultrasonic mist-direct analysis in real-time time-of-flight mass spectrometry, J. Am. Soc. Mass Spectrom. 28 (2017) 2393–2400, https://doi.org/10.1007/s13361-017-1746-4. Copyright © 2017 American Society for Mass Spectrometry.)*

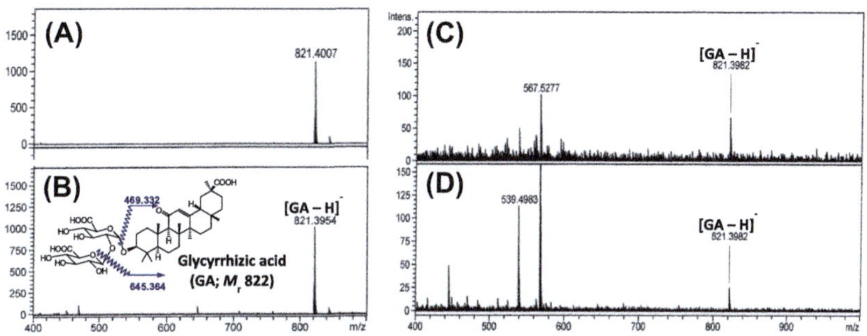

Figure 2.13 Effect of the nonthermal desorption system using ultrasonic atomization. (A) Negative ion mass spectrum of glycyrrhizic acid (500 μg/mL) obtained by conventional He-Direct Analysis in Real Time (helium gas temperature = 450°C). (B)–(D) Negative ion mass spectra of glycyrrhizic acid (500 μg/mL), cream sample (2.5 mg/mL in water), and lotion sample (2.5 mg/mL in water) obtained by nonthermal He-DART (50°C) combined with the ultrasonic atomization system. The cream and lotion samples contained glycyrrhizic acid of 0.3% (in weight) and other compounds of 99.7% (a large part of which were glycerol and water). All of the mass spectra were acquired with a Brucker MicrOTOF-Q II mass spectrometer. *(Reproduced with permission from H. Shimada, K. Maeno, K. Kinoshita, Y. Shida, Rapid analysis of ingredients in cream using ultrasonic mist-direct analysis in real-time time-of-flight mass spectrometry, J. Am. Soc. Mass Spectrom. 28 (2017) 2393–2400, https://doi.org/10.1007/s13361-017-1746-4. Copyright © 2017 American Society for Mass Spectrometry.)*

2.4.3 Corona^{++} [63]

Corona^{++} (AMR Inc.; Fig. 2.14A) is a corona discharge device to improve ionization efficiency of He-DART. This discharge system is very simple: (1) a corona needle is placed in the sampling area and (2) DC voltage ($\approx \pm 3$ kV) is applied to this needle using the electrospray voltage source of the mass spectrometer. The He-DART ion source is operated as it is.

When positive and negative corona discharges are combined with conventional He-DART system (denoted as Corona^{++}-DART), a visible glow occurs along the helium gas flow between the DART source exit and the corona needle (Fig. 2.14B). A single corona is also localized at the needle tip. According to spectroscopic analysis, this glow includes helium atoms in various excited states (He*), e.g., the high Rydberg states 3^1D

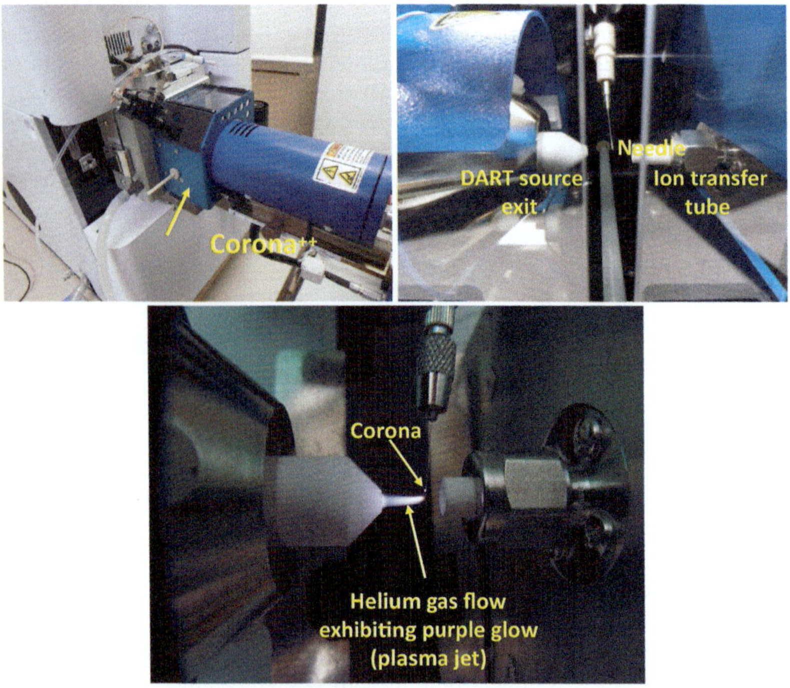

Figure 2.14 (A) External and (B) internal appearance of Corona^{++} system. Photographs by courtesy of AMR Inc. (C) Photograph of a helium plasma jet occurring between the DART source exit and the corona needle. *(Reproduced with permission from K. Sekimoto, M. Sakakura, T. Kawamukai, H. Hike, T. Shiota, F. Usui, Y. Bando, M. Takayama, Improvement in ionization efficiency of direct analysis in real time-mass spectrometry (DART-MS) by corona discharge, Analyst 141 (2016) 4879–4892, https://doi.org/10.1039/c6an00779a. Copyright © 2016 Royal Society of Chemistry.)*

(internal energy $Q = 23.1$ eV), 3^3D ($Q = 23.1$ eV), and 3^3S ($Q = 22.7$ eV), a resonance state 2^1P ($Q = 21.2$ eV), an optically allowed state 2^3P ($Q = 21.0$ eV), and metastable states 2^1S ($Q = 20.62$ eV) and 2^3S ($Q = 19.8$ eV). In contrast, the temperature of the helium gas within the glow is almost identical (+1.5°C) to that without a glow due only to DART. This means that the glow corresponds to an atmospheric pressure nonequilibrium plasma jet. Formation mechanism of excited helium in the plasma jet is described at the end of this section.

Corona^{++}-DART ionization, in which analytes are inserted into the plasma jet, results in quite a high (de)protonation efficiency compared to conventional He-DART ionization (Fig. 2.15A). This phenomenon can be interpreted as follows: as already mentioned, (de)protonation occurring in

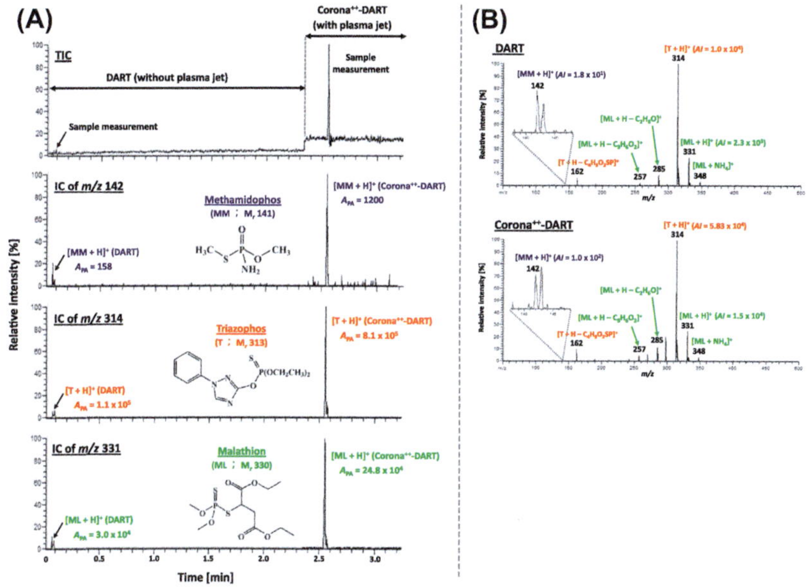

Figure 2.15 Comparison in analyte ionization characteristics between Corona^{++}-Direct Analysis in Real Time (DART) and DART, by measuring a mixture including methamidophos, triazophos, and malathion at each concentration of 100 ppm in water. (A) Total ion chromatogram (TIC) and ion chromatogram (IC) of protonated analytes. A_{PA} represents the peak area value [arbitrary unit] of a given ion. (B) DART and Corona^{++}-DART mass spectra. *AI* represents the absolute intensity [arbitrary unit] of a given ion. These data were obtained with a Thermo Fisher Scientific LCQ ion-trap mass spectrometer. *(Reprinted with permission from K. Sekimoto, Ionization characteristics of direct analysis in real time (DART) and improvement in its ionization efficiency by corona discharge, J. Mass Spectrom. Soc. Jpn. 65 (2017) 102–106. Copyright © 2017 Mass Spectrometry Society of Japan.)*

He-DART is attributable to proton transfer reactions with the reagent ions H_3O^+ and $O_2^{\cdot-}$ (Eqs. 2.10 and 2.11), and/or resonant capture of slow electrons e^-_{slow} (Eq. 2.12). These reagents (i.e., H_3O^+, $O_2^{\cdot-}$, and e^-_{slow}) originate from Penning ionization of $N_2/O_2/H_2O$ with He* (Eq. 2.2). In Corona^{++}-DART with a plasma jet, the analyte ionization area contains abundant He* as well as N_2, O_2, and H_2O. These conditions promote Penning ionization to form the reagents and subsequent (de)protonation of analytes.

Notably, the mass spectral pattern obtained in Corona^{++}-DART is quite similar to that observed in He-DART only. The type and number of analyte-related ions detected in Corona^{++}-DART mass spectra are essentially identical to those in He-DART mass spectra (Fig. 2.15B). These results suggest that the variety of ionization reactions occurring in corona-DART is almost the same as that in He-DART (i.e., (de)protonation, molecular ion formation, fragmentation, oxidation, and background ion attachment as described in Section 2.2). It is concluded, therefore, that the combination of He-DART with corona discharge can easily improve the ionization efficiency of He-DART without side reactions and any effect of He* in the plasma jet on analyte fragmentation processes.

Corona^{++}-DART is very useful for the highly sensitive and simple analysis of *n*-alkanes in positive-ion mode. When *n*-alkanes have been supplied into He-DART, the resulting analyte ions have varied significantly depending on the type of *n*-alkanes and mass spectrometer used (see Fig. 2.16A). In case of Corona^{++}-DART, *n*-alkanes (Alk) can be ionized as $[\text{Alk} + \text{O} - 3\text{H}]^+$ (m/z Alk + 13) and/or $[\text{Alk} + 2\text{O} - \text{H}]^+$ (m/z Alk + 31), independent of the alkane or mass spectrometer used (Fig. 2.16B). Furthermore, the absolute intensities of the alkane-related ions obtained in Corona^{++}-DART are 5−1000 times greater than those observed in He-DART. These results demonstrate that Corona^{++}-DART can be applied to the analyses of petroleum and diesel/gasoline fuels with high sensitivity.

Formation mechanism of excited helium in plasma jet: Ignition of the plasma jet, i.e., the origin of the He* formation, is attributed to electrons which are accelerated on the corona needle tip surface and possess kinetic energies above 19.8 eV, corresponding to the lowest energy of a metastable helium atoms ($He(2^3S)$). Those electrons collide with helium atoms flowing in the vicinity of the needle tip surface through the DART source exit, a large portion of which can be found in the ground state (1^1S), resulting in the excitation and ionization of $He(1^1S)$ (Eq. 2.26). The He*

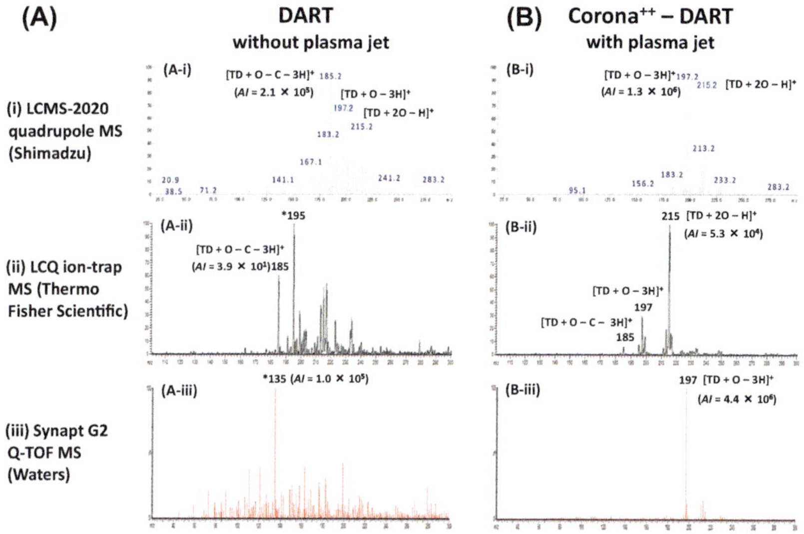

Figure 2.16 Positive-ion mass spectra of *n*-tridecane ($C_{13}H_{28}$; M_r 184) obtained with (A) Direct Analysis in Real Time (DART) and (B) Corona^{++}-DART coupled with three different mass spectrometers (i)–(iii). *AI* represents the absolute intensity [arbitrary unit] of a given ion. *(Reprinted with permission from K. Sekimoto, Ionization characteristics of direct analysis in real time (DART) and improvement in its ionization efficiency by corona discharge, J. Mass Spectrom. Soc. Jpn. 65 (2017) 102–106. Copyright © 2017 Mass Spectrometry Society of Japan.)*

atoms formed here include species in various different energy levels such as the high Rydberg state, the resonance state, the optically allowed states, and the metastable states.

$$He(1^1S) + e^-(\geqq 19.8\ eV) \rightarrow He^*/He^+ + e^-_{slow} \tag{2.26}$$

The He^+ ion can alter the He^* atoms via dimerization reactions to form He_2^+, subsequent generation of superexcited state due to recombination with an electron, and finally, radiative de-excitation.

The high Rydberg atoms with 3^1D, 3^3D, and 3^3S states are de-excited to lower energy levels such as the resonance (2^1P) and optically allowed states (2^3P) with visible radiations of 667.8, 587.6, and 706.5 nm. The transition from the resonance-state atom $He(2^1P)$ to the ground 1^1S state by emission of UV radiation (58.43 nm, Eq. 2.27) is allowed, whereas at atmospheric pressure, this radiation is efficiency absorbed by other $He(1^1S)$ to regenerate $He(2^1P)$ (Eq. 2.28).

$$\mathrm{He}(2^1P) \rightarrow \mathrm{He}(1^1S) + h\nu(58.43\ \mathrm{nm}) \tag{2.27}$$

$$\mathrm{He}(1^1S) + h\nu(58.43\ \mathrm{nm}) \rightarrow \mathrm{He}(2^1P) \tag{2.28}$$

The energy corresponding to 58.43 nm (21.2 eV) is possessed by a helium atom for quite a long time (as compared to that in which this energy behaves as UV radiation), indicating that $\mathrm{He}(2^1P)$ has a long effective lifetime in ambient air [83]. In contrast, the optically allowed state $\mathrm{He}(2^3P)$ becomes the metastable $\mathrm{He}(2^3S)$ via de-excitation with radiation of 1008.3 nm:

$$\mathrm{He}(2^3P) \rightarrow \mathrm{He}(2^3S) + h\nu(1008.3\ \mathrm{nm}) \tag{2.29}$$

The natural lifetimes of metastable atoms are significantly long (on the order of ms to min) because the transition from the metastable state to the ground state is optically inhibited. Taking into account the lifetimes of individual excited states, the resonance state atom $\mathrm{He}(2^1P)$ and the metastable atoms $\mathrm{He}(2^3S)$ and $\mathrm{He}(2^1S)$ can exist dominantly in the plasma jet occurring in Corona^{++}-DART.

2.5 Applications using argon gas: atmospheric pressure dark current argon discharge ionization with comparable performance of helium DART

Helium DART has been performed with a great amount of success as shown in Table 2.1. However, the use of helium is limited by its in situ and analytical utility in continuous analysis, as well as experimental cost. As mentioned earlier, an additional vacuum pumping system such as Vapur Interface is required for general mass spectrometers to eliminate helium gas with a low atomic mass of 4. This additional component results in an increase in the cost of mass spectrometers and prevents the miniaturization of instruments that is required for in situ analysis. Furthermore, helium gas is expensive and difficult to obtain, which leads to an increase in experimental cost and makes it hard to sustain its use.

Argon is a possible alternative gas that can be used to overcome the issue described above. It has an atomic mass of 40 and can be obtained much more easily at a lower cost than helium gas. However, argon has not been widely used in DART because of the lower internal energies of excited state argon. Excited argon stably exists in discharges (including DART glow discharge) in metastable states, such as the $4S\ ^3P_2$ and $4S\ ^3P_0$ states, with internal energies of 11.6 and 11.7 eV, respectively. These energies are lower than the IE of H_2O, which results in the formation of fewer H_3O^+ ions and give rise to quite low analyte ionization efficient in argon DART (denoted as Ar-DART) [57,65]. Thus, dopant-assisted protonation based on atmospheric pressure photoionization has been used for the effective operation of Ar-DART [59,65].

Recently, a novel atmospheric pressure argon discharge ionization technique has been developed in which the analyte ionization efficiency and mechanism are comparable to those of conventional He-DART [66]. The present discharge system is easily established by modifying the conventional DART source: (1) a needle, which has a tip end radius with a curvature of ca. 1 μm and includes a tip end formed into a hyperboloid of revolution (Fig. 2.17B), is placed in the sampling area, (2) heated ($\approx$ 500°C) but *ground-state* argon is flowed through the sampling area, and

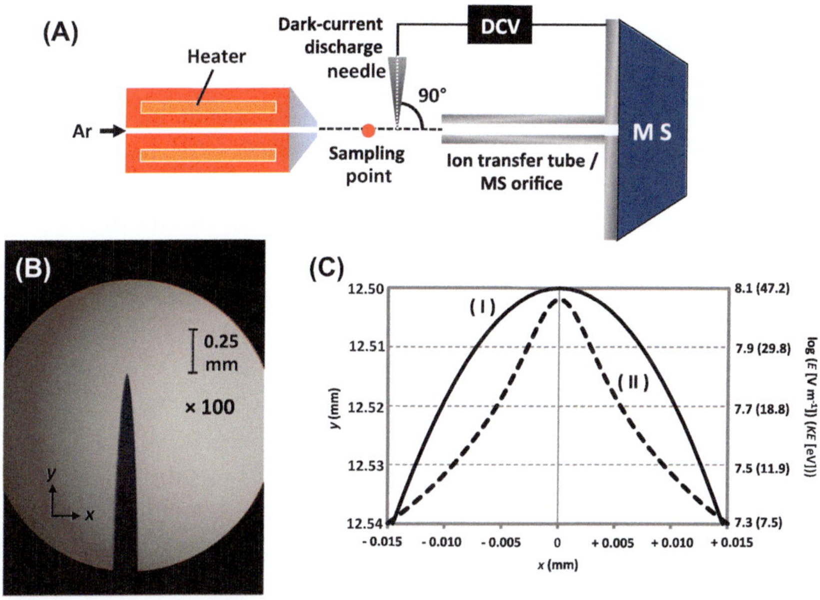

Figure 2.17 (A) Schematic illustration of the dark current argon discharge ionization source. (B) Optical micrograph of the dark current discharge needle. (C) I: The hyperbola approximating the contour of the cross-section needle tip used here. II: Logarithm of the electric field strength on the needle tip surface (log *E*) and the corresponding electron kinetic energy (*KE*) for a typical dark current discharge condition (discharge gap of 12.5 mm and dark current discharge voltage of 1.5 kV) as a function of the *x* coordinate of the tip surface. A detailed description of how to calculate *E* and *KE* can be found in Ref. [85].

(3) low DC voltage (<±2 kV) is applied to this needle using the electrospray voltage source of the mass spectrometer (Fig. 2.17A). Notably, the use of dopants and a DART glow discharge is not required. The resulting discharge state in the sampling area is referred to as a "dark current," a very low electric current (0.2–1 μA) compared to the DART glow discharge (≈ 3 mA). Excited state argon formed makes it possible to efficiently generate H_3O^+ and $O_2^{\cdot-}$ reagent ions. These reagent ions lead to the formation of (de)protonated analytes, the abundances of which are equal to those generated in He-DART. A detailed description of the dark current argon discharge ionization is present below.

2.5.1 Ionization characteristics

No ions are observed in background mass spectra when performing only Ar-DART. However, when the dark current discharge needle is used, a fairly low electric current of 0.2–1 μA is recorded, and background ions are readily observed regardless of whether the DART glow discharge is turned on or not. Notably, the resulting total ion intensities are very similar to those observed in the conventional He-DART. The turn-on needle voltage of 0.2–1 μA results in *no visible* light on the needle tip or between the DART source exit and the mass spectrometer orifice (or the ion transfer tube of Vapur Interface). This nonself-sustaining discharge state can be described as "dark current (or Townsend dark)". Dark current discharge is energetically much lower than self-sustaining discharge with visible light and electric current higher than 1 μA, e.g., glow and corona discharges.

The dark current argon discharge can lead to analyte ionization efficiencies and characteristics comparable to those of conventional He-DART (see Figs. 2.18 and 2.6). Typical polar compounds desorbed by heated argon (M_{gas}) are dominantly ionized as (de)protonated molecules $[M_{gas} \pm H]^{\pm}$. The other ion species formed are molecular ions $M_{gas}^{\cdot\pm}$, fragment ions $[M_{gas} \pm H - \text{Fragment}]^{\pm}$, oxygenated (de)protonated molecules $[(M_{gas} + nO) \pm H]^{\pm}$, dehydrogenated deprotonated molecules $[M_{gas} - 2H^{\cdot} - H]^{-}$, and background ion adducts $[M_{gas} + R]^{\pm}$. Nonpolar compounds such as *n*-alkanes (Alk) are detected as $[Alk + O - 3H]^+$ and $[Alk + 2O - H]^+$.

An interesting ionization characteristic of the dark current argon discharge is the high efficiency of (de)protonation. Furthermore, the background mass spectra (Fig. 2.19) indicate that the present (de)protonation are attributable to hydronium ion H_3O^+ and superoxide $O_2^{\cdot-}$ in a similar manner as He-DART. Taking into account the IEs of air

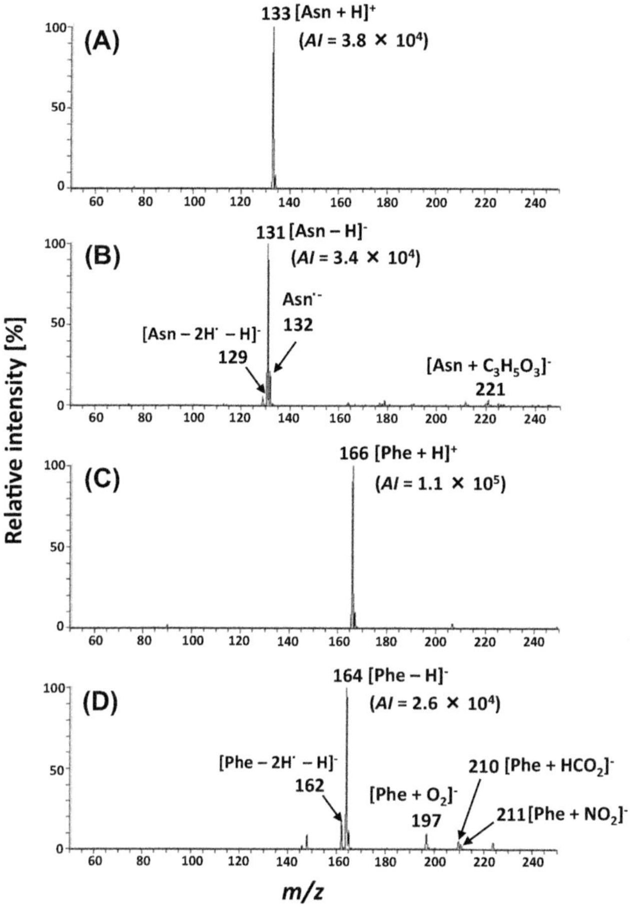

Figure 2.18 Dark current argon discharge ionization mass spectra of (A) L-asparagine in positive-ion mode, (B) L-asparagine in negative-ion mode, (C) L-phenylalanine in positive-ion mode, and (D) L-phenylalanine in negative-ion mode, obtained with a Thermo Fisher Scientific LCQ ion-trap instrument.

constituents, it is most likely that the dark current argon discharge efficiently generates excited state argon with an internal energy higher than those of metastable states, e.g., the resonance state $5S^3P_1$ at an internal energy of 14.1 eV and $5S^1P_1$ at 14.3 eV. It is known that these resonance states of argon form $O_2^{\cdot +}$ and $H_2O^{\cdot +}$ via the Penning ionization of O_2 and H_2O [84]. Therefore, the formation pathways of the reagent ions H_3O^+ and $O_2^{\cdot -}$ and resulting analyte ionization mechanism are comparable to He-DART (see Sections 2.1 and 2.2).

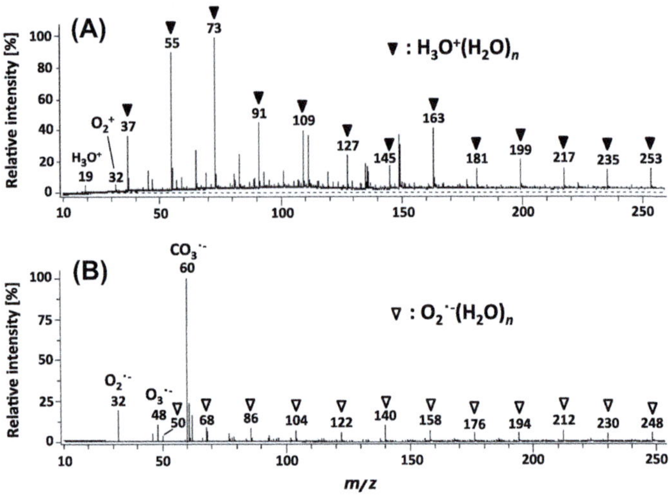

Figure 2.19 (A) Positive and (B) negative background ion mass spectra obtained by dark current argon discharge combined with a JEOL AccuTOF mass spectrometer.

2.5.2 Excitation of argon in atmospheric pressure dark current discharge

In the case of argon, it is known that there are a number of excited states up to the first ionization potential (15.8 eV). Of these, the metastable states at $\approx$ 11.6 eV (e.g., $4S^3P_2$ and $4S^3P_0$) and the resonance states at $\approx$ 14.2 eV (e.g., $5S^3P_1$ and $5S^1P_1$) can occur via Penning ionization. As mentioned above, the present dark current argon discharge efficiently generates H_3O^+. This indicates that resonance-state argon is dominantly formed in the present discharge. The resonance state is allowed to transition to a ground state via the emissions of UV radiation, however, under atmospheric pressure, this radiation is efficiently absorbed by other ground-state argon species to regenerate the resonance state. This is similar to the case of helium (Eqs. 2.27 and 2.28).

The formation of resonance-state argon can be attributed to electrons that are accelerated on the dark current discharge needle tip surface and possess a kinetic energy of over $\approx$ 14.2 eV. The kinetic energy (KE) of an electron determined by the product of the electric field strength (E) and the mean free path of electron ($\lambda_e = 375$ nm at 760 torr), i.e., KE [eV] $= E$ [V/m] $\times \lambda_e$ [m]. Furthermore, the field strength depends on the local curvature of each needle tip position, in addition to the applied voltage (Fig. 2.17C)

[85]. As electrons accelerated on the needle tip apex have kinetic energies of higher than 16 eV, these can form high Rydberg state argon with an internal energy of ≈ 15.6 eV, as well as argon ion (Ar^{+}). The high Rydberg argon easily converts to dimer ions (Ar_2^{+}) upon the loss of electrons. However, neither Ar^{+} nor Ar_2^{+} is detected in the positive background ion mass spectrum (Fig. 2.19A). This indicates that the present discharge conditions result in the formation of fewer high Rydberg state atoms and argon-related ions.

The needle tip shape plays a crucial role in enhancing and sustaining the dark current argon discharge ionization efficiency. When the dark current discharge is created by the present needle (Fig. 2.17B), the ion intensities remain constant even after a time lapse of 30 min. However, if a needle with a larger radius of curvature (>30 μm) is used, ions are not detected in the dark current discharge state. If the tip end radius of curvature is too large, and hence in the dark current discharge state, it is impossible to ensure a region of the tip end position at which sufficient amounts of high-energy electrons and resonance-state argon are formed. In contrast, if a needle, including a tip end formed into a reversed curved surface is used, the ion intensities significantly reduce right after the application of the dark current discharge voltage. This is because the tip end radius of the curvature is too small, and hence the shape of the tip end surface changes with the passage of time, meaning that the electric field established on the tip end surface cannot be kept constant. Therefore, the dark current argon discharge needle should have the following conditions: (1) the tip end has a radius of curvature that results in the generation of a sufficient number of high-energy electrons, a sufficient amount of resonance-state argon, and reagent ions H_3O^{+} and $O_2^{\cdot -}$ and (2) the tip shape must be kept constant during the application of voltage.

References

[1] R.B. Cody, J.A. Laramée, H.D. Durst, Versatile new ion source for the analysis of materials in open air under ambient conditions, Anal. Chem. 77 (2005) 2297–2302, https://doi.org/10.1021/ac050162j.

[2] Z. Takats, J.M. Wiseman, B. Gologan, R.G. Cooks, Mass spectrometry sampling under ambient conditions with desorption electrospray ionization, Science 306 (2004) 471–473, https://doi.org/10.1126/science.1104404.

[3] S.E. Edison, L.A. Lin, L. Parrales, Practical considerations for the rapid screening for pesticides using ambient pressure desorption ionisation with high-resolution mass spectrometry, Food Addit. Contam. Part A Chem. Anal. Control Expo Rist Assess. 28 (2011) 1393–1404, https://doi.org/10.1080/19440049.2011.596165.

[4] E. Crawford, B. Musselman, Evaluating a direct swabbing method for screening pesticides on fruit and vegetable surfaces using Direct Analysis in Real Time (DART) coupled to an Exactive benchtop orbitrap mass spectrometer, Anal. Bioanal. Chem. 403 (2012) 2807–2812, https://doi.org/10.1007/s00216-012-5853-6.
[5] L. Song, A.B. Dykstra, H. Yao, J.E. Bartmess, Ionization mechanism of negative ion-direct analysis in real time: a comparative study with negative ion-atmospheric pressure photoionization, J. Am. Soc. Mass Spectrom. 20 (2009) 42–50, https://doi.org/10.1016/j.jasms.2008.09.016.
[6] F. Rowell, J. Seviour, A.Y. Lim, C.G. Elumbaring-Salazar, J. Loke, J. Ma, Detection of nitro-organic and peroxide explosives in latent fingermarks by DART- and SALDI-TOF-mass spectrometry, Forensic Sci. Int. 221 (2012) 84–91, https://doi.org/10.1016/j.forsciint.2012.04.007.
[7] J.M. Nilles, T.R. Connell, H.D. Durst, Quantification of chemical warfare agents using the direct analysis in real time (DART) technique, Anal. Chem. 81 (2009) 6744–6749, https://doi.org/10.1021/ac900682f.
[8] R.R. Steiner, R.L. Larson, Validation of the direct analysis in real time source for use in forensic drug screening, J. Forensic Sci. 54 (2009) 617–622, https://doi.org/10.1111/j.1556-4029.2009.01006.x.
[9] A.H. Grange, G.W. Sovocool, Detection of illicit drugs on surfaces using direct analysis in real time (DART) time-of-flight mass spectrometry, Rapid Commun. Mass Spectrom. 25 (2011) 1271–1281, https://doi.org/10.1002/rcm.5009.
[10] R.A. Musah, M.A. Domin, R.B. Cody, A.D. Lesiak, A.J. Dane, J.R.E. Shepard, Direct analysis in real time mass spectrometry with collision-induced dissociation for structural analysis of synthetic cannabinoids, Rapid Commun. Mass Spectrom. 26 (2012) 2335–2342, https://doi.org/10.1002/rcm.6354.
[11] R.A. Musah, M.A. Domin, M.A. Walling, J.R.E. Shepard, Rapid identification of synthetic cannabinoids in herbal samples via direct analysis in real time mass spectrometry, Rapid Commun. Mass Spectrom. 26 (2012) 1109–1114, https://doi.org/10.1002/rcm.6205.
[12] A.D. Lesiak, R.A. Musah, M.A. Domin, J.R.E. Shepard, DART-MS as a preliminary screening method for "herbal incense": chemical analysis of synthetic cannabinoids, J. Forensic Sci. 59 (2014) 337–343, https://doi.org/10.1111/1556-4029.12354.
[13] S.J.B. Dunham, P.D. Hooker, R.M. Hyde, Identification, extraction and quantification of the synthetic cannabinoid JWH-018 from commercially available herbal marijuana alternatives, Forensic Sci. Int. 223 (2012) 241–244, https://doi.org/10.1016/j.forsciint.2012.09.010.
[14] A.D. Lesiak, R.A. Musah, R.B. Cody, M.A. Domin, A.J. Dane, J.R.E. Shepard, Direct analysis in real time mass spectrometry (DART-MS) of "bath salt" cathinone drug mixtures, Analyst 138 (2013) 3424–3432, https://doi.org/10.1039/C3AN00360D.
[15] A. Rodriguez-Lafuente, F.S. Mirnaghi, J. Pawliszyn, Determination of cocaine and methadone in urine samples by thin-film solid-phase microextraction and direct analysis in real time (DART) coupled with tandem mass spectrometry, Anal. Bioanal. Chem. 405 (2013) 9723–9727, https://doi.org/10.1007/s00216-013-6993-z.
[16] W.F. Duvivier, T.A. van Beek, E.J. Pennings, M.W. Nielen, Rapid analysis of Δ-9-tetrahydrocannabinol in hair using direct analysis in real time ambient ionization orbitrap mass spectrometry, Rapid Commun. Mass Spectrom. 28 (2014) 682–690, https://doi.org/10.1002/rcm.6831.
[17] C.Y. Pierce, J.R. Barr, R.B. Cody, R.F. Massung, A.R. Woolfitt, H. Moura, H.A. Thompson, F.M. Fernández, Ambient generation of fatty acid methyl ester ions from bacterial whole cells by direct analysis in real time (DART) mass spectrometry, Chem. Commun. 28 (2007) 807–809, https://doi.org/10.1039/b613200f.

[18] Y. Zhao, M. Lam, D. Wu, R. Mak, Quantification of small molecules in plasma with direct analysis in real time tandem mass spectrometry, without sample preparation and liquid chromatographic separation, Rapid Commun. Mass Spectrom. 22 (2008) 3217–3224, https://doi.org/10.1002/rcm.3726.
[19] E. Jagerdeo, M. Abdel-Rehim, Screening of cocaine and its metabolites in human urine samples by direct analysis in real-time source coupled to time-of-flight mass spectrometry after online preconcentration utilizing microextraction by packed sorbent, J. Am. Soc. Mass Spectrom. 20 (5) (2009) 891–899, https://doi.org/10.1016/j.jasms.2009.01.010.
[20] J.Y. Yew, R.B. Cody, E.A. Kravitz, Cuticular hydrocarbon analysis of an awake behaving fly using direct analysis in real-time time-of-flight mass spectrometry, Proc. Natl. Acad. Sci. U. S. A 105 (20) (2008) 7135–7140, https://doi.org/10.1073/pnas.0802692105.
[21] T. Cajka, K. Riddellova, M. Tomaniova, J. Hajslova, Ambient mass spectrometry employing a DART ion source for metabolomic fingerprinting/profiling: a powerful tool for beer origin recognition, Metabolomics 7 (2011) 500–508, https://doi.org/10.1007/s11306-010-0266-z.
[22] C.M. Jones, F.M. Fernandez, Transmission mode direct analysis in real time mass spectrometry for fast untargeted metabolic fingerprinting, Rapid Commun. Mass Spectrom. 27 (2013) 1311–1318, https://doi.org/10.1002/rcm.6566.
[23] T. Nah, M. Chan, S.R. Leone, K.R. Wilson, Real time in situ chemical characterization of submicrometer organic particles using direct analysis in real time-mass spectrometry, Anal. Chem. 85 (2013) 2087–2095, https://doi.org/10.1021/ac302560c.
[24] M.N. Chan, T. Nah, K.R. Wilson, Real time in situ chemical characterization of submicron organic aerosols using direct analysis in real time mass spectrometry (DART-MS): the effect of aerosol size and volatility, Analyst 138 (2013) 3749–3757, https://doi.org/10.1039/C3AN00168G.
[25] O.P. Haefliger, N. Jeckelmann, Direct mass spectrometric analysis of flavors and fragrances in real applications using DART, Rapid Commun. Mass Spectrom. 21 (2007) 1361–1366, https://doi.org/10.1002/rcm.2969.
[26] T. Sagawa, Flavor release measurement of spices and herbs, J. Mass Spectrom. Soc. Jpn. 66 (2018) 75–81.
[27] R.W. Jones, R.B. Cody, J.F. McClelland, Differentiating writing inks using direct analysis in real time mass spectrometry, J. Forensic Sci. 51 (2006) 915–918, https://doi.org/10.1111/j.1556-4029.2006.00162.x.
[28] S. Houlgrave, G.M. LaPorte, J.C. Stephens, J.L. Wilson, The classification of inkjet inks using AcuTOF DART (direct analysis in real time) mass spectrometry – a preliminary study, J. Forensic Sci. 58 (2013) 813–821, https://doi.org/10.1111/1556-4029.12048.
[29] M. Maric, J. Marano, R.B. Cody, C. Bridge, DART-MS: a new analytical technique for forensic paint analysis, Anal. Chem. 90 (2018) 6877–6884, https://doi.org/10.1012/acs.analchem.8b01067.
[30] A.T. Navare, J.G. Mayoral, M. Nouzova, F.G. Noriega, F.M. Fernández, Rapid direct analysis in real time (DART) mass spectrometric detection of juvenile hormone III and its terpene precursors, Anal. Bioanal. Chem. 398 (2010) 3005–3013, https://doi.org/10.1007/s00216-010-4269-4.
[31] M. Haunschmidt, W. Buchberger, C.W. Klampfl, R. Hertsens, Identification and semi-quantitative analysis of parabens and UV filters in cosmetic products by direct-analysis-in-real-time mass spectrometry and gas chromatography with mass spectrometric detection, Anal. Methods 3 (1) (2011) 99–104, https://doi.org/10.1039/C0AY00588F.

[32] M. Haunschmidt, C.W. Klampfl, W. Buchberger, R. Hertsens, Determination of organic UV filters in water by stir bar sorptive extraction and direct analysis in real-time mass spectrometry, Anal. Bioanal. Chem. 397 (2010) 269–275, https://doi.org/10.1007/s00216-009-3438-9.

[33] C. Wang, H. Zhu, Z. Cai, F. Song, Z. Liu, S. Liu, Newborn screening of phenylketonuria using direct analysis in real time (DART) mass spectrometry, Anal. Bioanal. Chem. 405 (2013) 3159–3164, https://doi.org/10.1007/s00216-013-6713-8.

[34] S. Zeng, T. Chen, L. Wang, H. Qu, Monitoring batch-to-batch reproducibility using direct analysis in real time mass spectrometry and multivariate analysis: a case study on precipitation, J. Pharm. Biomed. Anal. 76 (2013) 87–95, https://doi.org/10.1016/j.jpba.2012.12.01.

[35] R.L. Self, Direct analysis in real time-mass spectrometry (DART-MS) for rapid qualitative screening of toxic glycols in glycerin-containing products, J. Pharm. Biomed. Anal. 80 (2013) 155–158, https://doi.org/10.1016/j/jpba.2013.02.037.

[36] H. Shimada, K. Maeno, K. Kinoshita, Y. Shida, Rapid analysis of ingredients in cream using ultrasonic mist-direct analysis in real-time time-of-flight mass spectrometry, J. Am. Soc. Mass Spectrom. 28 (2017) 2393–2400, https://doi.org/10.1007/s13361-017-1746-4.

[37] M. Haunschmidt, C.W. Klamofl, W. Buchberger, R. Herstens, Rapid identification of stabilizers in polypropylene using time-of-flight mass spectrometry and DART as ion source, Analyst 135 (1) (2010) 80–85, https://doi.org/10.1039/b911040b.

[38] J.J. Pérez, G.A. Harris, J.E. Chipuk, J.S. Brodbelt, M.D. Green, C.Y. Hampton, F.M. Fernéndez, Transmission-mode direct analysis in real time and desorption electrospray ionization mass spectrometry of insecticide-treated bednets for malaria control, Analyst 135 (2010) 712–719, https://doi.org/10.1039/b924533b.

[39] A. Mess, J.P. Vietzke, C. Rapp, W. Francke, Qualitative analysis of tackifier resins in pressure sensitive adhesives using direct analysis in real time time-of-flight mass spectrometry, Anal. Chem. 83 (2011) 7323–7330, https://doi.org/10.1021/ac2011608.

[40] A. Kuki, L. Nagy, M. Zsuga, S. Kéki, Fast identification of phthalic acid esters in poly(vinyl chloride) samples by direct analysis in real time (DART) mass spectrometry, Int. J. Mass Spectrom. 303 (2011) 225–228, https://doi.org/10.1016/j.ij,s.2011.02.011.

[41] X. Li, J. Xing, C. Chang, X. Wang, Y. Bai, X. Yan, H. Liu, Solid-phase extraction with the metal-organic framework MIL-101(Cr) combined with direct analysis in real time mass spectrometry for the fast analysis of triazine herbicides, J. Sep. Sci. 37 (2014) 1489–1495, https://doi.org/10.1002/jsac.201400151.

[42] X. Wang, X. Li, Z. Li, Y. Zhang, Y. Bai, H. Liu, Online coupling of in-tube solid-phase microextraction with direct analysis in real time mass spectrometry for rapid determination of triazine herbicides in water using carbon-nanotubes-incorporated polymer monolith, Anal. Chem. 86 (2014) 4739–4747, https://doi.org/10.1021/ac500382x.

[43] L. Vaclavik, T. Cajka, V. Hrbek, J. Hajslova, Ambient mass spectrometry employing direct analysis in real time (DART) ion source for olive oil quality and authenticity assessment, Anal. Chim. Acta 645 (2009) 56–63, https://doi.org/10.1016/j.aca.2009.04.043.

[44] M. Zachariasova, T. Cajka, M. Godula, A. Malachova, Z. Veprikova, J. Hajslova, Analysis of multiple mycotoxins in beer employing (ultra)-high-resolution mass spectrometry, Rapid Commun. Mass Spectrom. 24 (2010) 3357–3367, https://doi.org/10.1002/rcm.4746.

[45] E. Block, R.B. Cody, A.J. Dane, R. Sheridan, A. Vattekkatte, K. Wang, Allium chemistry: use of new instrumental techniques to "see" reactive organosulfur species

formed upon crushing garlic and onion, Pure Appl. Chem. 82 (2010) 535–539, https://doi.org/10.1351/PAC-CON-09-08-12.
[46] E. Block, A.J. Dane, S. Thomas, R.B. Cody, Applications of direct analysis in real time mass spectrometry (DART-MS) in allium chemistry. 2-propenesulfenic and 2-propenesufinic acids, diallyl trisulfane S-oxide, and other reactive sulfur compounds from crushed garlic and other alliums, J. Agric. Food Chem. 58 (8) (2010) 4617–4625, https://doi.org/10.1021/jf000106.
[47] L. Vaclavik, V. Hrbek, T. Cajka, B.A. Rohlik, P. Pipek, J. Hajslova, Authentication of animal fats using direct analysis in real time (DART) ionization-mass spectrometry and chemometric tools, J. Agric. Food Chem. 59 (2011) 5919–5926, https://doi.org/10.1021/jf200734x.
[48] H. Danhelova, J. Hradecky, S. Prinosilova, T. Cajka, K. Riddellova, L. Vaclavik, J. Hajslova, Rapid analysis of caffeine in various coffee samples employing direct analysis in real-time ionization-high-resolution mass spectrometry, Anal. Bioanal. Chem. 403 (2012) 2883–2889, https://doi.org/10.1007/s00216-012-5820-2.
[49] E.S. Chernetsova, M. Bromirski, O. Scheibner, G.E. Morlock, DART-Orbitrap MS: a novel mass spectrometric approach for the identification of phenolic compounds in propolis, Anal. Bioanal. Chem. 403 (2012) 2859–2867, https://doi.org/10.1007/s00216-012-5800-6.
[50] L. Vaclavik, B. Belkova, Z. Reblova, K. Riddellova, J. Hajslova, Rapid monitoring of heat-accelerated reactions in vegetable oils using direct analysis in real time ionization coupled with high resolution mass spectrometry, Food Chem. 138 (2013) 2312–2320, https://doi.org/10.1016/j.foodchem.2012.12.019.
[51] M. Busman, J. Liu, H. Zhong, J.R. Bobell, C.M. Maragos, Determination of the aflatoxin AFB1 from corn by direct analysis in real time-mass spectrometry (DART-MS), Food Addit. Contam. Part A. Chem. Anal. Control Expo Risk. Assess. 31 (2014) 932–939, https://doi.org/10.1080/19440049.2014.900572.
[52] V. Krtova, V. Schlzova, O. Lacina, V. Hrbek, M. Tomaniova, J. Hajslova, Analytical strategies for controlling polysorbate-based nanomicelles in fruit juice, Anal. Bioanal. Chem. 406 (2014) 3909–3918, https://doi.org/10.1007/s00216-014-7823-7.
[53] G.A. Gómez-Ríos, T. Vasiljevic, E. Gionfriddo, M. Yu, J. Pawliszyn, Towards on-site analysis of complex matrices by solid-phase microextraction-transmission mode coupled to a portable mass spectrometer: via direct analysis in real time, Analyst 142 (2017) 2928–2935, https://doi.org/10.1039/C7AN00718C.
[54] D. Saang'onyo, G. Selby, D.L. Smith, Validation of a direct analysis in real time mass spectrometry (DART-MS) method for the quantitation of six carbon sugars in a saccharification matrix, Anal. Methods 4 (2012) 3460–3465, https://doi.org/10.1039/C2AY25337B.
[55] A.H. Grange, Semi-quantitative analysis of contaminants in soils by direct analysis in real time (DART) mass spectrometry, Rapid Commun. Mass Spectrom. 27 (2013) 305–318, https://doi.org/10.1002/rcm.6450.
[56] R.B. Cody, Observation of molecular ions and analysis of nonpolar compounds with the direct analysis in real time ion source, Anal. Chem. 81 (2009) 1101–1107, https://doi.org/10.1021/ac8022108.
[57] J. Kratzer, Z. Mester, R.E. Sturgeon, Comparison of dielectric barrier discharge, atmospheric pressure radiofrequency-driven glow discharge and direct analysis in real time sources for ambient mass spectrometry of acetaminophen, Spectrochim. Acta Part B 66 (2011) 594–603, https://doi.org/10.1016/j.sab.2011.06.005.
[58] R.B. Cody, A.J. Dane, Soft ionization of saturated hydrocarbons, alcohols and nonpolar compounds by negative-ion direct analysis in real-time mass spectrometry, J. Am. Soc. Mass Spectrom. 24 (3) (2013) 329–334, https://doi.org/10.1007/s13361-012-0569-6.

[59] H. Yang, D. Wan, F. Song, Z. Liu, S. Liu, Argon direct analysis in real time mass spectrometry in conjunction with makeup solvents: a method for analysis of labile compounds, Anal. Chem. 85 (2013) 1305–1309, https://doi.org/10.1021/ac3026543.

[60] R.V. Bennett, E.M. Morzan, J.O. Huckaby, M.E. Monge, H.I. Christensen, F.M. Fernández, Robotic plasma probe ionization mass spectrometry (RoPPI-MS) of non-planar surfaces, Analyst 139 (2014) 2658–2662, https://doi.org/10.1039/c4an00277f.

[61] J.D. Keelor, P. Dwivedi, F.M. Fernández, An effective approach for coupling direct analysis in real time with atmospheric pressure drift tube ion mobility spectrometry, J. Am. Soc. Mass Spectrom. 25 (2014) 1538–1548, https://doi.org/10.1007/s13361-014-0926-8.

[62] K. Sekimoto, M. Sakakura, T. Kawamukai, H. Hike, T. Shiota, F. Usui, Y. Bando, M. Takayama, Ionization characteristics of amino acids in direct analysis in real time mass spectrometry, Analyst 139 (2014) 2589–2599, https://doi.org/10.1039/c3an02193a.

[63] K. Sekimoto, M. Sakakura, T. Kawamukai, H. Hike, T. Shiota, F. Usui, Y. Bando, M. Takayama, Improvement in ionization efficiency of direct analysis in real time-mass spectrometry (DART-MS) by corona discharge, Analyst 141 (2016) 4879–4892, https://doi.org/10.1039/c6an00779a. K. Sekimoto, M. Takayama, D. Okumura, Ionizer and mass spectrometer including first section for ionizing sample under atmospheric pressure while vaporizing or desorbing the sample component and second section for generating corona discharge 2017, Japanese Patent No. 6091620, United States Patent No. US 9,691,598 B2, Chinese Patent No. ZL 201380078641.2.

[64] K. Sekimoto, Ionization characteristics of direct analysis in real time (DART) and improvement in its ionization efficiency by corona discharge, J. Mass Spectrom. Soc. Jpn. 65 (2017) 102–106.

[65] R.B. Cody, A.J. Dane, Dopant-assisted direct analysis in real time mass spectrometry with argon gas, Rapid Commun. Mass Spectrom. 30 (2016) 1181–1189, https://doi.org/10.1002/rcm.7552.

[66] K. Sekimoto, T. Takayama, H. Hike, M. Sakakura, Atmospheric Pressure Ionization Method, Japanese Patent No. 6382166, 2018, United States Patent No. US 10,262,852 B2, 2019.

[67] J. Hajslova, T. Cajka, L. Vaclavik, Challenging applications offered by direct analysis in real time (DART) in food-quality and safety analysis, Trends Anal. Chem. 30 (2011) 204–218, https://doi.org/10.1016/j.trac.2010.11.001.

[68] J.H. Gross, Direct analysis in real time – a critical review on DART-MS, Anal. Bioanal. Chem. 406 (2014) 63–80, https://doi.org/10.1007/s00216-013-7316-0.

[69] A.D. Lesiak, J.R. Shepard, Recent advances in forensic drug analysis by DART-MS, Bioanalysis 6 (6) (2014) 819–842, https://doi.org/10.4155/bio.14.31.

[70] M. Smoluch, P. Mielczarek, J. Silberring, Plasma-based ambient ionization mass spectrometry in bioanalytical sciences, Mass Spectrom. Rev. 35 (2016) 22–34, https://doi.org/10.1002/mas.21460.

[71] M. Domin, R.B. Cody (Eds.), Ambient Ionization Mass Spectrometry, The Royal Society of Chemistry, Cambridge, UK, 2015.

[72] Y. Dong (Ed.), Direct Analysis in Real Time Mass Spectrometry: Principles and Practices of DART-MS, Wiley, Hoboken, NJ, USA, 2018.

[73] G.A. Harris, F.M. Fernández, Simulations and experimental investigation of atmospheric transport in an ambient metastable-induced chemical ionization source, Anal. Chem. 81 (2009) 322–329, https://doi.org/10.1021/ac802117u.

[74] J.L. Pack, A.V. Phelps, Electron attachment and detachment. I. Pure O_2 at low energy, J. Chem. Phys. 44 (5) (1966) 1870–1883, https://doi.org/10.1063/1.1726956.

[75] L.M. Chanin, A.V. Phelps, M.A. Biondi, Measurements of the attachment of low-energy electrons to oxygen molecules, Phys. Rev. 128 (1) (1962) 219–230.
[76] M. Pavlik, J.D. Skalny, Generation of $[H_3O]^+ \cdot (H_2O)_n$ clusters by positive corona discharge in air, Rapid Commun. Mass Spectrom. 11 (1997) 1757–1766.
[77] Y. Ikezoe, S. Matsuoka, M. Takebe, A. Viggiano, Gas Phase Ion-Molecule Reaction Rate Constants through 1986, Maruzen, Tokyo, 1987, p. 154.
[78] V.G. Anicich, Evaluated bimolecular ion-molecule gas phase kinetics of positive ions for use in modeling planetary atmospheres, cometary comae, and interstellar clouds, J. Phys. Chem. Ref. Data 22 (1993) 1469–1569, https://doi.org/10.1063/1.555940.
[79] D.L. Albritton, Ion-neutral reaction-rate constants measured in flow reactors through 1977, Atomic Data Nucl. Data Tables 22 (1) (1978) 1–89.
[80] D.J. Jacob, Introduction to Atmospheric Chemistry, first ed., Princeton University Press, Princeton, New Jersey, 1999, pp. 163–171.
[81] T. Kotiaho, M.N. Eberlin, P. Vainiotalo, R. Kostiainen, Electrospray mass and tandem mass spectrometry identification of ozone oxidation products of amino acids and small peptides, J. Am. Soc. Mass Spectrom. 11 (2000) 526–535.
[82] R.M. Le Lacheur, W.H. Glaze, Reactions of ozone and hydroxyl radicals with serine, Environ. Sci. Technol. 30 (1996) 1072–1080.
[83] M. Ukai, Fundamentals of mass spectrometry – basic understanding of Penning ionization –, J. Mass Spectrom. Soc. Jpn. 57 (6) (2009) 393–404.
[84] G.S. Hurst, T.E. Bortner, R.E. Glick, Ionization and excitation of argon with alpha particles, J. Chem. Phys. 42 (1965) 713–719, https://doi.org/10.1063/1.1695995.
[85] K. Sekimoto, M. Takayama, Negative ion formation and evolution in atmospheric pressure corona discharges between point-to-plane electrodes with arbitrary needle angle, Eur. Phys. J. D 60 (2010) 589–599, https://doi.org/10.1140/epjd/e2010-10449-7.

CHAPTER 3

Desorption corona beam ionization

Wenjian Sun
Shimadzu Research Laboratory (Shanghai) Co., Ltd., Pudong New District, Shanghai, China

Contents

3.1 Introduction

Since the recent introduction of desorption electrospray ionization (DESI) by R. Graham Cooks and coworkers [1], many kinds of direct analysis approaches have been reported. These direct analysis ionization sources all have minimal sample pretreatment steps as a common feature, which has facilitated the rapid and high-throughput mass analysis of samples.

Many techniques with the capability of directly ionizing samples under atmospheric pressure have appeared since. Two review articles covering ambient ionization methods have been published [2−4]. In these reviews, the various techniques were divided into two groups. One group contained electrospray ionization (ESI)-related techniques such as DESI and electrospray-assisted laser desorption ionization (ELDI) [5], and the other contained atmospheric pressure chemical ionization (APCI)-related techniques such as direct

Ambient Ionization Mass Spectrometry in Life Sciences
ISBN 978-0-12-817220-9
https://doi.org/10.1016/B978-0-12-817220-9.00003-5

analysis in real time (DART) [6], atmospheric solid analysis probe (ASAP) [7], and desorption atmospheric pressure chemical ionization (DAPCI) [8].

In common with ASAP and DAPCI, our desorption corona beam ionization (DCBI) source [9] uses a low-current, high-voltage DC power supply, making it convenient to couple with commercially available mass spectrometry (MS) instruments that are normally equipped with ESI and APCI sources. Furthermore, using helium as the discharge gas can enhance the ionization efficiency of the source. When compared with its most closely related counterpart, DART, our DCBI source has a slightly different mechanism but a similar range of applications for most of the compounds tested. The visibility of the ionization area and the adaptability of the probe to the current Shimadzu liquid chromatography/mass spectrometry (LC-MS) platforms make DCBI a more seamless integrated part for direct analysis with higher efficiency. One example of these advantages is that such a gas-tight structure enables a pure nitrogen environment and thus the ionization efficiency for compounds with high electron affinity can be increased by one order of magnitude in the negative-ion mode.

3.2 Principles of DCBI

As shown in Fig. 3.1A, the DCBI probe mainly comprises a stainless-steel thin-walled gas heating tube, a needle electrode (discharge electrode), a metal ring electrode (counter electrode), and a stainless-steel enclosure with a glass inner tube for insulation. The other end of the gas heating tube is connected to a gas transfer line through which the helium gas is transported to the heating region. The gas stream flowing out of the heating tube passes through the discharge needle and reaches the sample surface lying near the MS inlet.

The ends of the gas heating tube are connected to a high-current, low-voltage DC power supply (25 A, 9 V). The gas flow can be up to 2 L/minutes at a maximum temperature of 400°C after heating. A high DC voltage (up to 5 kV) is superimposed on the gas heating tube and the discharge needle (the two are connected) to generate the corona discharge. The discharge current is normally kept around 5 μA, and it can reach 20 μA when the voltage is set high enough. Fig. 3.1B shows the appearance of the corona beam. The bright beam can help to easily identify the sampling area, and its low current ensures the safety of the device.

The DCBI probe is controlled by a stand-alone control box, as shown in Fig. 3.1C. The control box has three main functions: (1) providing a high

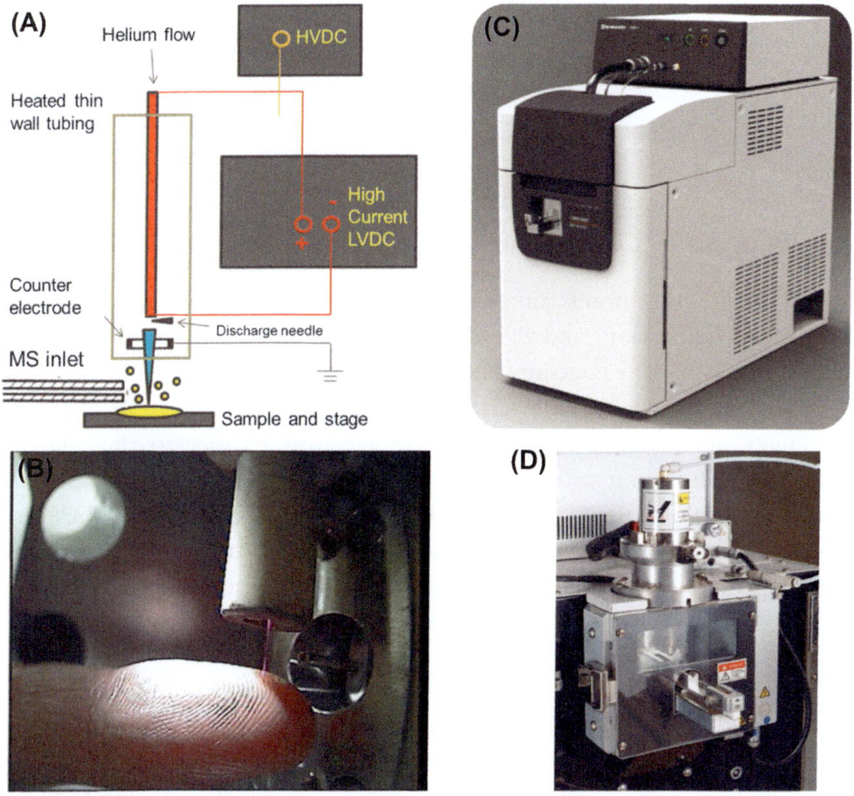

Figure 3.1 (A) Schematic drawing of the desorption corona beam ionization (DCBI) source probe; (B) photograph showing the bright discharge beam; (C) photograph of the DCBI probe control box on top of a Shimadzu Single Quadrupole Mass Spectrometer (LCMS-2020); (D) photograph of the sample introduction stage.

voltage for the corona discharge, (2) modulating a high DC current for control of the heating temperature of the discharge gas, and (3) providing an accurate discharge gas flow. Fig. 3.1D shows a photograph of the sample introduction stage upon which the sample holder is mounted. The sample holder can be smoothly slid into the sampling zone.

The mechanism of DCBI is similar to that of DART, where helium molecules are first excited to a metastable state and then undergo subsequent reactions in either positive- or negative-ion mode, as shown in Fig. 3.2. In the positive-ion mode, the metastable helium atoms can either directly eject one electron from the analyte molecule to form the radical cation or they can ionize water molecules to form hydronium ions for subsequent proton transfer reactions. In the negative-ion mode, the

Positive mode:

$$He^* + M \rightarrow He + M^{+\cdot} + e^-$$

$$He^* + nH_2O \rightarrow H_3O^+(H_2O)_{n-2} + OH^- + He,\ n > 1$$

$$H_3O^+(H_2O)_n + M \rightarrow [M+H]^+ + (n+1)\ H_2O\ \text{(indirect)}$$

Negative mode:

$$He^* + 2N_2 \rightarrow He + N_4^+ + e^-$$

$$M + e^- \rightarrow M^{-\cdot}$$

$$M + OH^- \rightarrow [M\text{-}H]^- + H_2O$$

$$O_2 + e^- \rightarrow O_2^{-\cdot}$$

Figure 3.2 Proposed mechanism of desorption corona beam ionization.

metastable helium atoms can release electrons from N_2 molecules when interacting with them, and the electrons can then directly react with the analyte molecules to generate the radical anions. Note that O_2 molecules have a very strong electron affinity and may play a role as an electron scavenger, suppressing the efficiency of electron transfer to the analyte molecules. This point will be brought up again in a later section where we discuss the use of pure N_2 to replace air in the ion source region in order to boost the sensitivity of the DCBI source in the negative-ion mode (by reducing the competition with O_2 molecules for electrons).

3.3 Features of DCBI

The performance of the DCBI source was evaluated with regard to the following: (1) sensitivity, (2) reproducibility, (3) mass range, (4) analyte types, and (5) helium consumption. Table 3.1 shows the specifications of the source. To test the robustness of the DCBI source, a continuous test lasting 10 hours was performed for both discharge current and temperature of the discharge gas, which are the two basic factors determining the stability of the desorption/ionization processes. Both factors did not change significantly over 10 hours of testing, during which the gas temperature and discharge current were maintained within $\pm 5°C$ and $\pm 1\ \mu A$, respectively.

The upper mass limit of the DCBI source was determined in the normal way by the volatility of the molecules being tested. In most of the cases, the DCBI can desorb molecules with masses of less than 800 Da. However, certain high-mass species with high volatility, such as perfluorinated nonyl 1,3,5-triazine ($C_{30}F_{57}N_3$, MW 1485.26 Da) in Table 3.1, can still be thermally desorbed from the sample surface. In both positive- and negative-ion modes, the precursor ions of the triazine can easily be identified as acetone adduct ions. Some fragment ions of such high-mass molecules arising due to thermal decomposition can also be detected when the temperature used for desorption is high.

Table 3.1 Specifications of the desorption corona beam ionization source.

Item	Specs
Limit of detection of α-hexyl cinnamic aldehyde (on LCMS-2020)	≥1 pg
Analyte types	Semivolatile
Gas temperature range	Room to 400°C
Upper mass limit	1485 Da
Helium flow rate	0−2 L/minutes
Discharge voltage	0 to ±5 kV
Probe mode	Positive and negative
Sample stage control	Manual and automatic

In addition to the basic specifications, the DCBI source also possesses some special advantages due to its unique engineering features. One example of such a feature is the use of N_2 protection in the source chamber to enhance the sensitivity in negative-ion mode [10]. Fig. 3.3 illustrates that the S/N was boosted by about one order of magnitude when ionizing RDX and trinitrophenol. This can mainly be attributed to the lack of the strong electron scavenger oxygen in the ionization chamber, which makes it easier for the analyte to obtain electrons for ionization.

3.4 Applications of DCBI

3.4.1 Food and drug safety

3.4.1.1 Fake drug screening

The fake drug issue is quite widespread in many South and East Asian countries. The main challenge in fake drug detection is to identify large amounts of drugs in a limited time. Usually, the requirement of high speed is more important than quantitation. Therefore, DCBI is very suitable for such an application. Collaborating with the Shanghai Institute of Food and Drug Control, we studied a series of fake drugs that had been confiscated from the market. Fig. 3.4 shows the DCBI mass spectra of one type of antitumor drug (Xeloda brand, capecitabine m/z 359.55) from four different markets [10]. The first three (A−C) were authentic and the m/z 359.55 peak was observed. The last spectrum (D) was obtained from a fake drug and did not contain the peak at m/z 359.55. All these tests were completed within 10 seconds, which is much faster than the minutes or hours required by optical methods or the LC-MS method.

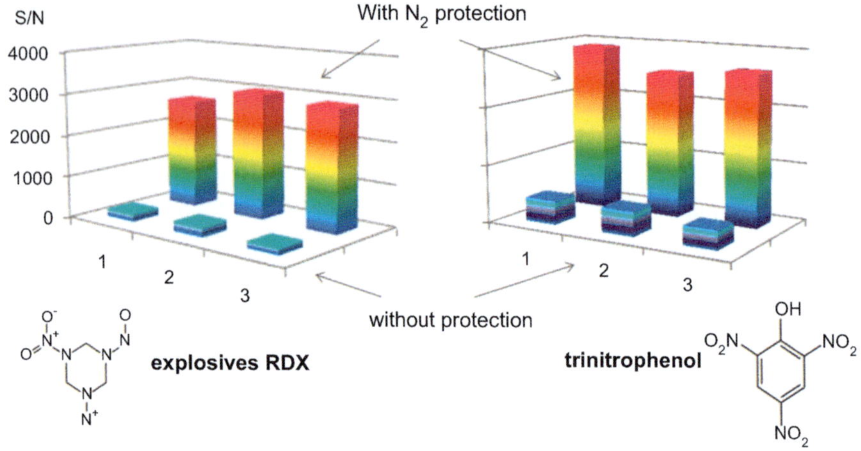

Figure 3.3 Signal enhancement of explosives detected by desorption corona beam ionization–mass spectrometry with N_2 gas–filled source enclosure. The numbers 1 to 3 in the lower part of the figure represent the three replicates.

When analyzing many Chinese traditional herbal medicines, the complex matrix of these medicines may significantly suppress ionization of the analytes or make the spectra far more complicated, which may cause false-negative or false-positive results. For instance, in the case of direct desorption and ionization of a real sample (Gan mao ling pill) using DCBI-MS, the ion signal of chlorpheniramine was not observed in the mass spectrum (Fig. 3.5A) [11]. After ethanol extraction, chlorpheniramine was detected when the ethanol extract of the pill was spotted onto filter paper without development, but the matrix interference was still obvious because the signal intensity of the analyte ion was low (Fig. 3.5B). Only several high ion signals such as *m/z* 195, 218, 302, 285, and 326 could be observed in the mass spectrum using the positive-ion mode. To tackle this issue, a simple separation using frontal elution paper chromatography was developed to couple with DCBI-MS. When the solution containing the analyte elutes, the analyte will gradually reach to the tip of the tapered paper (Fig. 3.6). Using the corona beam to probe the tip will release the analytes with much less matrix interference, and the signal intensity of chlorpheniramine can be increased 30-fold (Fig. 3.5C).

3.4.1.2 Pesticide detection

Food safety is another important issue in China where many cases have been disclosed in recent years. The analysis of pesticide residues on

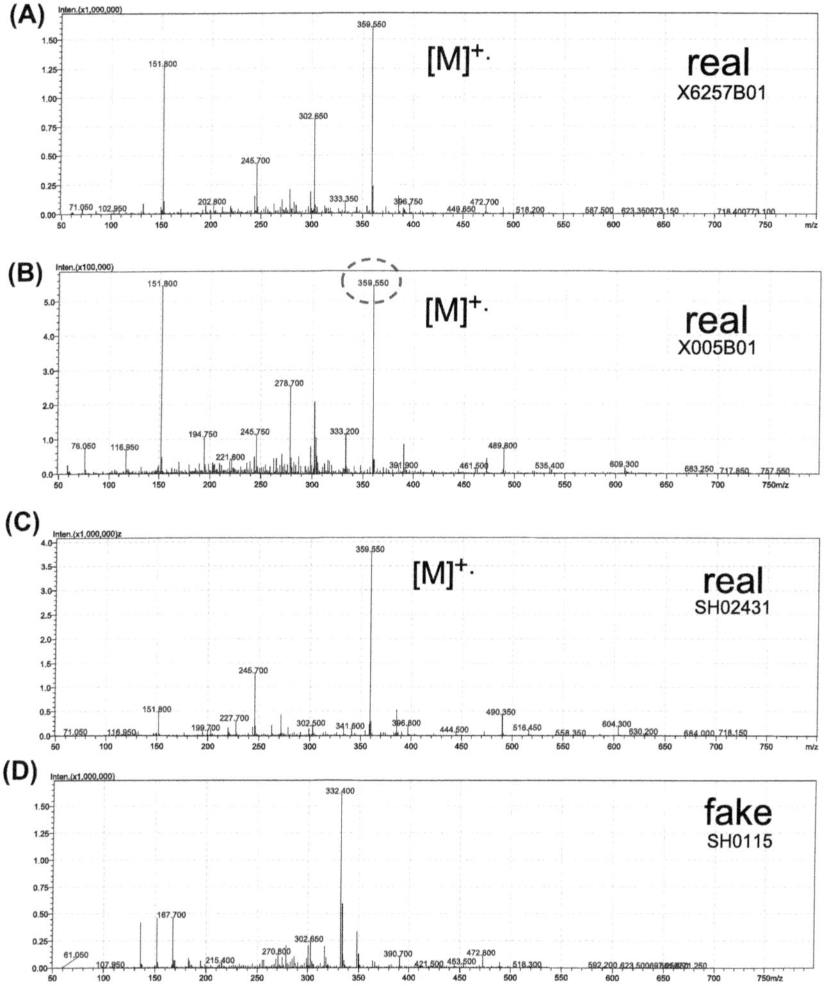

Figure 3.4 Mass spectra of various samples of Xeloda obtained from Shanghai Food and Drug Control and measured using desorption corona beam ionization as the ion source.

vegetables is one area that presents a heavy burden to the government and third-party test agencies. DCBI was used for the rapid determination of methamidophos residue directly from vegetables without any sample pretreatment. Pak choi leaves onto which methamidophos was sprayed (1 μg/mL) were irradiated directly with a DCBI beam and measured by our collaborator at Hunan Normal University with a 3D ion trap (Thermo Fisher). As shown in Fig. 3.7, the methamidophos (*m/z* 141.92) ions were

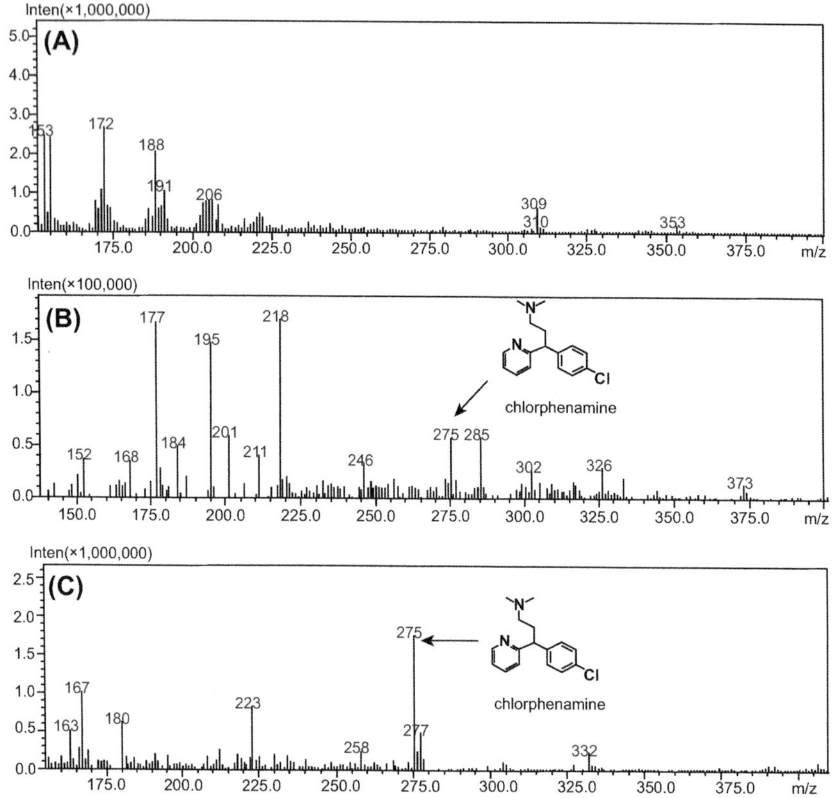

Figure 3.5 Mass spectra of chlorpheniramine obtained using (A) direct desorption ionization of the pills, (B) ethanol extract on filter paper without development, and (C) ethanol extract on filter paper after development. *(Reprinted with permission from Y. Huang, J. You, J. Zhang, W. Sun, L. Ding, Y. Feng, Coupling frontal elution paper chromatography with desorption corona beam ionization mass spectrometry for rapid analysis of chlorphenamine in herbal medicines and dietary supplements, J. Chrom. A 1218 (2011) 7371–7376.)*

clearly identified. Note that methamidophos has been banned from use on any type of vegetable in many countries; therefore, a positive result for its existence on the samples clearly demonstrates the importance and utility of this method.

3.4.1.3 Determination of usage period of gutter oil

The issue of using gutter oil (old/overcooked oil or purified waste oil from sewers and drains) for cooking is another severe food safety issue in China. Cooking oil is a complex mixture of chemicals, in which diglycerides

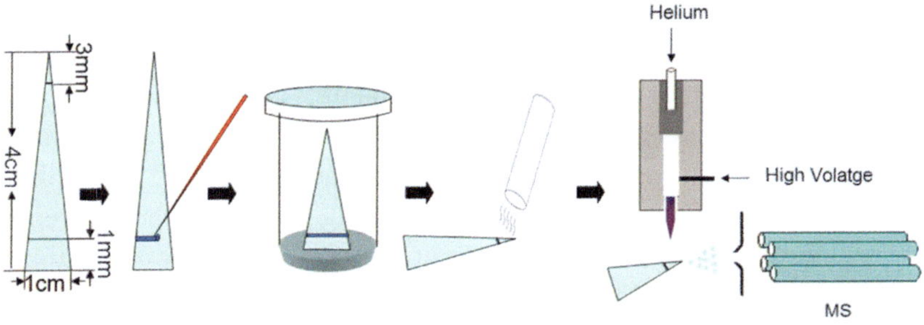

Figure 3.6 Schematics showing the use of frontal elution paper chromatography for desorption corona beam ionization–mass spectrometry analysis. *(Reprinted with permission from Y. Huang, J. You, J. Zhang, W. Sun, L. Ding, Y. Feng, Coupling frontal elution paper chromatography with desorption corona beam ionization mass spectrometry for rapid analysis of chlorphenamine in herbal medicines and dietary supplements, J. Chrom. A 1218 (2011) 7371–7376.)*

(DAGs), triglycerides (TAGs), and free fatty acids are the main components. Besides these major components, a series of minor polar compounds are also present, and their distribution is characteristic of different types of oil.

The term "gutter oil" is quite general and could be used to describe multiple types of used oil such as disposed oil and oil that has been in use for a long time in restaurants. The latter type (oil in use for a long time) is a

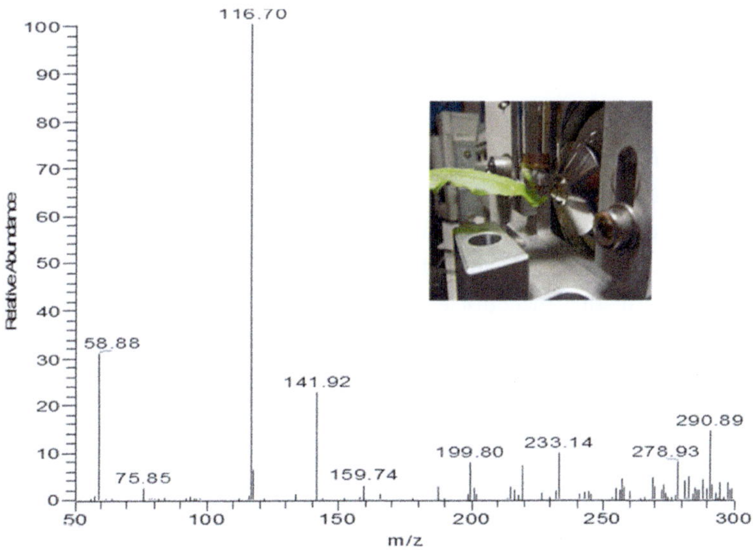

Figure 3.7 MS spectrum of methamidophos (*m/z* 141.92) residue on a vegetable sample (pak choi).

common issue for many restaurants. Using oil for an extended period can tremendously reduce costs but the oil quality will continuously degrade over time and eventually cause health problems. The traditional way an inspection agency identifies how long the oil has been used is to visually examine its color and viscosity. To more quantitatively identify the cooking period, a series of oil samples were analyzed using DCBI and the differences between fresh and used oil were quantified for various cooking times, cooking temperatures, oil types, and origins of the oil [12]. One example involved analyzing bean oil that had been cooked at 180 °C for various time periods. The heated oils were collected at 4, 8, 12, 24, and 36 hours and then left at room temperature to cool. The fresh (0 hour) and heated oil (4, 8, 12, 24, and 36 hours) samples were diluted with toluene (oil:toluene = 1:5, v/v) and then the diluted solutions (5 μL) were characterized by DCBI-MS. Fig. 3.8A shows the appearance of the oil samples at different cooking times, where it can be seen that distinct colors exist for each cooking time. Fig. 3.8B–G show the mass spectra corresponding to the oil samples shown in Fig. 3.8A. It can be seen that the fresh bean oil contains both DAGs (such as *m/z* 575, 577, 601, 603, and 617) and TAGs (such as *m/z* 856, 879, and 881) at similar levels. As the cooking progressed, the relative intensity of the TAGs became lower and ions in the lower mass range such as *m/z* 439, 397, and 463 started to be observed (Fig. 3.8F and G). The ions in the lower mass range might be generated by thermal decomposition of the oil. This result shows that DCBI-MS can be used as a good indicator for oils that have been overused in restaurants.

3.4.1.4 Detection of illicit additives in weight-loss dietary supplements

Weight-loss foods and proprietary sliming products are often natural foods or fortified foods, which are intended to facilitate weight reduction and health enhancement. However, many of these products have been found to be adulterated with pharmaceutical chemicals (e.g., sibutramine) and even forbidden substances (e.g., fenfluramine). Sibutramine used to be approved by the USFDA for the treatment of obesity [13] but was withdrawn in October 2010. If the recommended dose is exceeded, it may cause a series of side effects such as palpitations, chest pain, insomnia, diabetes, anorexia, and abnormal liver function [14–16]. As for fenfluramine, due to heart valve disease and pulmonary hypertension [17,18], it has been banned by the USFDA since 1997.

Six drug molecules commonly used to adulterate traditional Chinese medicine for weight loss have been analyzed with DCBI-MS [19]. The six

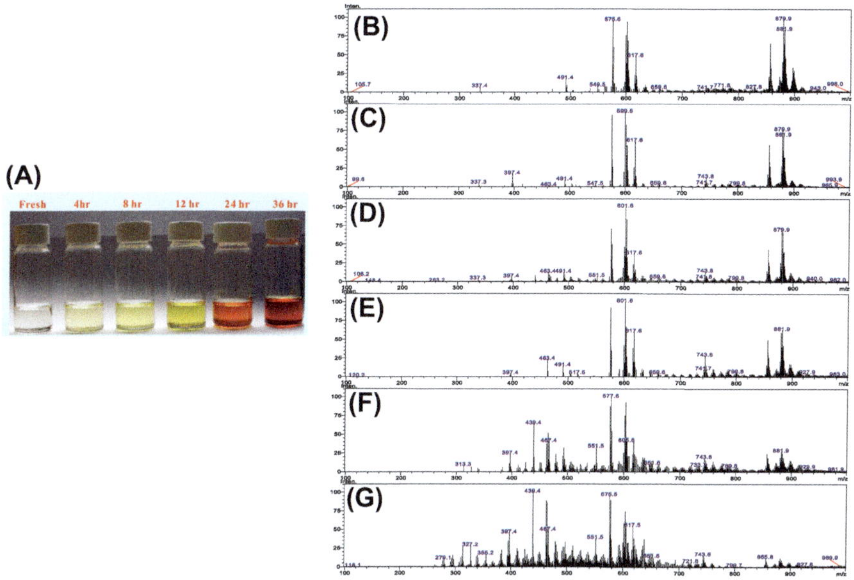

Figure 3.8 (A) Appearance of used oil after different cooking periods; (B–G) desorption corona beam ionization mass spectra of bean oils detected at 300°C. The bean oil was continuously heated at 180°C and then collected at different times (B) 0, (C) 4, (D) 8, (E) 12, (F) 24, and (G) 36 hours for detection.

compounds are M1: fenfluramine, M2: N, N-didesmethyl sibutramine, M3: N-mono-desmethyl sibutramine, M4: sibutramine, M5: phenolphthalein, and M6: sildenafil. Two sets of experiments were conducted. In the first set, a mixture of pure compounds was directly analyzed with DCBI-MS and the spectrum is shown in Fig. 3.9A. All six compounds can be detected without much difficulty except for M5 and M6. In another set of experiments, the same mixture of pure compounds was embedded in a matrix (tea leaves) and directly analyzed with DCBI-MS (Fig. 3.9B). In this case, the peak intensity for M5 is particularly strong but the peaks associated with the other compounds are suppressed to some extent. Nevertheless, all compounds are still detectable.

To determine the amount of the drugs in the real sample on a semiquantitative basis, some simple sample pretreatment steps are still needed. Solvent extraction and concentration were applied to different matrices such as drug capsules and tea bags. After this pretreatment, the same sample solution was tested with both DCBI-MS and high performance liquid chromatography (HPLC)-ESI-MS. The final results are compared in Table 3.2, and similar concentrations are observed for both

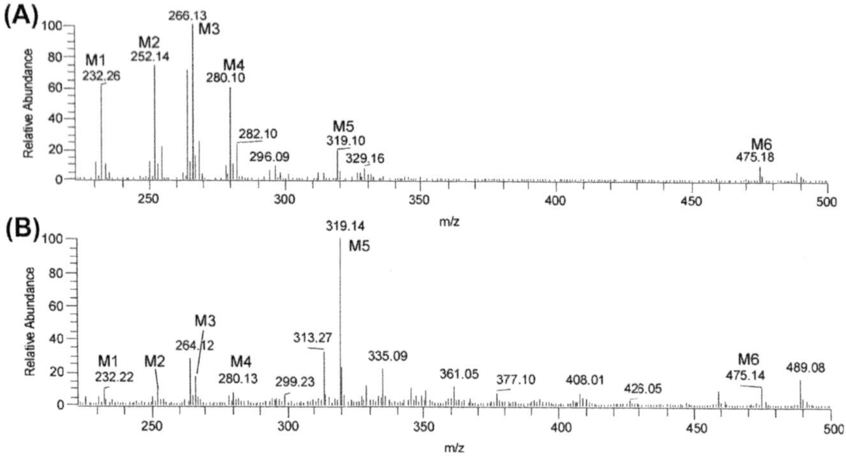

Figure 3.9 Mass spectra of six different drug molecules commonly used to adulterate traditional Chinese medicine for weight loss. The six compounds are M1: fenfluramine, M2: *N, N*-didesmethyl sibutramine, M3: *N*-mono-desmethyl sibutramine, M4: sibutramine, M5: phenolphthalein, and M6: sildenafil. (A) standard compounds without matrix (direct analysis from the real sample) and (B) compounds in the matrix. *(Reprinted with permission from H. Wang, Y. Wu, Y. Zhao, W. Sun, L. Ding, B. Guo, B. Chen, Rapid screening of illicit additives in weight loss dietary supplements with desorption corona beam ionisation (DCBI) mass spectrometry, Food Addit. Contam. 29 (2012) 1194–1201.)*

methods. As also seen in Table 3.2, the signal variation when using DCBI is larger than that of HPLC-ESI; however, such a DCBI screening process can save more than half an hour for each sample because the pretreatment process is not sophisticated and can be performed very quickly. Once the existence and approximate concentration range of the compounds have been determined, one can still use the HPLC-ESI method to determine the concentration more accurately and more reproducibly.

3.4.1.5 Rapid determination of bisphenol-A in food packaging materials

The food packing materials industry is a billion dollar market in China, and the demand for paper, plastic, and metal-based packing materials is still increasing every year. Plastic among all others is the most frequently used material due to its lower cost and higher durability. However, there have been serious issues associated with plastic material usage in the food industry over the past 10 years in China. One important example is residual bisphenol-A (BPA) being over the prescribed limit. BPA is a photoinitiator

Table 3.2 Calculated concentrations of illicit additives in real products.

		DCBI		HPLC	
Sample	**Adulteration chemical**	**Concentration (mg/g)**	**RSD (n = 6) (%)**	**Concentration (mg/g)**	**RSD (n = 6) (%)**
Capsule A	Sibutramine	16.32	10.62	20.11	6.72
Capsule B	N-mono-desmethyl sibutramine	18.20	14.16	23.08	7.60
Capsule C	Sibutramine	25.79	13.38	32.72	4.30
Capsule D	N-mono-desmethyl sibutramine	1.86	13.93	5.13	2.90
	Phenolphthalein	14.14	25.05	13.83	4.80
	Sibutramine	16.05	15.63	20.42	4.20
Capsule E	N-di-desmethyl sibutramine	1.38	22.83	1.73	6.86
Capsule F	N-di-desmethyl sibutramine	0.18	14.33	0.23	3.66
Capsule G	Sibutramine	0.16	22.29	0.39	5.81
Capsule H	Sibutramine	0.11	13.32	0.21	5.30
	Phenolphthalein	44.1	25.22	43.9	5.80
Tea bag A	Sibutramine	1.56	8.40	2.17	3.32
Tea bag B	N-di-desmethyl sibutramine	0.81	14.33	2.04	2.17
Tea bag C	N-di-desmethyl sibutramine	1.09	9.92	3.17	6.04

DCBI, desorption corona beam ionization; *HPLC*, high performance liquid chromatography.

used during the polymerization process of certain plastic materials. High concentrations of BPA in the human body, especially in babies, are known to cause severe health problems. Supposedly, all BPA should decompose after the production process; however, large amounts of residue may remain in the final product if the production stage is not well controlled.

Analysis of BPA content is typically a burdensome task for a testing company due to the long pretreatment and LC separation times. Similar to the case of detecting illegal drug adulteration described in the previous session, the total time for one sample can be shortened by more than 30 minutes. Fig. 3.10 illustrates the pretreatment process for DCBI-MS. Using the same pretreatment method, a calibration curve can be constructed, as shown in Fig. 3.11. The dynamic range is about three orders of magnitude and the overall linearity is relatively good for a direct analysis method ($R^2 = 0.9863$). Using this calibration curve, the concentration of BPA in unknown samples can be determined and the results are in close agreement with those obtained using the HPLC-ESI method (271 ppb from DCBI-MS vs. 200 ppb from HPLC-ESI), and no positive result was found for the other two samples using either method (Table 3.3).

3.4.2 Public safety

The detection of explosives and illicit drugs at the security checkpoints of large public facilities such as airports is currently achieved using ion mobility spectrometers. However, the high false-positive rate makes ion mobility

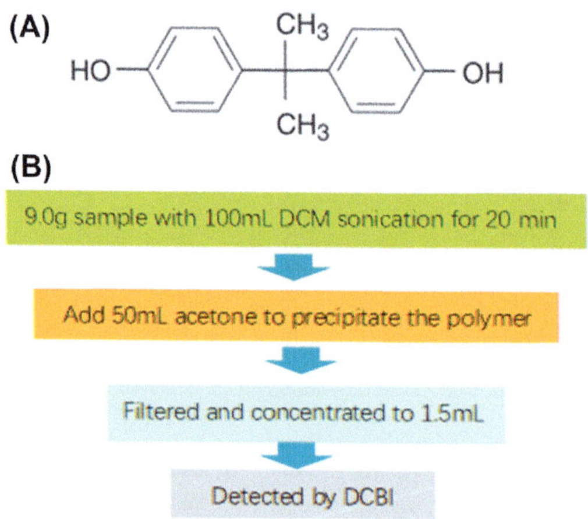

Figure 3.10 (A) Structure of bisphenol-A (BPA); (B) pretreatment process before using desorption corona beam ionization for testing BPA.

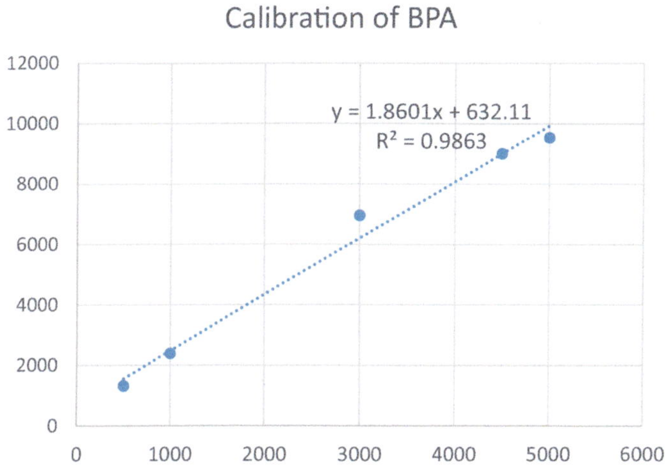

Figure 3.11 Calibration curve of bisphenol-A detected using desorption corona beam ionization–mass spectrometry.

devices difficult to rely on, and demand for an alternative method with high accuracy and high speed is increasing. Motivated by this, we tested DCBI coupled with a single quadrupole MS for the same purpose of analyzing explosives and illicit drug molecules [10].

Fig. 3.12 shows three explosives (TNT, CE, and hexanitrostilbene standard solutions) detected with DCBI-MS. Normally all explosives are thermally unstable chemicals, and thermal decomposition is not uncommon for such a thermal desorption process. Nevertheless, as shown in Fig. 3.12, the observed major peaks were mainly parent ions or parent ions with the loss of nitro groups when we use DCBI-MS, which indicates that DCBI is a very soft ionization source. In order to obtain more complete information about the explosives, we developed an accurate temperature scan process using the DCBI source, as shown in Fig. 3.13. The chromatogram in Fig. 3.13A shows that the three explosives (nitroguanidine [MW 104];

Table 3.3 Measured concentrations of bisphenol-A in different plastic samples with both LC-MS and DCBI-MS methods.

Sample #	Plastic #1	Plastic #2 (ppb)	Plastic #3
LC-MS	Not detected	200	Not detected
DCBI-MS	Not detected	271	Not detected

DCBI-MS, desorption corona beam ionization–mass spectrometry; *LC-MS*, liquid chromatography–mass spectrometry.

PETN [MW 315], and hexanitrostilbene [MW 450]) were sequentially desorbed and ionized within 5 minutes. Fig. 3.13B shows the same process but over a shorter time scale (3 minutes) with a higher heating rate. This programmed temperature scan is a powerful method for the identification of various thermally unstable compounds while maintaining a short analysis time.

Illicit drug molecules were also tested using DCBI-MS. Fig. 3.14 shows the mass spectra of two drugs (heroin and ketamine) in the positive-ion mode. The radical cations of the precursors give rise to the major peaks in both spectra, which again proves how soft this DCBI source is and that it does not rely on proton transfer.

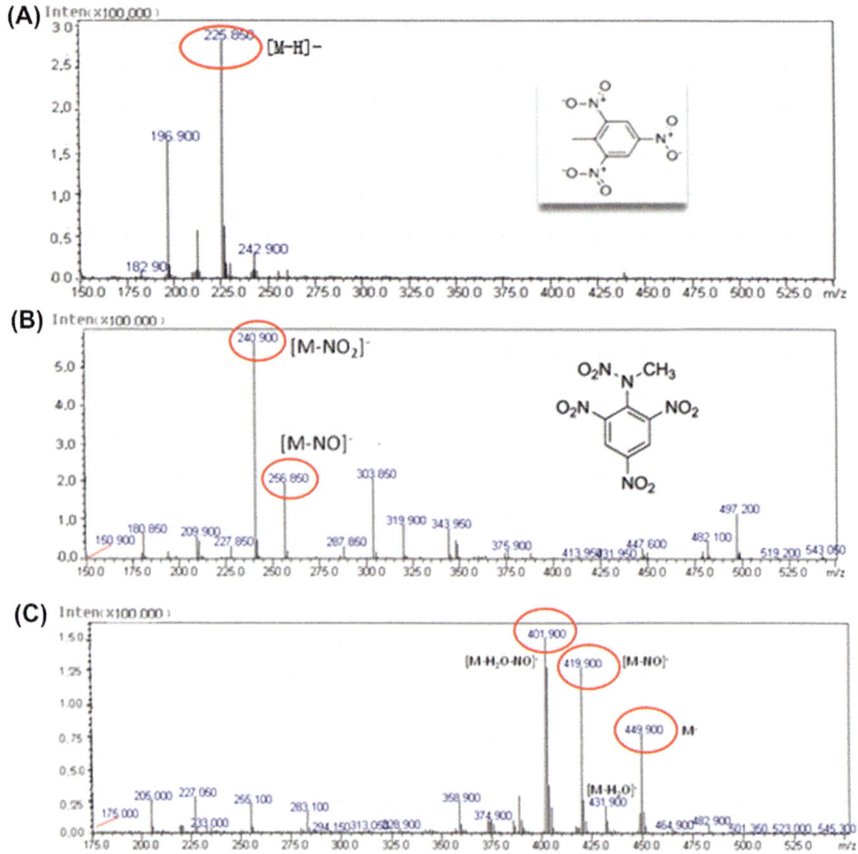

Figure 3.12 Mass spectra of (A) TNT (MW: 227.1), (B) CE (MW: 288.2), and (C) hexanitrostilbene (MW: 450.2) acquired with desorption corona beam ionization–mass spectrometry.

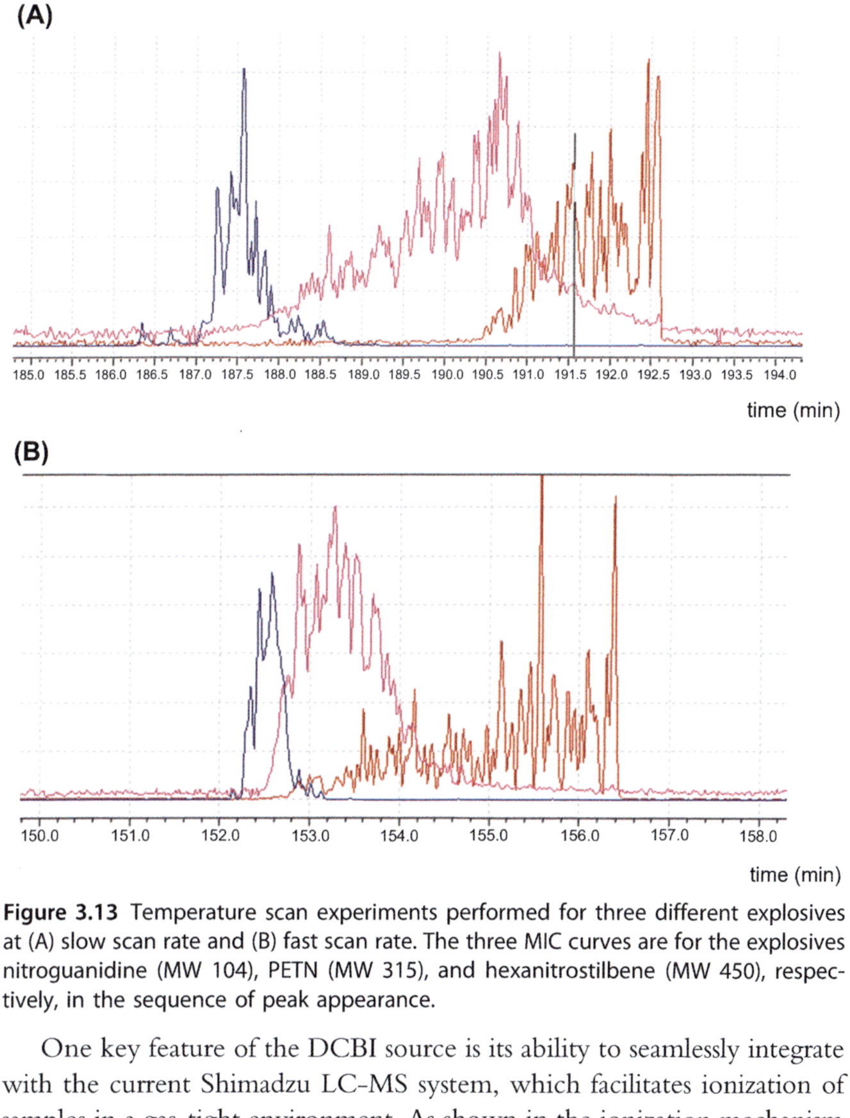

Figure 3.13 Temperature scan experiments performed for three different explosives at (A) slow scan rate and (B) fast scan rate. The three MIC curves are for the explosives nitroguanidine (MW 104), PETN (MW 315), and hexanitrostilbene (MW 450), respectively, in the sequence of peak appearance.

One key feature of the DCBI source is its ability to seamlessly integrate with the current Shimadzu LC-MS system, which facilitates ionization of samples in a gas-tight environment. As shown in the ionization mechanism for DCBI (Fig. 3.2), electrons in the negative-ion mode may be captured by O_2. If the electron affinity of the analyte is low, the chance that it is ionized by obtaining electrons will be very low. Therefore, removing O_2 from the environment by replacing with N_2 is a viable way to further improve sensitivity. Fig. 3.3 demonstrates such advantage for explosive molecules, and similar results have also been obtained with other compounds such as BPA.

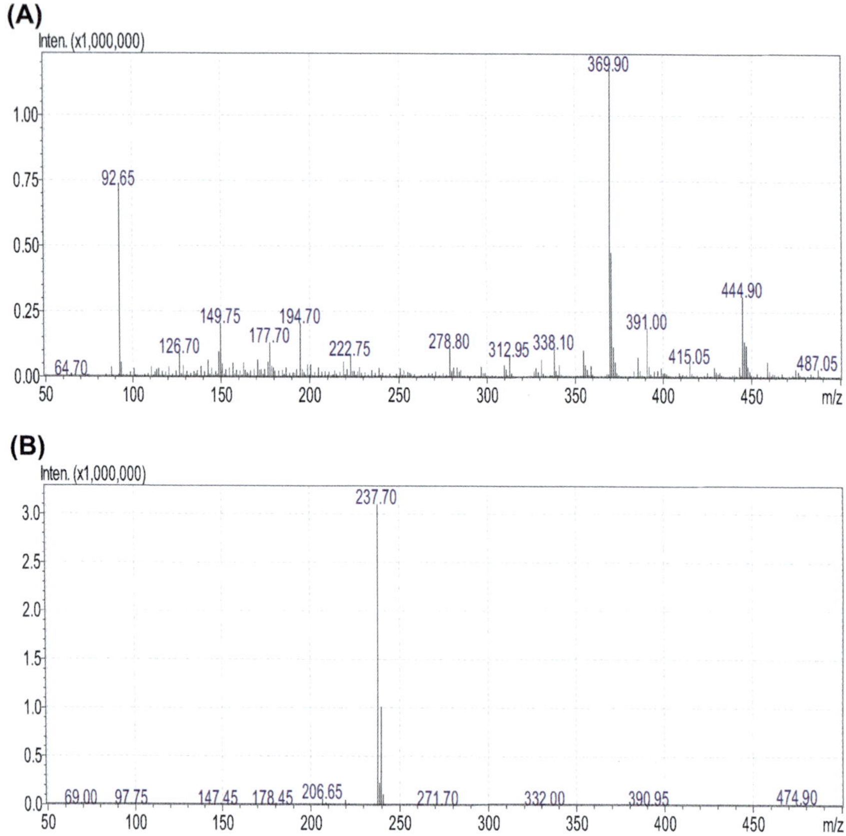

Figure 3.14 Mass spectra for two illicit drugs: (A) heroin, *m/z* 369.4 and (B) ketamine, *m/z* 237.7.

3.4.3 Direct detection of compounds in body fluids for life science

Direct analysis of drug compounds and their metabolites in body fluids such as blood, urine, and saliva is always desirable, especially considering the huge annual expenditure of the in vitro diagnosis industry globally and in China. One challenge for body fluid analysis is to detect low concentrations of analytes in a complex matrix. MS has become the main driving force in this area owing to its high sensitivity and both qualitative and quantitative results. Similar to the case described earlier in which BPA was analyzed using DCBI-MS, the ability of DCBI-MS to analyze samples directly without going through flow injection, where clogging often happens, makes much simpler sample pretreatments possible. For example, You and coworkers have reported a simple and fast sample preparation method for analyzing drug levels in blood samples with DCBI-MS [20].

In this method, a micropipette tip-based microextraction was coupled to DCBI-MS. A 10 μL micropipette tip filled with C18 ZipTip column material was used. The main procedure is as follows (also shown in Fig. 3.15): (1) activation of the C18 material with 10 μL of acetonitrile three times (wetting), (2) rinsing with 10% acetonitrile three times (equilibration), (3) taking a blood sample from a fingertip with a pipette (releasing and drawing back the blood sample three times to ensure full interaction of the target molecules with the column; sampling), (4) using 20% acetonitrile for rinsing in order to release interfering materials such blood cells, proteins, and other large polar molecules (rinsing), (5) rinsing the tip again with 2 μL of 100% methanol in order to release the analytes (drug molecules) of interest onto a test paper (eluting), and (6) loading the test paper for DCBI-MS analysis after the solvent has evaporated. By following this procedure, the drug molecules can be significantly enriched and interference from the other components of the blood can be removed. The whole process takes less than 3 minutes including the extraction lasting for 2 minutes, drying for half a minute, and DCBI-MS for half a minute.

This method was successfully used for the analysis of antihypertension drugs such as nifedipine, nitrendipine, nimodipine, and illegal drugs in body fluids. The pyridine cycle for each compound means that it is easy for hydrogen to be lost under illumination from DCBI and, therefore, two peaks are observed for each compound with a mass difference of 2 Th

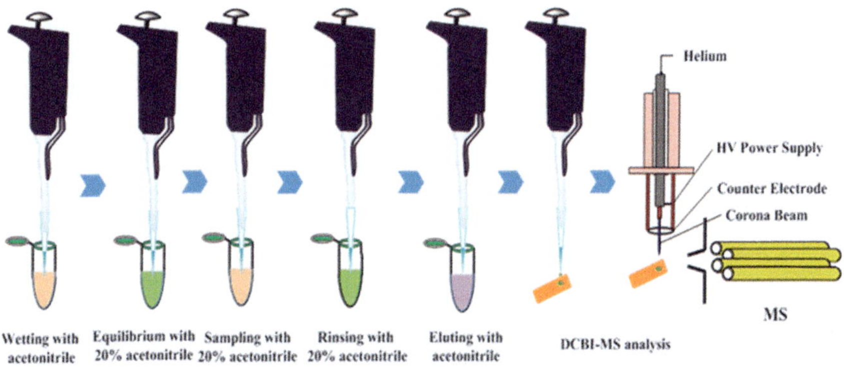

Figure 3.15 Sample pretreatment procedure for ZipTip desorption corona beam ionization–mass spectrometry. *(Reprinted with permission from K.M. Walsh, E. Leen, M.E.J. Lean, The effect of sibutramine on resting energy expenditure and adrenaline-induced thermogenesis in obese females, Int. J. Obes. 23 (1999) 1009–1015.)*

(Nifedipine *m/z* 347, 345; Nitrendipine *m/z* 361, 359; Nimodipine 419, 417). The sensitivities of these compounds to DCBI ionization were examined first by varying the concentration and upload quantity of the three compounds, as shown in Fig. 3.16. When the concentration of the sample was reduced from 500 to 200 ug/L, the signal was obviously reduced. However, if the upload quantity was increased to 50 μL, the signal can be recovered by the compensating amount of sample, as shown in Fig. 3.16C. This indicates that the enrichment process on the ZipTip is as effective when using a higher volume of sample.

After testing the pure compounds, experiments were conducted to analyze the antihypertension drugs in blood plasma. Fig. 3.17 shows the results for different conditions. Plasma was spiked with 10 μL of each compound in Fig. 3.17A and 50 μL of each compound in Fig. 3.17B. In Fig. 3.17C, plasma spiked with 50 uL of each compound was re-extracted and analyzed. In Fig. 3.17D, plasma was analyzed directly without extraction. No positive signal was observed for direct analysis of the plasma, whereas a strong signal was observed after extraction, even from residual samples. The enrichment function of this method is therefore very effective.

A similar effect was also observed when analyzing ketamine in urine samples. Fig. 3.18A shows the DCBI-MS analysis results after extraction; the major $[M]^{\cdot+}$ peak has a good S/N ratio, whereas no peak due to ketamine can be observed if only analysis without extraction is performed, as shown in Fig. 3.18B.

Similar to the concept of using ZipTip for sample preconcentration, other types of extraction materials were also tested in order to reduce matrix effects and enrich the targeted analytes in body fluids. In the following two examples, either a magnetic solid-phase extraction (MSPE) [21] or a thin-film microextraction (TFME) [22] method was used for the purposes mentioned above before desorbing and ionizing the samples with DCBI.

As a proof of concept, Chen and coworkers studied MSPE using pyrrole-coated Fe_3O_4 magnetic nanoparticles (Fe3O4@Ppy) for the extraction of antidepressants. As shown in Fig. 3.19, an Fe_3O_4@Ppy suspension (20 μL, 0.2 mg Fe_3O_4@Ppy) was added to urine or plasma spiked with antidepressants. After vortexing for 1 mintue, the antidepressant-coated Fe_3O_4@Ppy was magnetically gathered at the bottom of the vial with the assistance of an external magnet and then washed with H_2O. After the external magnet was removed, a magnetic glass capillary was inserted into the vial to collect the nanoparticles before being transferred to the DCBI source for detection. As a control without magnetic nanoparticle

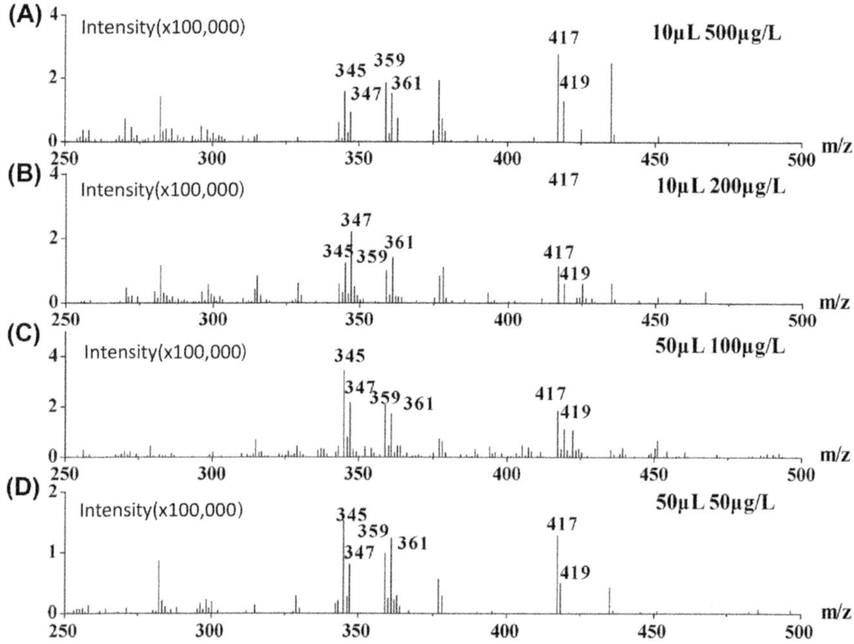

Figure 3.16 Dependence of signal intensity on the concentration and volume of the analyte solution: (A) 10 μL 500 μg/L, (B) 10 μL 200 μg/L, (C) 50 μL 100 μg/L, and (D) 50 μL 50 μg/L). *(Reprinted with permission from K.M. Walsh, E. Leen, M.E.J. Lean, The effect of sibutramine on resting energy expenditure and adrenaline-induced thermogenesis in obese females, Int. J. Obes. 23 (1999) 1009–1015.)*

enrichment, 2 μL of sample solution was deposited on the magnetic glass capillary directly and transferred to the DCBI source for analysis.

The analytes used in this study are three different kinds of antidepressants (citalopram, sertraline, and fluoxetine), all of which are selective serotonin reuptake inhibitors that have been marketed and widely introduced in depression therapy [21]. Therapeutic drug measurement of antidepressants in body fluids is important for the determination of an efficient and safe dose, and for the detection of adherence and compliance with the treatment by the patient. Fig. 3.20 shows the difference between using DCBI with and without MSPE pretreatment for detecting these antidepressants. Fig. 3.20A shows there was only one compound that could be identified when a urine sample was directly analyzed with DCBI-MS, whereas Fig. 3.20B shows that all three antidepressants can be identified after MSPE.

To further prove the adsorption efficiency of the pyrrole coating when using magnetic beads, experiments were conducted to compare the signal

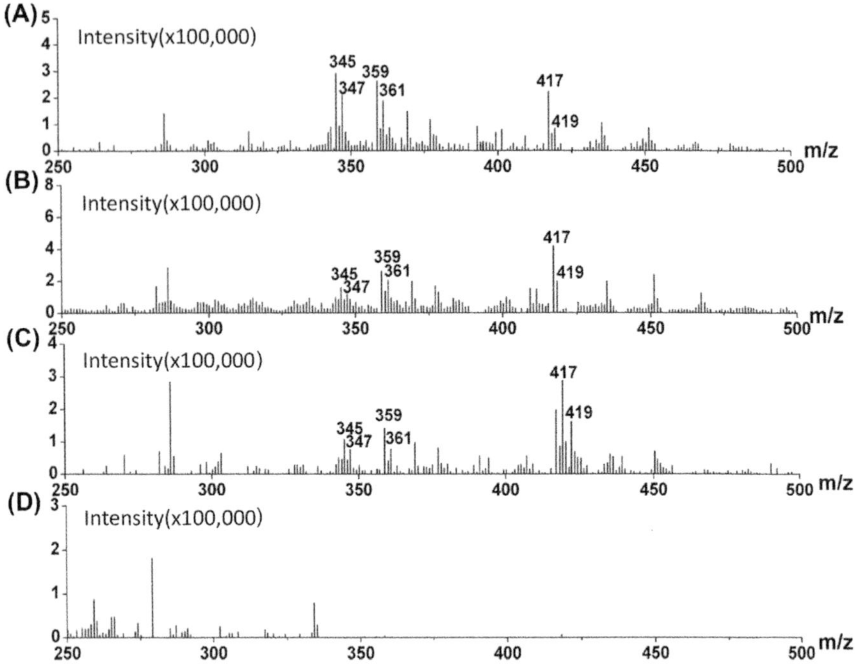

Figure 3.17 Different methods used for analysis of plasma spiked with antihypertensive drugs. (A) micropipette tip-based microextraction coupled with desorption corona beam ionization—mass spectrometry (DCBI-MS) for analysis of 10 μL of spiked plasma; (B) micropipette tip-based microextraction coupled with DCBI-MS for analysis of 50 μL of spiked plasma; (C) re-extraction and analysis of the residue of the 50 μL plasma sample; (D) direct analysis of the plasma sample without extraction. *(Reprinted with permission from K.M. Walsh, E. Leen, M.E.J. Lean, The effect of sibutramine on resting energy expenditure and adrenaline-induced thermogenesis in obese females, Int. J. Obes. 23 (1999) 1009—1015.)*

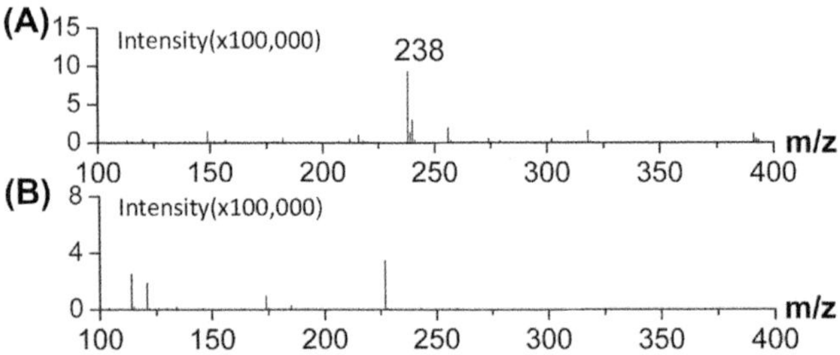

Figure 3.18 Analysis of ketamine (*m/z* 237.7) in urine samples. (A) analysis with extraction; (B) analysis without extraction. *(Reprinted with permission from K.M. Walsh, E. Leen, M.E.J. Lean, The effect of sibutramine on resting energy expenditure and adrenaline-induced thermogenesis in obese females, Int. J. Obes. 23 (1999) 1009—1015.)*

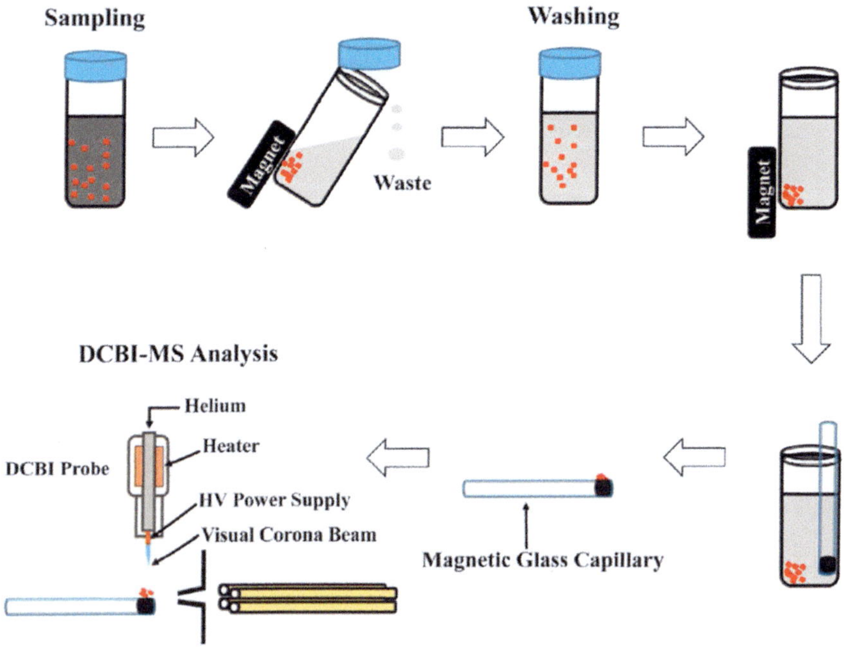

Figure 3.19 Experimental protocol and configuration of the magnetic solid-phase extraction–desorption corona beam ionization–mass spectrometry system. 213 × 159 mm (300 × 300 DPI). *(Reprinted with permission from D. Chen, H. Zheng, Y. Huang, Y. Hu, Q. Yu, B Yuan, Y. Feng, Magnetic solid phase extraction coupled with desorption corona beam ionization-mass spectrometry for rapid analysis of antidepressants in human body fluids, Analyst 140 (2015) 5662–5670.)*

intensity between three different conditions: direct analysis without MSPE, MSPE without pyrrole coating, and MSPE with pyrrole coating. Fig. 3.21 shows such a comparison for the three antidepressants without using MSPE (curve a) and using MSPE-DCBI-MS with bare Fe_3O_4 (curve b) or Fe_3O_4 with pyrrole as sorbents (curve c). It can be seen that the extraction effect of the pyrrole sorbent is very important for good extraction performance, with enrichment factors ranging from 20 to 60. These results suggest that the combination of MSPE with DCBI-MS provides an effective and sensitive method for the determination of antidepressants in body fluids.

Similar to the MSPE method, the three antidepressants were also enriched using TFME by the same group of authors [21]. In this work, the thin films used for extraction comprised submicron-sized highly ordered mesoporous silica-carbon composite fibers (OMSCFs), which were simply prepared by electrospinning and subsequent carbonization. Typically,

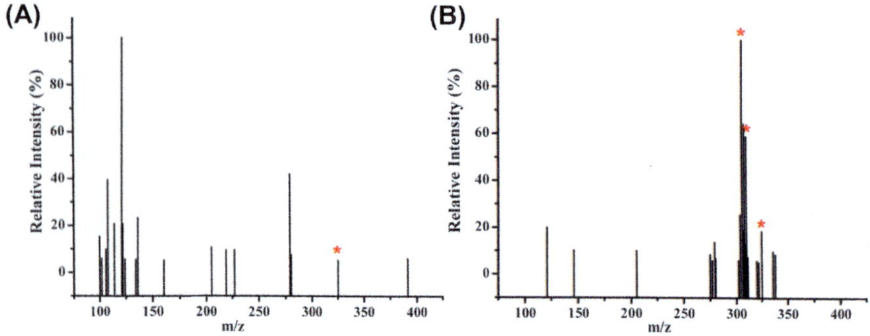

Figure 3.20 Analysis of three urine sample spiked with antidepressants (500 ng/mL) using full-scan positive-ion mode. Mass spectra obtained by (A) direct desorption corona beam ionization–mass spectrometry analysis of 2 μL spiked urine sample and (B) magnetic solid-phase extraction–desorption corona beam ionization–mass spectrometry analysis of spiked urine sample. 178 × 74 mm (300 × 300 DPI). *(Reprinted with permission from D. Chen, H. Zheng, Y. Huang, Y. Hu, Q. Yu, B Yuan, Y. Feng, Magnetic solid phase extraction coupled with desorption corona beam ionization-mass spectrometry for rapid analysis of antidepressants in human body fluids, Analyst 140 (2015) 5662–5670.)*

OMSCF thin films were immersed into the diluted plasma for extraction of the target analytes and then directly subjected to DCBI-MS for detection.

Compared with the MSPE method discussed above, this thin-film method is even simpler. Fig. 3.22A shows the basic procedure of this measurement process.

The mesopore structure of the thin-film has a size-exclusion effect, which can avoid protein precipitation and thus reduce interference from large molecules. Moreover, the OMSCFs provided mixed-mode hydrophobic/ion-exchange interactions toward target analytes, which can also greatly improve the sensitivity. Human plasma samples (2 μL) containing 1 mg/μL of each antidepressant were tested using DCBI-MS without (Fig. 3.22B) and with TFME enrichment (Fig. 3.22C), and it was obvious that enrichment by TFME is effective (analytes can only be observed after TFME).

Furthermore, coexisting interference from plasma would obviously affect the extraction efficiency as well as the desorption and ionization process. Fig. 3.23 indicates the matrix effect (the ratio of the signal intensity of the spiked plasma sample to the signal intensity obtained from a standard solution of the analyte with the same spiking concentration of 1 μg/mL). The absolute matrix effect values were calculated to be 0.375, 0.360, and 0.462 for citalopram, sertraline, and fluoxetine, respectively. Although these results show that TFME

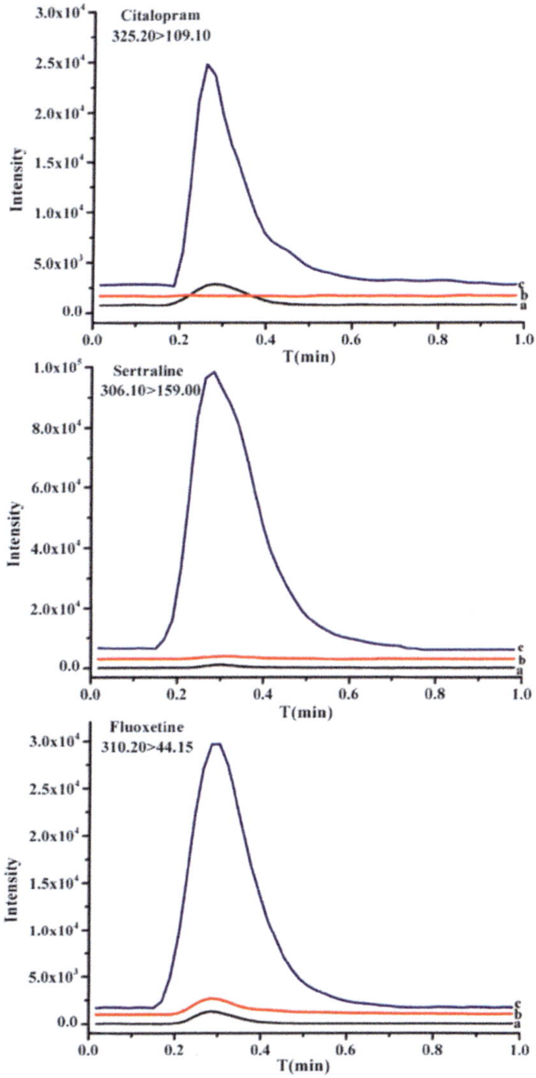

Figure 3.21 Mass spectrometry signals of three antidepressants obtained by direct desorption corona beam ionization—mass spectrometry (DCBI-MS) without magnetic solid-phase extraction (MSPE) (curve a), and MSPE-DCBI-MS using bare Fe_3O_4 (curve b) or Fe_3O_4@Ppy as sorbents (curve c). Urine was spiked with the analytes at 500 ng/mL 73 × 152 mm (300 × 300 DPI). *(Reprinted with permission from D. Chen, H. Zheng, Y. Huang, Y. Hu, Q. Yu, B Yuan, Y. Feng, Magnetic solid phase extraction coupled with desorption corona beam ionization-mass spectrometry for rapid analysis of antidepressants in human body fluids, Analyst 140 (2015) 5662–5670.)*

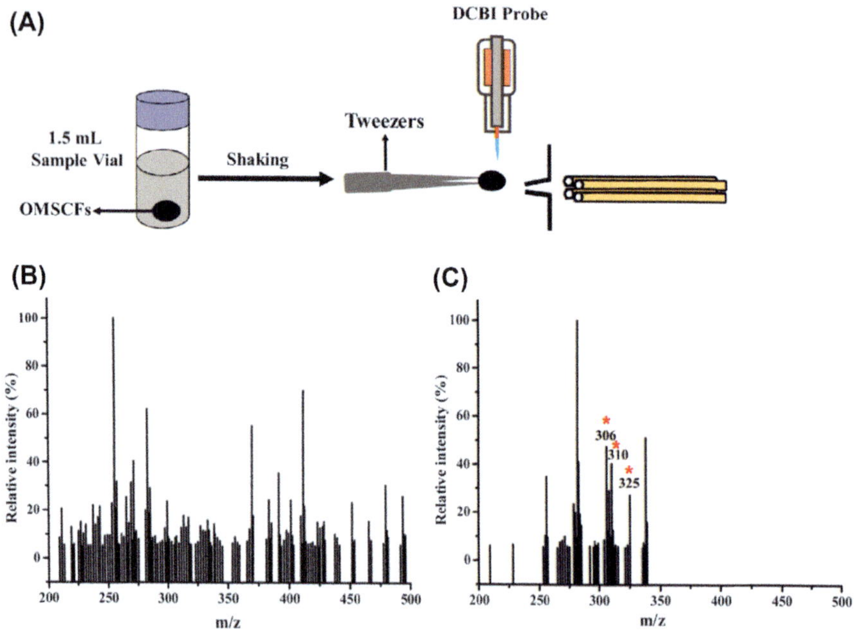

Figure 3.22 (A) Experimental protocol and configuration of the thin-film microextraction–desorption corona beam ionization–mass spectrometry (DCBI-MS) system, (B) mass spectra obtained by direct DCBI-MS analysis of 2 μL spiked plasma sample containing three antidepressants (1 μg/mL) in full-scan positive-ion mode, and (C) mass spectra obtained by TFME-DCBI-MS analysis for the same samples. *(Reprinted with permission from D. Chen, Y. Hu, D. Hussain, G. Zhu, Y. Huang and Y. Feng, Electrospun fibrous thin film microextraction coupled with desorption corona beam ionization-mass spectrometry for rapid analysis of antidepressants in human plasma, Talanta 152 (2016) 188–195.)*

cannot eliminate the matrix effect completely, the matrix interference after extraction with OMSCFs was less than reported for the materials (Fe3O4@Ppy) described in the previous session (0.17–0.25) [20].

3.5 Summary

DCBI has been developed as an ambient-pressure direct analysis source with a unique configuration. Helium gas was used to form a visible corona beam when a high DC voltage was applied to the source at low current (mA). For most of the samples to be desorbed/ionized, it was necessary to heat the helium gas for thermal desorption. The types of samples that can be analyzed are mainly small molecules, but molecules with molecular weight

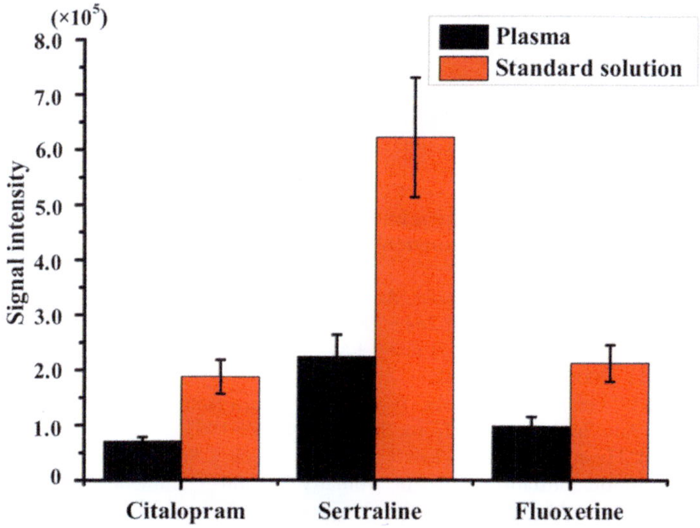

Figure 3.23 Matrix effect on thin-film microextraction–desorption corona beam ionization–mass spectrometry signals. Antidepressants were spiked at a concentration of 1 μg/mL. *(Reprinted with permission from D. Chen, H. Zheng, Y. Huang, Y. Hu, Q. Yu, B Yuan, Y. Feng, Magnetic solid phase extraction coupled with desorption corona beam ionization-mass spectrometry for rapid analysis of antidepressants in human body fluids, Analyst 140 (2015) 5662–5670.)*

of up to 1500 Th have also been detected at the highest heating temperature. The seamless integration of DCBI with LC-MS makes an oxygen-free environment possible, which can enhance the signal intensity by more than one order of magnitude in the negative-ion mode. In addition, the visibility of the corona beam makes the plasma easy to align with the sample and makes it possible for use in areas where lateral resolution is required.

DCBI has been used in a wide range of different applications. These areas include food and drug safety, explosives/illicit drugs, and medicine/metabolites in body fluids for life science. Two approaches to applying DCBI were used: direct analysis with and without pretreatment. For analytes in a very simple matrix (e.g., explosives directly sampled from a solid surface), the sampling glass tip can be directly introduced into the DCBI plasma and the matrix effect is not severe. On the other hand, particularly when using DCBI for life sciences where the matrix effect is most severe (e.g., body fluids contain cells, proteins, lipids, and other small molecules), special but simple pretreatments were developed to reduce such matrix effects and further enrich the analytes (examples include the MSPE and

TFME methods). The application of DCBI in clinical areas can be further broadened when newer and more efficient pretreatment techniques are developed.

References

[1] Z. Takats, J.M. Wiseman, B. Gologan, R.G. Cooks, Mass spectrometry sampling under ambient conditions with desorption electrospray ionization, Science 306 (2004) 471–473.
[2] A. Venter, M. Nefliu, R.G. Cooks, Ambient desorption ionization mass spectrometry, Trends Anal. Chem. 27 (2008) 284–290.
[3] R.G. Cooks, Z. Ouyang, Z. Takats, J.M. Wiseman, Ambient mass spectrometry, Science 311 (2006) 1566–1570.
[4] H. Chen, G. Gamez, R. Zenobi, What can we learn from ambient ionization techniques? J. Am. Soc. Mass Spectrom. 20 (2009) 1947–1963.
[5] J. Shiea, M. Huang, H. Hsu, C. Lee, C. Yuan, I. Beech, J. Sunner, Electrospray-assisted laser desorption/ionization mass spectrometry for direct ambient analysis of solids, Rapid Commun. Mass Spectrom. 19 (2005) 3701–3704.
[6] R.B. Cody, J.A. Laramee, H.D. Durst, Versatile new ion source for the analysis of materials in open air under ambient conditions, Anal. Chem. 77 (2005) 2297–2302.
[7] C.N. McEwen, R.G. McKay, B.S. Larsen, Analysis of solids, liquids, and biological tissues using solids probe introduction at atmospheric pressure on commercial LC/MS instruments, Anal. Chem. 77 (2005) 7826–7831.
[8] Z. Takats, I. Cotte-Rodriguez, N. Talaty, H. Chen, R.G. Cooks, Direct, trace level detection of explosives on ambient surfaces by desorption electrospray ionization mass spectrometry, Chem. Commun. (2005) 1950–1952.
[9] H. Wang, W. Sun, J. Zhang, X. Yang, T. Lin, L. Ding, Desorption corona beam ionization source for mass spectrometry, Analyst 135 (2010) 688–695.
[10] W. Sun, J. Zhang, J. Ting, L. Yang, J. Yan, X. Zhang, L. Ding, An optimized design of the desorption corona beam ionization source (DCBI) and its applications, in: Proceedings of 59th Annual Conference for Am. Soc. Mass Spectrom, 2011.
[11] Y. Huang, J. You, J. Zhang, W. Sun, L. Ding, Y. Feng, Coupling frontal elution paper chromatography with desorption corona beam ionization mass spectrometry for rapid analysis of chlorphenamine in herbal medicines and dietary supplements, J. Chrom. A 1218 (2011) 7371–7376.
[12] S. Cheng, Y. Cheng, W. Sun, J. Shiea, Differentiating cooking oil from gutter oil by desorption corona beam ionization mass spectrometry, in: Proceedings of 60th Annual Conference for Am. Soc. Mass Spectrom, 2012.
[13] S.D. Glick, R.E. Haskew, I.M. Maisonneuve, J.N. Carlson, T.P. Jerussi, Enantioselective behavioral effects of sibutramine metabolites, Eur. J. Pharmacol. 397 (2000) 93–102.
[14] I.D. Hind, J.E. Mangham, S.P. Ghani, R.E. Haddock, C.J. Garratt, R.W. Jones, Sibutramine pharmacokinetics in young and elderly healthy subjects, Eur. J. Clin. Pharmacol. 54 (1999) 847–849.
[15] K.M. Walsh, E. Leen, M.E.J. Lean, The effect of sibutramine on resting energy expenditure and adrenaline-induced thermogenesis in obese females, Int. J. Obes. 23 (1999) 1009–1015.
[16] C.A. Luque, J.A. Rey, The discovery and status of sibutramine as an anti-obesity drug, Eur. J. Pharmacol. 440 (2002) 119–128.

[17] H.M. Connolly, J.L. Crary, M.D. McGoon, D.D. Hensrud, B.S. Edwards, W.D. Edwards, H.V. Schaff, Valvular heart disease associated with fenfluramine–phentermine, N. Engl. J. Med. 337 (1997) 581–588.
[18] N.J. Weissman, Appetite suppressants and valvular heart disease, Am. J. Med. Sci. 321 (2001) 285–291.
[19] H. Wang, Y. Wu, Y. Zhao, W. Sun, L. Ding, B. Guo, B. Chen, Rapid screening of illicit additives in weight loss dietary supplements with desorption corona beam ionisation (DCBI) mass spectrometry, Food Addit. Contam. 29 (2012) 1194–1201.
[20] J. You, Y. Hao, Y. Huang, Y. Cheng, W. Sun, Q. Yu, B. Yuan, Y. Feng, Micropipette tip-based micro-extraction - desorption corona beam ionization mass spectrometry for rapid analysis of antihypertensive drugs in body fluids, Chin. J. Anal. Chem. 03 (2013) 319–322.
[21] D. Chen, H. Zheng, Y. Huang, Y. Hu, Q. Yu, B. Yuan, Y. Feng, Magnetic solid phase extraction coupled with desorption corona beam ionization-mass spectrometry for rapid analysis of antidepressants in human body fluids, Analyst 140 (2015) 5662–5670.
[22] D. Chen, Y. Hu, D. Hussain, G. Zhu, Y. Huang, Y. Feng, Electrospun fibrous thin film microextraction coupled with desorption corona beam ionization-mass spectrometry for rapid analysis of antidepressants in human plasma, Talanta 152 (2016) 188–195.

CHAPTER 4

DESI-based imaging mass spectrometry in forensic science and clinical diagnosis

Yuki Sugiura, Eiji Sugiyama, Makoto Suematsu
Department of Biochemistry, Keio University School of Medicine, Tokyo, Japan

Contents

4.1 Principle of DESI

Desorption electrospray ionization (DESI) serves as a source for the introduction of ionized molecules into the mass spectrometer (MS) inlet under atmospheric pressure [1,2] (Fig. 4.1A). A focused beam of charged ESI solvent droplets is aimed at the sample surface to desorb analyte molecules. The desorbed analytes are ionized in the process and carried into the inlet capillary, through which they are transferred to the MS. Due to the similarities between their ion formation mechanisms, DESI mass spectra generally resemble conventional ESI mass spectra [3]. The alkali metal adducts of small molecular ions and the multiply charged molecular ions of proteins are typically observed (Fig. 4.1B) [3].

DESI ionization parameters are flexible. For example, the composition of the spraying solution can be adjusted to selectively ionize target compounds [4], including small compounds, peptides and proteins present on metal, polymer, and mineral surfaces [3]. In some cases, DESI can ionize nonpolar compounds, such as cholesterol, carotene, and TNT(2,4,6-Trinitrotoluene), which are generally not ionized by electrospray ionization (ESI) or matrix-assisted laser desorption ionization (MALDI). The

Ambient Ionization Mass Spectrometry in Life Sciences
ISBN 978-0-12-817220-9
https://doi.org/10.1016/B978-0-12-817220-9.00004-7

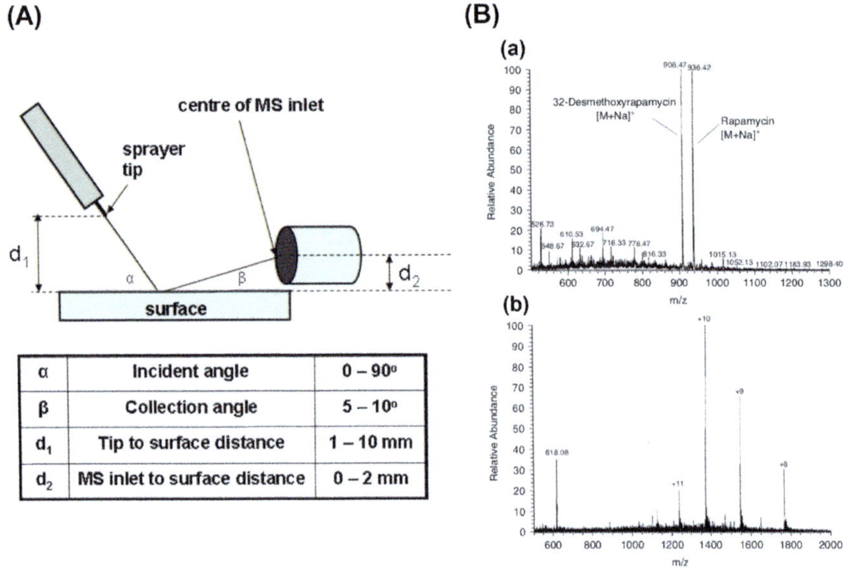

α	Incident angle	0 – 90°
β	Collection angle	5 – 10°
d_1	Tip to surface distance	1 – 10 mm
d_2	MS inlet to surface distance	0 – 2 mm

Figure 4.1 Instrumental setup and spectrum characteristics of desorption electrospray ionization (DESI). A Schematic diagram of a DESI source and MS inlet. B. (A) Positive ion DESI spectrum of human serum spiked with small molecular drug rapamycin and an equal amount of 32-desmethoxy rapamycin (60 μg/L). Ions at m/z 906 and 936 correspond to the sodiated molecular ions of each compound. (B) Positive ion DESI spectrum of protein, equine cytochrome *c* (10 ng),deposited on a poly(methyl methacrylate) (PMMA) surface showing multiple charged molecular ions. *(Adapted from Z. Takats, J.M. Wiseman, R.G. Cooks, Ambient mass spectrometry using desorption electrospray ionization (DESI): instrumentation, mechanisms and applications in forensics, chemistry, and biology, J. Mass Spectrom. 40 (10) (2005) 1261–1275.)*

spectral characteristics of DESI of such molecules are similar to those of corona discharge or atmospheric pressure chemical ionization [3]. Another attractive feature of DESI is that sample destruction is minimal, so samples can be analyzed multiple times, and tissue samples can undergo histochemical staining following IMS analysis.

4.2 Application I: forensic science

One of the most famous images generated by DESI is the "cocaine fingerprint" reported by Ifa et al. in 2008 [5] (Fig. 4.2). The authors performed chemical analysis and fingerprint identification with DESI-IMS. Such chemical fingerprints would make it possible for law enforcement to link evidence of illicit drug use to a specific person. This

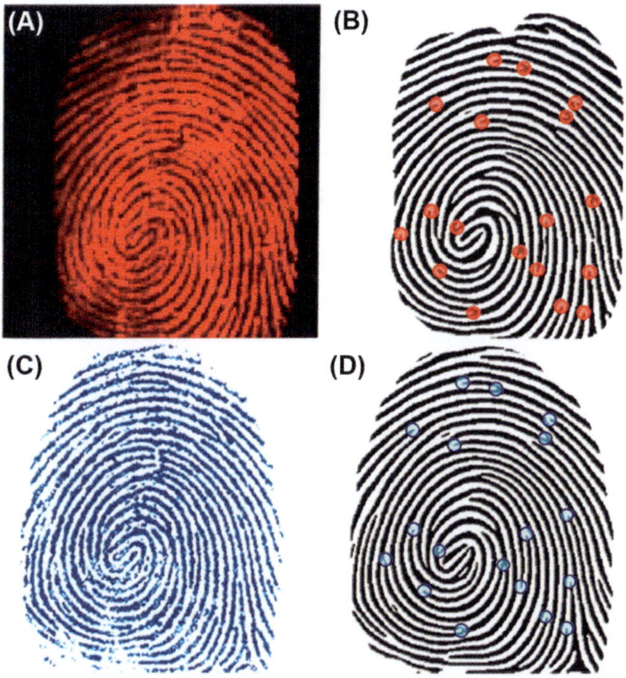

Figure 4.2 Fingerprint identification with chemical analysis by desorption electrospray ionization–imaging mass spectrometry (DESI-IMS). (A) DESI image of cocaine distribution on a live scan fingerprint blotted on glass. (B) Digital processed image of the fingerprint generated from the DESI image. (C) Ink fingerprint blotted on paper and (D) digital image of the fingerprint generated from (C). Several of the automatically detected points of interest (minutiae) are circled in B and D. *(Adapted from D.R. Ifa, N.E. Manicke, A.L. Dill, R.G. Cooks, Latent fingerprint chemical imaging by mass spectrometry, Science 321(5890) (2008) 805.)*

early study demonstrated two critical advantages of DESI that established the feasibility of its use in forensic analysis. First, nonpolar molecules, including lycopene, the alkaloid coniceine, and small molecule drugs were successfully detected by DESI. Second, surface analyses on both conductive and nonconductive materials were achieved. DESI was thus applied in subsequent studies for the detection of illicit drugs, explosives, chemical warfare agents, inks and documents, fingerprints, and gunshot residues. Surface analysis has been performed with a broad range of materials, including skin, glass, rubber and leather gloves, towels, medical gauze, and cotton [6]. DESI has even been used for the analysis of urine [7] and spotted blood specimens [8].

Again, another notable advantage of DESI is the ability to preserve a sample in optimal condition, as it remains nearly intact following analysis. This advantage was demonstrated in another study by Ifa et al. [9], in which DESI-IMS was used for forensic analysis of inks. Inks are commonly examined during document authentication when forgery is suspected. Analysis of a questionable document generally begins with a determination of whether different inks have been applied to it. Traditional analysis involves removing a piece of the document and extracting the inks, so valuable documents cannot be authenticated this way. In the study, the authors used DESI-IMS to identify different inks among several written characters on a document. The sample was preserved, thereby establishing the promise of DESI-IMS for this application. As Fig. 4.3 nicely illustrates, DESI-IMS provides sufficient sensitivity and spatial resolution for the analysis of ink on paper documents. Thus, the results reported by Ifa et al. are excellent examples of how DESI coupled with imaging technology could be employed for routine forensic analysis.

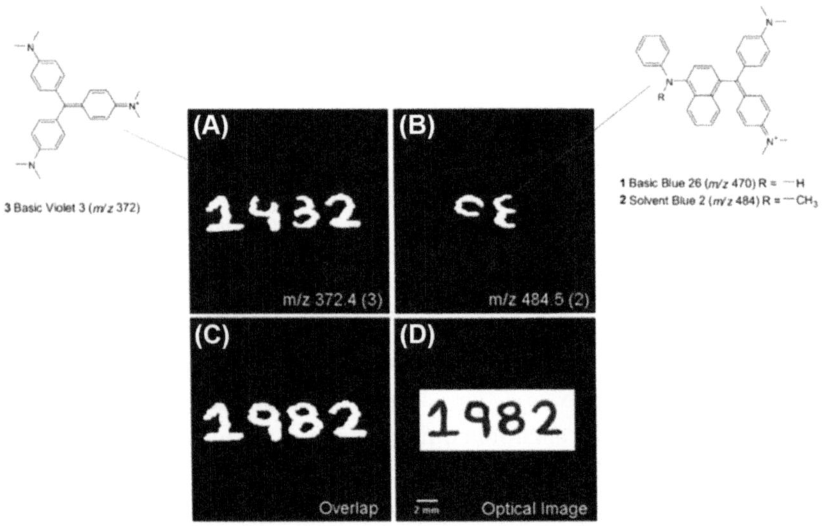

Figure 4.3 Desorption electrospray ionization images can discriminate forged numbers on printer paper demonstrated by Ifa et al. [9]. (A) Two-dimensional ion image of Basic Violet 3, m/z 372. (B) Two-dimensional ion image of Solvent Blue 2, m/z 484. (C) Overlay of A and B combining Basic Violet 3 and Basic Blue 2 chemical images. (D) Optical image of the surface. Identified structures are presented in Scheme 1. *(Adapted from D.R. Ifa, L.M. Gumaelius, L.S. Eberlin, N.E. Manicke, R.G. Cooks, Forensic analysis of inks by imaging desorption electrospray ionization (DESI) mass spectrometry, Analyst 132(5) (2007) 461–467.)*

4.3 Application II: metabolite imaging for clinical diagnosis

Cancer cells dynamically modify their protein expression, which enables them to alter energy and lipid metabolism to survive. Their abnormal molecular composition profiles contain information that is valuable for clinical diagnosis. This information enables clinicians to not only distinguish cancerous cells from surrounding normal cells, but also to recognize specific cell types, their stages, malignancy, and to develop prognoses for patients. There are numerous reports in the clinical literature of applying IMS to study surgically dissected tumor specimens [10−14]. These researchers have successfully visualized and characterized significant differences between the cellular molecular compositions of normal, stroma, and cancerous tissues.

These advances illustrate the capability of IMS as a practical diagnostic tool, particularly as an intraoperative diagnostic method for identifying borders between normal and cancerous tissues [15]. MALDI-IMS is the most frequently used clinical imaging technique, as it enables sensitive detection for the analysis of a wide range of molecular species. However, sample preparation requires skill and long-term experience, especially for the matrix coating process. This is a critical and practical problem when IMS is introduced to an in-hospital laboratory, where it will be operated by physicians and clinical laboratory technologists.

Obviously, the minimal and rapid DESI sample preparation process is a significant advantage for medical use. Several leading studies have shown the effectiveness of DESI for ionization and detection of diagnostic molecules and its usefulness for clinical diagnosis [16−19]. Eberlin et al. [20] provided an excellent demonstration of DESI profiling coupled with multivariate statistical analysis and machine learning, in which gliomas were rapidly classified according to their tumor subtypes (oligodendroglioma, astrocytoma, and oligoastrocytoma), histologic grade, and cell concentration. Through a statistical analysis of lipid compositions among the glioma types, they successfully characterized the tumors and achieved >97% cross-validation in sample sets from 36 patients (Fig. 4.4A).

The rapid determination of borders between cancerous and normal tissues based on a DESI dataset was demonstrated by Banerjee et al. [21]. Their rapid determinations are generally performed within minutes. The authors utilized glucose metabolites, including Krebs cycle intermediates, which were highly informative for distinguishing between cancerous and benign tissue. They developed a statistical model to distinguish between

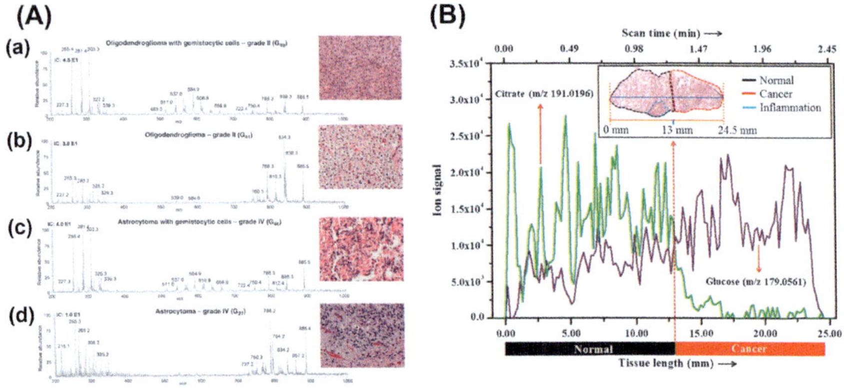

Figure 4.4 Feasibility studies of desorption electrospray ionization (DESI)-based clinical diagnosis to identify tumors and surrounding normal tissues. (A) Eberlin et al. [20] showed DESI profiling can rapidly classify according to tumor subtypes (oligodendroglioma, astrocytoma, and oligoastrocytoma). Negative ion mode mass spectra of (A) sample G49, oligodendroglioma grade II with gemistocytes; (B) sample G41, oligodendroglioma grade II; (C) sample G46, astrocytoma grade IV with gemistocytes; and (D) sample G27, astrocytoma grade IV, are shown. (B) Banerjee et al. [21] determined borders between cancerous and normal tissues based on a DESI dataset. Extracted ion chronogram of glucose and citrate over a line scan of a typical prostate tissue specimen that contains both benign (black outline) and cancer (red outline) areas are shown. *(Adapted from L.S. Eberlin, I. Norton, A.L. Dill, A.J. Golby, K.L. Ligon, S. Santagata, R.G. Cooks, N.Y. Agar, Classifying human brain tumors by lipid imaging with mass spectrometry, Cancer Res. 72 (3) (2012) 645–654; S. Banerjee, R.N. Zare, R.J. Tibshirani, C.A. Kunder, R. Nolley, R. Fan, J.D. Brooks, G.A. Sonn, Diagnosis of prostate cancer by desorption electrospray ionization mass spectrometric imaging of small metabolites and lipids, Proc. Natl. Acad. Sci. U. S. A. 114 (13) (2017) 3334–3339.)*

prostate cancer tissue and benign specimens and achieved nearly 90% accuracy per patient. Since prostate cancer tissue stored more glucose and less citrate than adjacent normal tissue, the ratio of the glucose and citrate ion signals accurately predicted cancerous regions. The cancerous regions exhibited glucose/citrate ratios exceeding 1.0, whereas the ratio in normal prostate regions was less than 0.5 (Fig. 4.4B).

Both of these examples show that using multivariate statistical methods to extract diagnostic data from the huge amount of chemical information in DESI data sets, is key for successful clinical application of DESI-IMS to cancer diagnosis. A recent study by Inglese et al. [22] advanced this concept (Fig. 4.5). They adopted a state-of-the-art, deep unsupervised neural network-based technique, parametric t-SNE, for 3D mapping of characteristic multivariate indices in a human colorectal adenocarcinoma biopsy.

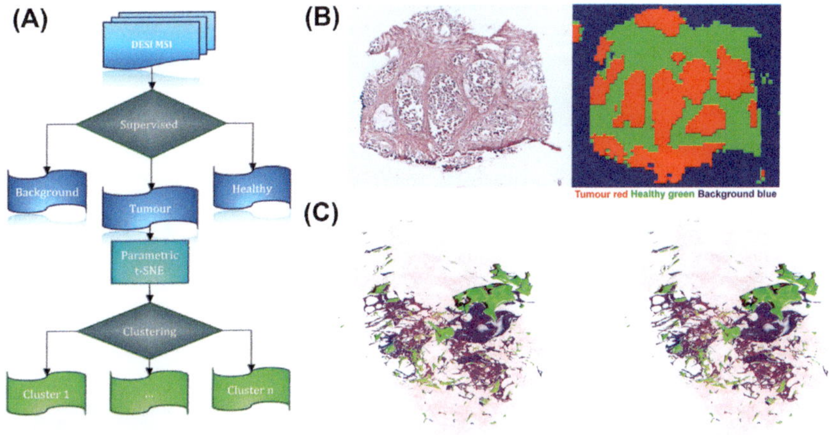

Figure 4.5 Deep learning and 3D-desorption electrospray ionization (DESI) imaging identify distinct metabolism of heterogeneous cancer subregions shown by Inglese et al. [22]. (A) Scheme of the DESI coupled with deep learning workflow. DESI imaging data are segmented into three classes—background, healthy, and tumor. A two-dimensional representation of the tumor spectrum is calculated by parametric t-SNE and clustered using OPTICS. (B) Generated statistical model was applied to whole obtained DESI-imaging mass spectrometry (IMS) spectra and classified into tumor (red), heathy (green) and back ground (blue). (C) Stereoscopic (cross-eyed) rendering of the 3D reconstruction of the three clusters [22]. *(Adapted from P. Inglese, J.S. McKenzie, A. Mroz, J. Kinross, K. Veselkov, E. Holmes, Z. Takats, J.K. Nicholson, R.C. Glen, Deep learning and 3D-DESI imaging reveal the hidden metabolic heterogeneity of cancer, Chem. Sci. 8(5) (2017) 3500–3511.)*

The results revealed the hidden metabolic heterogeneity of cancer tissue and shed light on the complex metabolic changes that differentiate cancer tissue and its subtypes from healthy tissues.

4.4 Application III: reactive DESI

Due to the nature of direct MS with complex tissue samples, targeting nonpolar molecules with low ionization efficiencies is always challenging. The potential of DESI-IMS was reported early [23], and DESI-IMS has since been applied for the analysis of various types of clinical specimens [19,24–27]. However, the analytes measured in many published studies were abundant lipids. For detection of a particular molecular species by MALDI-IMS, researchers can optimize matrix compounds and solvents to improve ionization of the target compound. On the other hand, an analyst

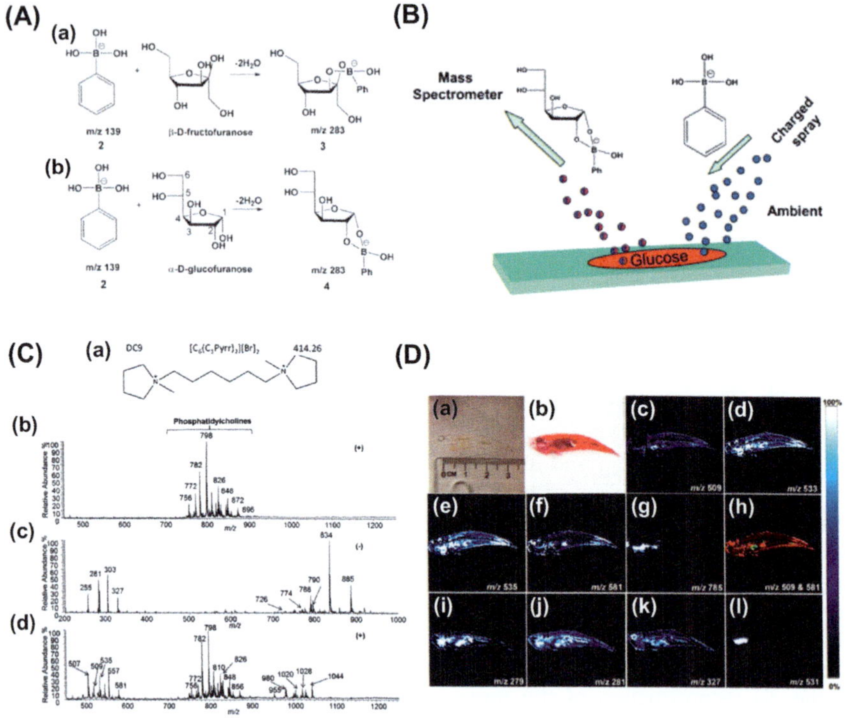

Figure 4.6 Use of heterogeneous reactions at a solution-solid interface in desorption electrospray ionization (DESI) to identify/enhance targeted molecular detection. (A–B) In Chen et al. [32], a cyclic boronate is used to detect sugars via in-situ derivatization with phenylboronic acid [PhB(OH)2]. (C–D) In Lostun et al. [35], a stable dicationic DC9 reagent (a) was used to form lipid-dication pairs to enhance targeted lipid signals. DESI-MS mass spectra of lipids in rat brain in positive mode (a) and negative mode (b) and reactive DESI-MS spectrum of enhanced intensity DC9-lipid ion pairs in rat brain in the positive ion mode(c) are shown. Application of DC9-using reactive DESI allowed enhanced detection of compounds both commonly observed in the positive and negative ion mode, in zebra fish tissues.

can place a chemical additive in the DESI-IMS spraying solvent that reacts specifically with the targeted molecule to increase its ionization efficiency. In this unique DESI method, called reactive DESI [7,28–31], specific reagents are added to the spray solution that are intended to facilitate particular ionizing reactions during the sampling process. In an early study by Chen et al. [32], the authors used reactive DESI to detect non-polar sugars via *in situ* derivatization with phenylboronic acid [$PhB(OH)_2$] (Fig. 4.6A–B). Other phenylboronic acids, including 3-nitrophenylboronic acid and *N*-methyl-4-pyridineboronic acid iodide, were employed in subsequent studies [33].

Reactive DESI involving simple adduct formation, such as complexation of trinitrohexahydro-1,3,5-triazine (RDX) with the CF_3COO- anion or TNT with the methoxide anion [34], has also been reported. In a study by Lostun et al. [35], a stable dicationic DC9 reagent was used to form lipid-dication pairs for which the signal intensity was higher than that obtained in a conventional DESI-MS lipid analysis (Fig. 4.6C–D). Their reactive DESI-IMS method was more sensitive, and they used it to visualize fatty acids in rat brain and zebra fish whole bodies. Fatty acid ion artifacts can be produced via postsource phospholipid decay reactions, particularly during analysis of lipid-abundant tissues; however, their reactive DESI can suppress formation of such artifacts.

Sensitivity enhancement by reactive DESI even enabled us to detect natural products directly on algal tissue surfaces under ambient conditions without any disruptive sample processing. Nyadong et al. [36] achieved bromophycolide detection by the addition of various anions, including Cl^-, Br^-, and CF_3COO^-, to the DESI spray solvent and optimized ionization conditions with Cl^-.

4.5 Conclusion and perspective

Here we have briefly reviewed applications of DESI with a focus on direct detection of various analytes, including nonpolar molecules that are sometimes not detectable with other ionization methods. As we have shown, DESI-IMS has been successfully adopted in the forensic and life sciences owing to its minimal sample preparation requirements and short analysis times. The simplicity of DESI protocols will make DESI-based chemical detection and imaging routine in forensic testing as well as clinical diagnosis. Further technical developments that facilitate use by nonexperts will expand opportunities to utilize DESI as a practical inspection tool that can be operated by technicians at inspection agencies. For example, optimization of measurement parameters like spraying conditions, the sprayer angle, and stage movement, could be automated. The integration of ambient ionization MS with other imaging modalities for medical use, such as MRI and positron emission tomography (PET), will yield valuable information that can assist doctors in making critical medical decisions.

References

[1] Z. Takats, J.M. Wiseman, B. Gologan, R.G. Cooks, Mass spectrometry sampling under ambient conditions with desorption electrospray ionization, Science 306 (5695) (2004) 471–473.
[2] R.G. Cooks, Z. Ouyang, Z. Takats, J.M. Wiseman, Ambient mass spectrometry, Science 311 (5767) (2006).
[3] Z. Takats, J.M. Wiseman, R.G. Cooks, Ambient mass spectrometry using desorption electrospray ionization (DESI): instrumentation, mechanisms and applications in forensics, chemistry, and biology, J. Mass. Spectrom. 40 (10) (2005) 1261–1275.
[4] T.J. Kauppila, N. Talaty, T. Kuuranne, T. Kotiaho, R. Kostiainen, R.G. Cooks, Rapid analysis of metabolites and drugs of abuse from urine samples by desorption electrospray ionization-mass spectrometry, Analyst 132 (9) (2007) 868–875.
[5] D.R. Ifa, N.E. Manicke, A.L. Dill, R.G. Cooks, Latent fingerprint chemical imaging by mass spectrometry, Science 321 (5890) (2008) 805.
[6] M. Zhao, S. Zhang, C. Yang, Y. Xu, Y. Wen, L. Sun, X. Zhang, Desorption electrospray tandem MS (DESI-MSMS) analysis of methyl centralite and ethyl centralite as gunshot residues on skin and other surfaces, J. Forensic Sci. 53 (4) (2008) 807–811.
[7] G. Huang, H. Chen, X. Zhang, R.G. Cooks, Z.J. Ouyang, Rapid screening of anabolic steroids in urine by reactive desorption electrospray ionization, Anal. Chem. 79 (21) (2007) 8327–8332.
[8] J.M. Wiseman, C.A. Evans, C.L. Bowen, J.H. Kennedy, Direct analysis of dried blood spots utilizing desorption electrospray ionization (DESI) mass spectrometry, Analyst 135 (4) (2010).
[9] D.R. Ifa, L.M. Gumaelius, L.S. Eberlin, N.E. Manicke, R.G. Cooks, Forensic analysis of inks by imaging desorption electrospray ionization (DESI) mass spectrometry, Analyst 132 (5) (2007) 461–467.
[10] K. Yanagisawa, Y. Shyr, B.J. Xu, P.P. Massion, P.H. Larsen, B.C. White, J.R. Roberts, M. Edgerton, A. Gonzalez, S. Nadaf, J.H. Moore, R.M. Caprioli, D.P. Carbone, Proteomic patterns of tumour subsets in non-small-cell lung cancer, Lancet 362 (9382) (2003) 433–439.
[11] S. Rauser, C. Marquardt, B. Balluff, S.R.O. Deininger, C. Albers, E. Belau, R. Hartmer, D. Suckau, K. Specht, M.P. Ebert, Classification of HER2 receptor status in breast cancer tissues by MALDI imaging mass spectrometry, J. Proteome Res. 9 (4) (2010) 1854–1863.
[12] L.H. Cazares, D.A. Troyer, B. Wang, R.R. Drake, O.J. Semmes, MALDI tissue imaging: from biomarker discovery to clinical applications, Anal. Bioanal. Chem. 401 (1) (2011) 17–27.
[13] M.R. Groseclose, P.P. Massion, P. Chaurand, R.M. Caprioli, High-throughput proteomic analysis of formalin-fixed paraffin-embedded tissue microarrays using MALDI imaging mass spectrometry, Proteomics 8 (18) (2008) 3715–3724.
[14] T.W. Powers, B.A. Neely, Y. Shao, H. Tang, D.A. Troyer, A.S. Mehta, B.B. Haab, R.R. Drake, MALDI imaging mass spectrometry profiling of N-glycans in formalin-fixed paraffin embedded clinical tissue blocks and tissue microarrays, PLoS One 9 (9) (2014) e106255.
[15] T. Alexandrov, M. Becker, O. Guntinas-Lichius, G. Ernst, F. von Eggeling, MALDI-imaging segmentation is a powerful tool for spatial functional proteomic analysis of human larynx carcinoma, J. Cancer Res. Clin. Oncol. 139 (1) (2013) 85–95.
[16] A.L. Dill, L.S. Eberlin, A.B. Costa, C. Zheng, D.R. Ifa, L. Cheng, T.A. Masterson, M.O. Koch, O. Vitek, R.G. Cooks, Multivariate statistical identification of human bladder carcinomas using ambient ionization imaging mass spectrometry, Chemistry 17 (10) (2011) 2897–2902.

[17] D. Calligaris, D. Caragacianu, X. Liu, I. Norton, C.J. Thompson, A.L. Richardson, M. Golshan, M.L. Easterling, S. Santagata, D.A. Dillon, F.A. Jolesz, N.Y. Agar, Application of desorption electrospray ionization mass spectrometry imaging in breast cancer margin analysis, Proc. Natl. Acad. Sci. U. S. A. 111 (42) (2014) 15184–15189.
[18] A.L. Dill, D.R. Ifa, N.E. Manicke, A.B. Costa, J.A. Ramos-Vara, D.W. Knapp, R.G. Cooks, Lipid profiles of canine invasive transitional cell carcinoma of the urinary bladder and adjacent normal tissue by desorption electrospray ionization imaging mass spectrometry, Anal. Chem. 81 (21) (2009) 8758–8764.
[19] A.L. Dill, L.S. Eberlin, C. Zheng, A.B. Costa, D.R. Ifa, L. Cheng, T.A. Masterson, M.O. Koch, O. Vitek, R.G. Cooks, Multivariate statistical differentiation of renal cell carcinomas based on lipidomic analysis by ambient ionization imaging mass spectrometry, Anal. Bioanal. Chem. 398 (7–8) (2010) 2969–2978.
[20] L.S. Eberlin, I. Norton, A.L. Dill, A.J. Golby, K.L. Ligon, S. Santagata, R.G. Cooks, N.Y. Agar, Classifying human brain tumors by lipid imaging with mass spectrometry, Cancer Res. 72 (3) (2012) 645–654.
[21] S. Banerjee, R.N. Zare, R.J. Tibshirani, C.A. Kunder, R. Nolley, R. Fan, J.D. Brooks, G.A. Sonn, Diagnosis of prostate cancer by desorption electrospray ionization mass spectrometric imaging of small metabolites and lipids, Proc. Natl. Acad. Sci. U. S. A. 114 (13) (2017) 3334–3339.
[22] P. Inglese, J.S. McKenzie, A. Mroz, J. Kinross, K. Veselkov, E. Holmes, Z. Takats, J.K. Nicholson, R.C. Glen, Deep learning and 3D-DESI imaging reveal the hidden metabolic heterogeneity of cancer, Chem. Sci. 8 (5) (2017) 3500–3511.
[23] J.M. Wiseman, S.M. Puolitaival, Z. Takats, R.G. Cooks, R.M. Caprioli, Mass spectrometric profiling of intact biological tissue by using desorption electrospray ionization, Angew Chem. Int. Ed. Engl. 44 (43) (2005) 7094–7097.
[24] K.S. Kerian, A.K. Jarmusch, V. Pirro, M.O. Koch, T.A. Masterson, L. Cheng, R.G. Cooks, Differentiation of prostate cancer from normal tissue in radical prostatectomy specimens by desorption electrospray ionization and touch spray ionization mass spectrometry, Analyst 140 (4) (2015) 1090–1098.
[25] L.S. Eberlin, R.J. Tibshirani, J. Zhang, T.A. Longacre, G.J. Berry, D.B. Bingham, J.A. Norton, R.N. Zare, G.A. Poultsides, Molecular assessment of surgical-resection margins of gastric cancer by mass-spectrometric imaging, Proc. Natl. Acad. Sci. U. S. A. 111 (7) (2014) 2436–2441.
[26] A.K. Jarmusch, V. Pirro, Z. Baird, E.M. Hattab, A.A. Cohen-Gadol, R.G. Cooks, Lipid and metabolite profiles of human brain tumors by desorption electrospray ionization-MS, Proc. Natl. Acad. Sci. U. S. A. 113 (6) (2016) 1486–1491.
[27] A.K. Jarmusch, K.S. Kerian, V. Pirro, T. Peat, C.A. Thompson, J.A. Ramos-Vara, M.O. Childress, R.G. Cooks, Characteristic lipid profiles of canine non-Hodgkin's lymphoma from surgical biopsy tissue sections and fine needle aspirate smears by desorption electrospray ionization–mass spectrometry, Analyst 140 (18) (2015) 6321–6329.
[28] H. Chen, N.N. Talaty, Z. Takáts, R.G. Cooks, Desorption electrospray ionization mass spectrometry for high-throughput analysis of pharmaceutical samples in the ambient environment, Anal. Chem. 77 (21) (2005) 6915–6927.
[29] I. Cotte-Rodriguez, H. Chen, R.G. Cooks, Rapid trace detection of triacetone triperoxide (TATP) by complexation reactions during desorption electrospray ionization, Chem. Commun. (9) (2006) 953–955.
[30] I. Cotte-Rodriguez, H. Hernandez-Soto, H. Chen, R.G. Cooks, In situ trace detection of peroxide explosives by desorption electrospray ionization and desorption atmospheric pressure chemical ionization, Anal. Chem. 80 (5) (2008) 1512–1519.
[31] M.D. Green, H. Nettey, O. Villalva Rojas, C. Pamanivong, L. Khounsaknalath, M. Grande Ortiz, P.N. Newton, F.M. Fernandez, L. Vongsack, O. Manolin, Use of

refractometry and colorimetry as field methods to rapidly assess antimalarial drug quality, J. Pharm. Biomed. Anal. 43 (1) (2007) 105–110.
[32] H. Chen, I. Cotte-Rodriguez, R.G. Cooks, cis-Diol functional group recognition by reactive desorption electrospray ionization (DESI), Chem. Commun. (6) (2006) 597–599.
[33] Y. Zhang, H. Chen, Detection of saccharides by reactive desorption electrospray ionization (DESI) using modified phenylboronic acids, Int. J. Mass Spectrom. 289 (2–3) (2010) 98–107.
[34] I. Cotte-Rodríguez, Z. Takáts, N. Talaty, H. Chen, R.G. Cooks, Desorption electrospray ionization of explosives on surfaces: sensitivity and selectivity enhancement by reactive desorption electrospray ionization, Anal. Chem. 77 (21) (2005) 6755–6764.
[35] D. Lostun, C.J. Perez, P. Licence, D.A. Barrett, D.R. Ifa, Reactive DESI-MS imaging of biological tissues with dicationic ion-pairing compounds, Anal. Chem. 87 (6) (2015) 3286–3293.
[36] L. Nyadong, E.G. Hohenstein, A. Galhena, A.L. Lane, J. Kubanek, C.D. Sherrill, F.M. Fernandez, Reactive desorption electrospray ionization mass spectrometry (DESI-MS) of natural products of a marine alga, Anal. Bioanal. Chem. 394 (1) (2009) 245–254.

CHAPTER 5

Ambient laser-based mass spectrometry analysis methods: a survey of core technologies and reported applications

Alessandra Tata[1], Michael Woolman[1,2] Emma Bluemke[1,2,*], Arash Zarrine-Afsar[1,2,3,4]

[1]Techna Institute for the Advancement of Technology for Health, University Health Network, Toronto, ON, Canada; [2]Department of Medical Biophysics, University of Toronto, Toronto, ON, Canada; [3]Department of Surgery, University of Toronto, Toronto, ON, Canada; [4]Keenan Research Center for Biomedical Science & the Li Ka Shing Knowledge Institute, St. Michael's Hospital, Toronto, ON, Canada; [*]Present address: Department of Engineering Science, Institute of Biomedical Engineering, University of Oxford, Old Road Campus Research Building, Oxford, United Kingdom

Contents

Ambient Ionization Mass Spectrometry in Life Sciences
ISBN 978-0-12-817220-9
https://doi.org/10.1016/B978-0-12-817220-9.00005-9

5.1 Introduction

Following the seminal discovery of the stimulated emission of radiation phenomena and the works of Charles Towns, lasers have paved the way for many other discoveries and advancements in science and technology. The current applications of lasers range from "sensing" to "machining" where the absorptive properties of target media in the context of the wavelength, pulse duration, pulse shape, and the energy of the laser light have made a wide variety of applications possible. While to those outside the realm of science, laser applications may be limited to surgery or advanced weaponry, the field of analytical chemistry has greatly benefited from the advent of lasers. The fields of "laser photochemistry" and "pump-probe spectroscopy" made it possible to capture the first frames of ultrafast molecular reactions, with free-electron lasers pushing the boundaries of temporal resolution in reaction monitoring. The field of Mass Spectrometry has not been an exception and also greatly benefited from the advent of lasers. This chapter aims to review the utility of lasers as part of the most commonly used ambient ion sources for detection as well as for spatially resolved imaging. Lasers are used in modern ion sources to either desorb the analyte for postionization, promote concurrent desorption or ionization, or to extract precharged molecules to the gas phase for capture and analysis by mass analyzers. Like with many other mass spectrometry techniques and technologies, a variety of acronyms exist for laser-based ambient ion sources, and slightly modifying the source geometry has been taken by the mass spectrometry community to indicate novelty. While a discussion with respect to how much variance in ion source property constitutes a legitimate regime to warrant a "new technique" designation, it must be noted that not all lasers are created equal. While few lasers offer pure desorption (defined herein as placement of analytes from a condensed phase to the gas phase), almost all lasers offer some ablative properties, creating aerosolized particles of said material present within condensed phase. These aerosols may contain chunks of solid material or smaller, somewhat solvated microparticles that could be intercepted by a postablation ionization mechanism for detection. The mechanisms of ablation could be assisted, thermal, nonthermal or even photoacoustic. The requirement of postionization is heavily tied it to the "state" at which the laser in question extracts the material from the condensed phase and could be modulated by laser properties such as wavelength, pulse duration, or energy. For instance, infrared (IR) radiation is strongly absorbed by water and water-rich media such as biological tissues. An IR laser with microsecond pulse duration is likely to offer more thermal ablation than its counterpart systems operating on the shorter pulse durations of pico- or femtosecond.

In the sections below, we have reviewed major laser-based ambient ion sources for mass spectrometry and provided original citations to exemplary applications for each source as demonstrated in the literature. Laser applications of Rapid Evaporative Ionization Mass Spectrometry (REIMS) are discussed in Chapter 8. At the end of this chapter we have included a comparative table that highlights key features, advantages, and proven application domains of each method alongside illustrations describing the reported optimal setup geometry. To ensure this review remains as faithful to the interpretations intended by original authors, in many instances statements have been borrowed *en bloc* or presented with minimal rewording. This is deliberate to ensure an unbiased review of the original literature provided in this chapter.

5.2 ELDI-MS

Electrospray-assisted **L**aser **D**esorption/**I**onization **M**ass **S**pectrometry (ELDI-MS) combines laser desorption with postionization by electrospray. This pioneering method was used to rapidly analyze a variety of solid samples under ambient conditions more than a decade ago [1]. Fig. 5.1A illustrates the geometry of an ELDI setup.

The first report of ELDI-MS experiment in 2005 used a combination of ESI and a 337 nm pulsed (4 ns pulse duration, operating at 10 Hz) nitrogen ablative laser (20 μJ/pulse) where analysis under ambient conditions of organic molecules in solid samples without sample preparation was reported [1]. In this work, cytochrome *c* deposited as a neat compound on stainless steel substrate was desorbed from solid substrates without addition of "matrix material," capitalizing on the general ablative nature of the pulsed nitrogen laser systems operating at 337 nm. Here, postionization with the augmented ESI source produced high-quality mass spectra where multiple charging of protein molecules was seen. Notably, this pioneering work was the first to report mass spectra of intact proteins via laser desorption in the absence of exogenous matrix material, extracted from solid metallic conductive and nonconductive insulating substrates, in stark contrast to its sister technology and vacuum counterpart MALDI-MS (Fig. 5.1B). This study used bovine cytochrome *c* and methaqualone (in the form of a tablet)

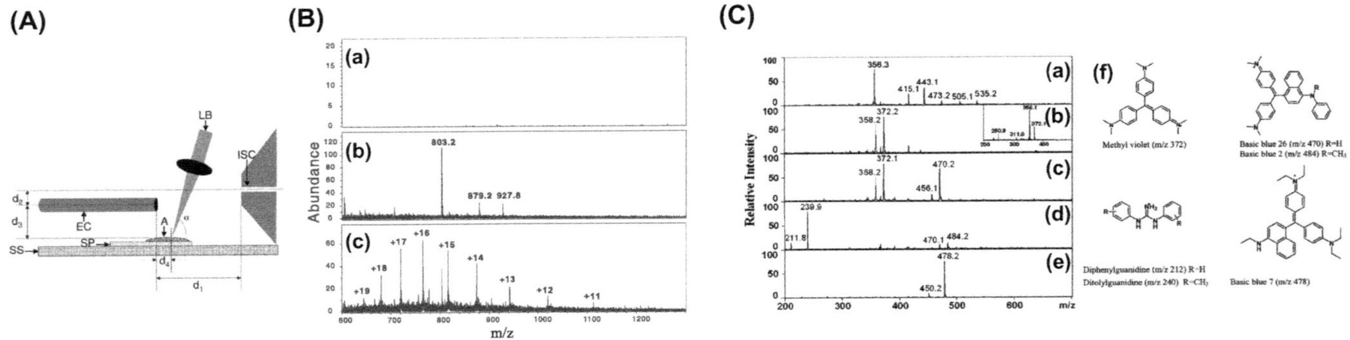

Figure 5.1 ***Electrospray-assisted Laser Desorption/Ionization Mass Spectrometry (ELDI-MS) schematics and reported applications.*** (A) The graphic representation of the ELDI-MS setup. The solid sample (labelled "A") deposited on a support plate made out of stainless steel is placed on a mobile sample stage that is then irradiated with a pulsed nitrogen laser. The laser-ablated materials are then ionized in the electrospray solvent plume. The solvent is delivered through an electrospray capillary, and the generated ions are transported to the mass analyzer via an ion sampling capillary. Distances d_1 to d_4 indicate the optimized geometry of the source: distance between the tip of the electrospray capillary and the ion transport capillary, $d_1 = 8$ mm; a lateral offset for the electrospray capillary from the sample plate, $d_2 = 2$ mm; distance between the electrospray capillary and the sample plate, $d_3 = 3$ mm; the distance between the optimal position of the laser light and the sampling point below ESI capillary tip, $d_4 = 2$ mm. The laser irradiates the sample at an angle $\alpha = 45$ degrees. (B) The ambient ELDI mass spectra of bovine cytochrome *c* (sample deposited on stainless steel surface) using the setup described above. The following control experiments of (a) laser desorption only; (b) ESI only; and (c) both laser desorption and ESI operational, are also included (C) ELDI mass spectra of ink lines in the positive ion mode drawn with: (a) SKB SB-2000 (in red ink), (b) SKB SB-2000 (in black ink), (c) SKB SB-2000 (in blue ink), (d) ZEBRA BNE1-BL (in blue ink), (e) PILOT SUPER GRIP (in blue ink) ballpoint pens, and (f) structures of methyl violet, basic blue 26, basic blue 2, diphenylguanidine, ditolylguanidine, as well as Basic Blue 7. The inset in (b) represents the MS/MS spectrum of the major component parent ion *m/z* 372. *(A and B), Reprinted (adapted) with permission from J. Shiea, M.Z. Huang, H.J. Hsu, C.Y. Lee, C.H. Yuan, I. Beech, J. Sunner, Electrospray-assisted laser desorption/ionization mass spectrometry for direct ambient analysis of solids, Rapid Commun. Mass Spectrom. 19 (2005) 3701–3704. Copyright (2005) John Wiley and Sons, Ltd. (C). Reprinted (adapted) with permission from S.C. Cheng, Y.S. Lin, M.Z. Huang, J. Shiea, Applications of electrospray laser desorption ionization mass spectrometry for document examination, Rapid Commun. Mass Spectrom. 24 (2010) 203–208. Copyright (2010) John Wiley and Sons, Ltd.)*

to showcase the applicability of matrix-free ELDI-MS for the analysis of both large protein molecules and small molecule compounds [1].

Shortly thereafter, the same ELDI-MS technique was applied to showcase rapid and sensitive detection of major proteins from dried biological fluids (e.g., blood, tear, saliva, serum), bacterial cultures, and tissues (e.g., porcine liver and heart), without the addition of exogenous matrix material and in the absence of sample preparations (pretreatment) under ambient conditions [2]. Here, ELDI-MS was shown to produce ESI-like multiply charged peptides and intact proteins up to 29 kDa (carbonic anhydrase) and 66 kDa (bovine albumin) both from single-protein solutions and complex sample mixtures subjected to digest. The ability to generate multiply charged peptides is a key attribute that enables tandem mass spectrometric (MS/MS)-based peptide sequencing, further conferring specificity to this ambient, matrix-free MS analysis technique. The ELDI-MS/MS analysis of protein digests and small intact proteins was showcased by collisionally activated dissociation (CAD) and by nozzle-skimmer dissociation (NSD). ELDI-MS/MS thus was predicted to be a useful tool for protein sequencing as well as for top-down proteomics approaches [2,3]. Further extending the application domain to aqueous material [3], ELDI-MS largely yielded protonated protein ions, also seen in various biological fluids (including tear from human subjects, cow milk, serum, as well as bacterial extracts). MS signals were not hindered by significant corresponding sodiated or potassiated adducts. It was also possible to rapidly quantify the levels of glycosylated hemoglobin protein from whole blood samples provided by diabetic patients; all in the absence of sample pretreatment and deposition of matrix material required for non-ambient laser-based methods such as MALDI [3].

Shiea et al. [4] also demonstrated that ELDI-MS could have utility in rapidly characterizing ink compounds deposited by inkjet printer on regular paper, also being able to detect chemical compounds from thermal papers. ELDI-MS analysis allowed ink and paper molecules to be distinguished in terms of their chemical compositions (Fig. 5.1C), where the molecules innate to the ink were desorbed through laser irradiation with a spatial resolution of $<100\ \mu m$; with the desorbed molecules being intercepted by the electrospray plume that was formed from an acidic methanol/water solution (50%). The mechanism of ionization was postulated to be ESI-like; facilitated through fusion with the charged solvent material present in plume droplets [4].

5.3 LAESI-MS

The second prominent laser-based two-step ablation/desorption with ionization technologies to be discussed are known as **L**aser **A**blation **E**lectro**S**pray **I**onization **M**ass **S**pectrometry (LAESI-MS) and **L**aser **A**ssisted **D**esorption **E**lectrospray **I**onization **M**ass **S**pectrometry (LA-DESI-MS). Both LAESI-MS and LA-DESI-MS use mid-IR lasers coupled to an ESI source [5–7]. Since this debut, LAESI-MS has become a commercially available ion source with application domains spanning from endogenous molecule mapping to drug and metabolite analysis.

LAESI-MS uses mid-IR lasers for the interrogation of water-rich samples (such as biological tissues or aqueous solutions) since the O–H bond in water molecules strongly absorbs the mid-IR radiation. This technique uses either an erbium-doped yttrium aluminum garnet Er:YAG laser (2.94 μm, pulse duration <100 ns, operating at 5–10 Hz, outputting 3.5 mJ/pulse) or a neodymium-doped yttrium aluminum garnet Nd:YAG laser augmented with optical parametric oscillator (OPO) outputting a pulsed 2.94 μm laser beam, with pulse duration of 4 ns, and repetition rate of 10 Hz.

Close to 0.3–3.5 mJ of laser energy is typically focused (via CaF_2 or ZnSe lenses) to irradiate the sample surface (incidence angle of 90 degrees) as shown in Fig. 5.2A. Fiber delivery of IR lasers is also reported where the coupling losses result in a reduction of effective average laser power. LAESI-MS and IR-**M**atrix **A**ssisted **L**aser **D**esorption **E**lectro**S**pray **I**onization (MALDESI) [8–17] (described in detail in the next section) are similar techniques: they both use a mid-IR laser as a desorption source with water molecules acting like a matrix material propelling the "dissolved" molecules into the gas phase. Here, fast imaging of the laser/ESI/substrate interaction area through shadow or dark field illumination may shed insights into the mechanism of LAESI-MS [5].

The response of LAESI-MS signal with respect to concentration of analyte is linear, and experiments using pure verapamil and reserpine (two easy-to-desorb molecules routinely used in MS for response curve determinations) exhibited linear calibration curves (from 1 to 10 mM) with correlation coefficients of approximately 0.99 reported. Further attempts to establish a limit of detection (LOD) for these compounds resulted in 8 and 25 fmol estimates, respectively [5]. These LOD values are sufficiently low to motivate semiquantitative detection of analytes of pharmaceutical interest.

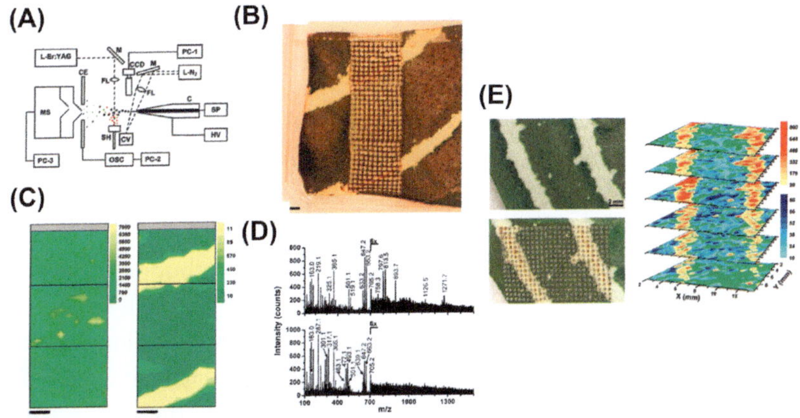

Figure 5.2 ***The schematic of Laser Ablation ElectroSpray Ionization Mass Spectrometry (LAESI-MS) setup for tissue analysis and imaging with results.*** (A) Schematic diagram of the LAESI instrumentation reported by works from the Vertes laboratory where the following abbreviations are used: *C*, capillary; *CCD*, CCD camera with short-distance microscope (for fast imaging); *CE*, counter electrode; *CV*, cuvette containing the dye for fast imaging; *FL*, focusing lenses; *HV*, high-voltage power supply; *L-Er:YAG*, Er:YAG laser for ablation; *L-N_2*, nitrogen laser (used for fast imaging); *M*, mirrors; *MS*, mass spectrometer; *OSC*, digital oscilloscope; *PC-1 to PC-3*, personal computers for control, data acquisition and processing; *SH*, sample holder; *SP*, syringe pump. (B) This panel shows a variegated *A. squarrosa* leaf probed with LAESI-MS through rastering the leaf surface with a focused infrared laser beam where three 4 mm by 4 mm areas have been probed separately. The tissue surface (fresh, live leaf) was ablated using a circular beam profile area (350-μm in diameter) along with a 400-μm step size. (C) Molecular images shown in the contour plots indicate the distribution of the ions detected at *m/z* 663 (diacetyl coumaryl-rhamnoside), present in both green and yellow regions, and *m/z* 493 (methoxykaempferol glucuronide), dominant in yellow regions. (D) Mass spectra obtained at varying regions within the leaf (top: green regions, bottom: yellow regions) suggesting different chemical compositions consistent with methoxykaempferol glucuronide distribution seen in the LAESI-MS molecular images included in panel (C). Scale bar is set to 1 mm (panels B and C). (E) Depth profiling with LAESI-MS illustrating chlorophyll (blue scale) and kaempferol distributions (red scale) that matched the expected localization alongside the leaf thickness with chlorophyll distribution being largely limited to mesophyll layers. *(Panel (A) is reprinted (adapted) with permission from P. Nemes, A. Vertes, Laser ablation electrospray ionization for atmospheric pressure, in vivo, and imaging mass spectrometry, Anal. Chem. 79 (2007) 8098–8106. Copyright (2007) American Chemical Society. Panels (B–D) are reprinted (adapted) with permission from P. Nemes, A.A. Barton, Y. Li, A. Vertes, Ambient molecular imaging and depth profiling of live tissue by infrared laser ablation electrospray ionization mass spectrometry, Anal. Chem. 80 (2008) 4575–4582. Copyright (2008) American Chemical Society. Panel (E) is reprinted (adapted) with permission from P. Nemes, A.A. Barton, A. Vertes, Three-dimensional imaging of metabolites in tissues under ambient conditions by laser ablation electrospray ionization mass spectrometry, Anal. Chem. 81 (2009) 6668–6675. Copyright (2009) American Chemical Society.)*

LAESI-MS analysis has been used for exploration of proteins, peptides, lipids, and metabolites in standard solutions, biological fluids, bacterial cultures, and untreated biological tissues [6,18–21]. By capitalizing on the high spatial resolution of the laser beam used to interrogate the tissue chemical content, preliminary analyses have demonstrated the potential of LAESI-MS for molecular imaging [6]. Subsequent studies have illustrated the ability to acquire 2D and 3D images (with lateral and depth resolutions of ~300 μm and 40–50 μm, respectively) of the distributions of metabolites and lipid molecules in a variety of biological tissues such as rat brain tissues and *A. squarrosa* and *S. lynise* leaves [19]. The depth "drilling" feature of the IR laser provides some level of three-dimensionality to LAESI-MS analysis compared to DESI-MS or MALDI-MS where the gentle nature of surface solvent extraction or laser-induced removal of nonpenetrating matrix material may prevent 3D imaging without serial section analysis [19,22]. In the *A. squarrosa* LAESI-MS study, the distributions of kaempferol (diacyl coumaryl-rhamnoside), *m/z* 493, present both in yellow and green regions of the leaf; methoxykaempferol glucuronide, *m/z* 493, present only in yellow sectors of the leaf; the unassigned ion at *m/z* 813, more abundant in the green sections, were all in good alignment with the optical image of the leaf pattern (Fig. 5.2B–D) [6]. The generation of 3D chemical content maps (i.e., molecular distribution images) based on depth profiling with the IR laser used in LAESI-MS analysis [6] was further evaluated in a follow-up study (Fig. 5.2E) where a more detailed metabolite imaging of the leaf content was presented [19]. These ion distribution images from LAESI-MS imaging are also utilized to study microbe–antibiotic interactions via interrogation of *in situ* microbial colonies [23]. Following this work, the use of LAESI-MS expanded to mass spectrometric analysis of rat brain slices [22].

The turning point for technology development of LAESI-MS was the incorporation of an optical fiber to deliver laser light to the sample surface without the need for open beam delivery. To improve resolution, the end of the GeO_2-based glass fiber was chemically etched with HNO_3 solution (1%) to form a sharp tip with a radius of 15 μm. This development improved the resolution, enhanced beam quality at the analysis site, and conferred additional stability, thus improving reproducibility. This setup offered the capability of desorbing individual plant cells such that the metabolic profiles of single cells could be obtained [18,24–26]. The results suggested that several metabolites were heterogeneously distributed in the pigment and nonpigment cells of the *A. cepa* [26]. Further mechanistic insights followed that during LA, the plume expanded hemispherically such that only a small portion of the neutral molecules could be ionized by interception with the ESI spray. It was also recognized that desorbing sample solutions within a capillary (of ~750 μm I.D.) limited the radial expansion of the plume, resulting in a marked reduction of the LOD for verapamil to 600 amol [25].

In addition, the performance has been improved by the application of heat via **H**eat **A**ssisted **LAESI-MS** (HA-LAESI-MS) as a variation of LAESI-MS that couples a heated nebulizer chip with the conventional LAESI source. The heated microchip is positioned in the front of the MS orifice, delivering a stream of hot nitrogen gas (180 or 360 mL/min flow rates) that is heated through the nozzle of the microchip (to 170–200°C). The ESI source for this setup is positioned 3–6 mm away from said microchip, and the angle between the ESI capillary and microchip is set to approximately 20–30 degrees. The heated gas is proposed to enhance the ionization of even less polar compounds present in the ESI plume allowing HA-LAESI to be used for simultaneous analysis of polar and nonpolar compounds, thus improving the application domain of the technique [27,28]. The heated technique also enhances the subcellular analysis of metabolites and lipids, as verified in the well-studied biological sample model *X. leaves* eggs [28].

In addition to these developments, LAESI-MS has also been combined with ion mobility separation [29] to allow further differentiation between isobaric species and isomers [30], with some success reported. The versatility of LAESI-MS is further shown via analysis of single cells using a variety of optimal fiber tip designs [18]: the cells with different stages of development or pigmentation showed different metabolic profiles. This ability to report molecular images at single cell level as pixels to mapping heterogeneity in biological tissues is a testament to the specificity of LAESI-MS in chemical analysis of biological entities at the extremes of spatial (lateral) resolution (on the order of tens of microns). This study carefully showcased high resolution analysis consistent with single cell requirements is possible with LAESI-MS [18].

5.4 IR-MALDESI-MS

IR-MALDESI-MS is a closely related variant to LAESI-MS. As in LAESI-MS, IR-MALDESI-MS uses a mid-IR laser; however, it may further utilize

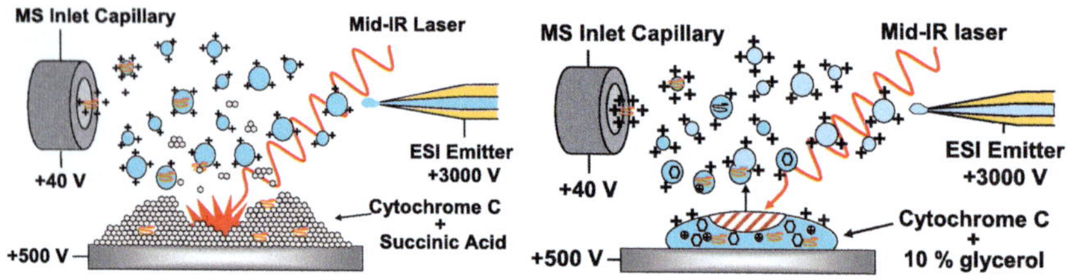

Figure 5.3 ***InfraRed-Matrix Assisted Laser Desorption ElectroSpray Ionization Mass Spectrometry (IR-MALDESI-MS) setup used for analysis of solid and liquid protein samples.*** Samples are first mixed with "additive" or matrix materials and deposited on a support plate. The IR laser beam induces desorption of particles carrying molecular signatures that are then intercepted with an ionization stream provided by charged ESI particles in a manner similar to **L**aser **A**blation **E**lectro**S**pray **I**onization **M**ass **S**pectrometry. The ionized materials are then taken up by the mass analyzer. *(Reprinted (adapted) with permission from J.S. Sampson, K.K. Murray, D.C. Muddiman, Intact and top-down characterization of biomolecules and direct analysis using infrared matrix-assisted laser desorption electrospray ionization coupled to FT-ICR mass spectrometry, J. Am. Soc. Mass Spectrom. 20 (2009) 667–673. Copyright (2009) American Society for Mass Spectrometry.)*

exogenous matrix materials to enhance the desorption of the analytes (Fig. 5.3). These matrix materials can include glycerol, an exogenous ice layer formed on the surface of sample [14], or succinic acid additives (which are used for the UV laser variant of MALDESI-MS [8]). Note that confusion exists with respect to the utility of correct terminology for "matrix," "additive," or "dopant." Matrix is a material that absorbs the laser energy (often resonantly) and catalyzes desorption by providing a transfer. Dopants are additives that enhance the process of desorption and additives are those that can facilitate desorption by addressing nonidealities such as cocrystallization or accelerated evaporation. In IR-MALDESI-MS, a pulsed IR laser is used to interrogate aqueous solutions of proteins mixed with appropriate additive [8] (an example IR laser is ER-YAG generating 2.8 mJ pulse energy at 2.94 μm operating 10-Hz repetition rate with a pulse width of 100 ns focused to a spot size of 200 μm using a calcium fluoride).

The application domain of IR-MALDESI method includes the top-down analysis of equine myoglobin in aqueous phase as well as direct detection of intact proteins and peptides [8]. An LOD on the order of 10 amol for cytochrome *c* has been reported. In addition, the method has been used for direct analysis of fabrics (textile) and trace fibers where an identification of dyes and fiber molecules with potential applications in forensic science has been demonstrated [17,31]. A variant of IR-MALDESI has also been reported for mass spectrometric imaging where a focused laser spot with diameter on the ranges of 45–250 μm has been used to desorb material from the sample [32]. In such studies, the sample stage was first cooled down to −9°C in order to facilitate capture of moisture on the tissue surface, forming a thin layer of ice that serves as additional matrix [14,32]. It was observed that the thin ice layer served efficiently as matrix absorbing thermal bath. The ice layer absorbs the excess laser energy and reduces the interference from organic matrix peaks at low mass range that would occur in MALDI-MS type experiments. Low temperature interrogation of specimens with IR-MALDESI-MS allows investigation of small molecule compounds [32]. In yet another published study, the influences of laser parameters including wavelength, fluence, the distance from the sampling stage, and the composition of matrix material on IR LD were examined. The study is based on cataloging IR-MALDESI-MS response to both pharmaceutical agents and endogenous lipids, present in animal tissue [33,34]. IR-MALDESI has also been applied in the imaging mode for spatially resolved study of hen ovaries, mouse liver, and a number of neonatal mouse tissues [12,35,36]. IR-MALDESI has been coupled to a high-resolution mass analyzer and applied in the imaging mode to examine the distribution of HIV drugs such as tenofovir, emtricitabine, and raltegravir in cervical tissues [12]. Furthermore, the analysis of small molecule distributions in high salt solution samples (an application which is problematic for ESI-MS) is possible with IR-MALDESI. This utility was well demonstrated in a study where triterpenoid lipids such as lupeol were found to be mainly localized to the skin (exocarp) of fermented cucumber in 1M salt brine, whereas β-sitosterol and stigmasterol were both detected in mesocarp (flesh) [36].

5.5 IR-LADESI-MS

Infra**R**ed **L**aser-**A**ssisted **D**esorption **E**lectrospray **I**onization **M**ass **S**pectrometry (IR-LADESI-MS), a closely related variant of IR-MALDESI [7], is a laser-based ambient MS method that uses IR lasers for desorption of materials, followed by ionization via interactions with electrospray solvent stream [7]. In this technique, an ion trap mass spectrometer interface was modified using a pulsed Er:YAG laser (2.94 μm wavelength, 100 ns pulse temporal width, at a repetition rate of 10 Hz) as shown in Fig. 5.4A. The

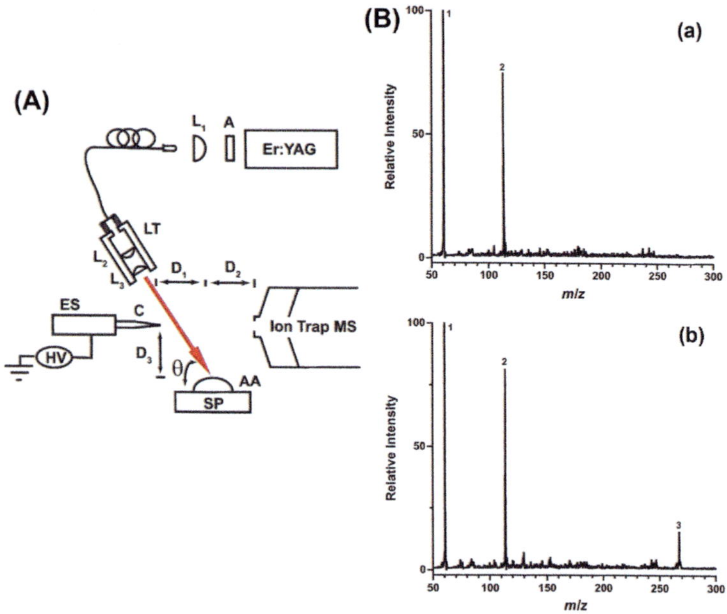

Figure 5.4 ***The schematic of an InfraRed Laser-Assisted Desorption Electrospray Ionization Mass Spectrometry (IR-LADESI-MS) ion source and a reported application.*** (A) Schematics of the setup. Here, ES is nanoelectrospray; C denotes fused-silica capillary; HV is the high voltage supply; L_1, L_2, and L_3 are planoconvex lenses used to focus the beam; A is an attenuator; LT indicates the lens tube; SP is the sample plate; AA denotes aqueous analyte; D_1 is the distance between the capillary tip and the ablation point; D_2 indicates the distance between the ablation point and the skimmer cone; D_3 is the distance between sample plate and the capillary tip; Ø is the laser incidence angle. (B) Direct analysis of human urine using IR-LADESI-MS. The spectrum of urine from (A) an unmedicated volunteer individual and (B) a volunteer sample medicated with atenolol, a drug used to treat hypertension. The identified mass to charge ratio peaks as 1 denotes urea; 2 is creatinine; and 3 is atenolol, respectively. Peak 2 indicates creatinine preset in both medicated and unmedicated patient urine samples. *(Reprinted (adapted) with permission from Y.H. Rezenom, J. Dong, K.K. Murray, Infrared laser-assisted desorption electrospray ionization mass spectrometry, Analyst 133 (2008) 226–232. Copyright (2008) The Royal Society of Chemistry.)*

laser was coupled into the instrument with a germanium oxide optical fiber (450 μm core diameter) for desorption at atmospheric pressure, and a nanoelectrospray source for postdesorption ionization of the laser extracted materials. These minor attributes, along with the use of nano-ESI, set apart this "fiber coupled" laser method from its closely related counterparts described previously.

IR-LADESI-MS was initially debuted for the direct analysis of water-containing samples under ambient conditions. Here, analytes contained in aqueous solutions are placed on a stainless-steel target support material and then irradiated with the IR laser. The material desorbed from the target specimen is ionized by the continuous stream of charged droplets from nano-ESI, in a similar manner as described in the previous paragraph. It was shown that peptides derived from bradykinin and cytochrome *c*, as well as intact large molecular weight protein samples (e.g., hemoglobin) investigated using this method yield mass spectra similar to those obtained by conventional electrospray analyses [7]. Further to this study, IR-LADESI-MS was shown to be suitable for direct analysis of whole biological fluids (blood and urine) without pretreatment of samples. In this work, the hypertension drug molecule atenolol was also observed in untreated urine samples from treated patients (Fig. 5.4B). Additionally, pharmaceutical products such as 1% hydrocortisone (in cream) and liquid cold medication containing acetaminophen, doxylamine succinatem, and dextromethorphan were also analyzed in a direct manner using IR-LADESI-MS [7].

5.6 LDSPI-MS

In a similar vein, a series of closely related works report on **L**aser **D**esorption **S**pray **I**onization **M**ass **S**pectrometry (LDSPI-MS) that combines LD and ESI for direct analysis of liquid and solid samples [37,38]. In LDSPI, the liquid specimen is placed on a support plate and desorbed by an Nd:YAG laser operating at a wavelength of 1064 nm (20 Hz, pulse duration of 10 ns, 11 mJ with an incident angle of 45 degrees) as shown in Scheme 4. Unlike previous methods described so far, this source does not use mid-IR radiation in the range of 2.94 μm. The relative positions of the ESI capillary, sample support plate, and the MS inlet are similar to those described in ELDI-MS and IR-MALDESI-MS except for slight changes in parameter space adjustments. The sample stage is grounded and the technique does not use any organic or inorganic matrix materials.

Food safety applications have been reported for LDSPI-MS. A notable study established a molecular signature fingerprint library for yogurt and yogurt-related products (various brands) and then used this information for test sample analysis via principal component analysis (PCA) of the most abundant peaks (Fig. 5.5) [37].

Furthermore, the method was amenable to LDSPI-MS/MS where a variety of carbohydrate dimers were identified. In another work, a low

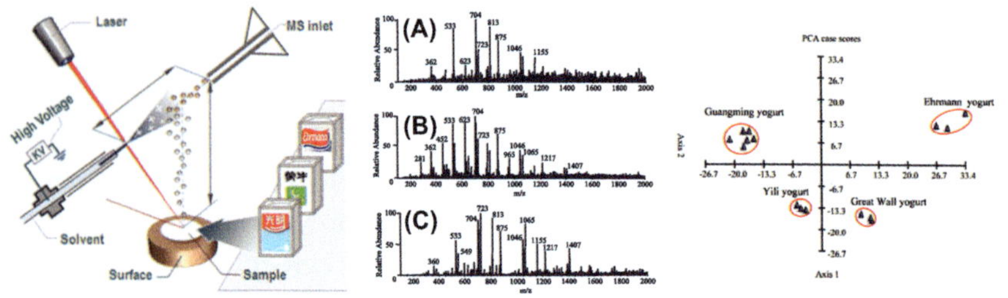

Figure 5.5 ***Laser Desorption Spray Ionization Mass Spectrometry source for the analysis of untreated samples.*** Samples (in the case shown here product from local yogurt brands) are deposited onto a sample plate subjected with ablation and aerosolization by a 1064 nm laser generating aerosols that are ionized by interception with ESI droplets. (A–C) spectra in the middle panel indicate fingerprints (positive ion mode) from three different local brands of yogurt, Yili, Ehrmann and Guangming, respectively. Principal component analysis of said three brands suggests that significant statistical difference exists that allows them to be distinguished from the less common brand Great Wall, in the rightmost panel. *(Reprinted (adapted) with permission from J. Liu, B. Qiu, H. Luo, Fingerprinting of yogurt products by laser desorption spray post-ionization mass spectrometry, Rapid Commun. Mass Spectrom. 24 (2010) 1365–1370. Copyright (2010) John Wiley and Sons, Ltd.)*

1.1 mJ laser energy was shown to be suitable for analyzing protein solutions deposited on metallic support targets. Here, interactions between protein and copper or iron ions generated by LD of the metal substrate induced unexpected metal–protein complex ions. It is likely that these ions arise due to general energy-driven ablative properties of 1064 nm laser light at the interface of metal protein solution surface (Fig. 5.6) [38].

5.7 AIRLAB-MS

Ambient **I**nfrared **L**aser **A**blation **M**ass **S**pectrometry (AIRLAB-MS) [39] is yet another variant that reports on the combination of desorption source adapted from an IR microscope and a continuous solvent probe. AIRLAB-MS uses an IR microscope (Continuum XL, 2.94 μm light, 12–14 mW, measured over 30 s with a pulse repetition rate of 20 Hz, corresponding to an energy density of 5.3 J/cm^2 per pulse) bearing an infinity-corrected reflective objective lens, and a continuous flow solvent probe coupled to a Fourier transform ion cyclotron resonance (FT-ICR) mass analyzer (Fig. 5.7). The transfer efficiency of AIRLAB-MS is significantly higher than those reported for similar techniques: a 50% efficiency for the transfer of material from the sample surface to the electrospray emitter was determined using glycerol/methanol droplets containing 1 mM nicotine as a

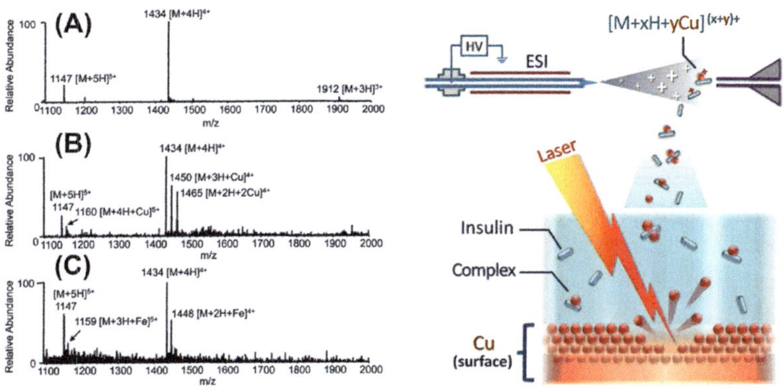

Figure 5.6 ***The Laser Desorption Spray Ionization Mass Spectrometry mass spectra of a 20 mM insulin solution from a variety of metal surfaces.*** Material desorbed from (A) gold, (B) copper, (C) stainless steel sample plates are shown. The **E**lectro**S**pray **I**onization spray solution used in this study was an acidified (0.1% FA) methanol/water (1:1 v/v). Unexpected complexations of protein with metal ions are seen. The experiment uses the setup shown in the panel to the right. *(Reprinted (adapted) with permission from J. Liu, C. Zhang, J. Sun, H. Luo, Unexpected complexation reaction during analysis of proteins using laser desorption spray post-ionization mass spectrometry, Analyst 137 (2012) 1764–1767. Copyright (2012) Royal Society of Chemistry.)*

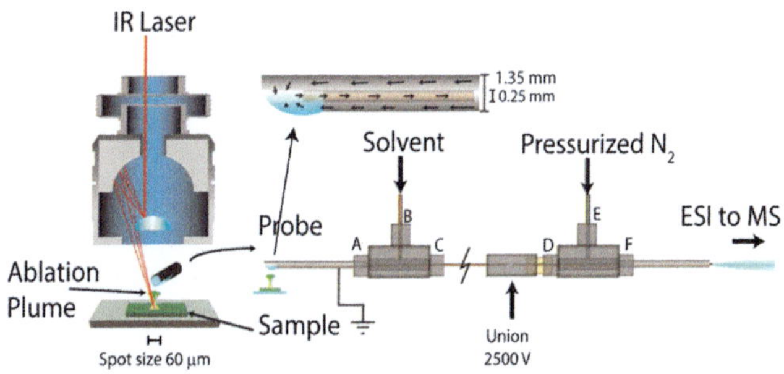

Figure 5.7 ***A schematic diagram (not drawn to scale) of the Ambient Infrared Laser Ablation Mass Spectrometry setup.*** The IR laser from the source is focused through a 15× objective mounted on a continuum XL infrared microscope. The laser ablation plume is captured by a liquid droplet at the tip of a stainless-steel capillary that is attached to a PEEK tee fitting denoted port A. The solvent material is pumped with a syringe into port B. A fused silica capillary is implemented to carry the solvent and the ablated material from the tip of the probe (enlarged in the view in order to better show the solvent flow) through port C and toward the electrospray emitter. Additionally, a conductive stainless-steel union (attached to port D) is utilized to apply the electrospray voltage to the solution. N_2 gas enters the PEEK emitter tee fitting located at port E. A fused silica capillary is the tasked with carrying the solvent and sample from the union, through the tee, and out of the port F where ions are generated by pneumatically assisted electrospray ionization technique. *(Reprinted (adapted) with permission from J.T. O'Brien, E.R. Williams, H.Y. Holman, Ambient infrared laser ablation mass spectrometry (AIRLAB-MS) of live plant tissue with plume capture by continuous flow solvent probe, Anal. Chem. 87 (2015) 2631–2638. Copyright (2015) American Chemical Society.)*

reported compound [39]. This transfer efficiency constitutes a major improvement compared to other IR laser couplings described so far, largely attributed to the continuous flow solvent probe implementation that allows efficient capture and transfer of the laser plume to MS.

Using solution droplets containing bradykinin, leucine enkephalin, and myoglobin it is further shown that the laser desorption does not induce fragmentation of biomolecules. Current reported applications of AIRLAB-MS include the analysis of liquid samples and plant leaves. For example, the chemical components of the tobacco leaf identified from the spatially resolved AIRLAB mass spectra of the ablated plant material included the expected compounds nicotine and uridine. The reproducibility of these measurements was demonstrated by the ablation of six closely spaced regions of interests (within 2 × 2 mm) on a young tobacco leaf, where a less than 10% standard deviation in the uridine signal was seen for each region of interest (ROI) analyzed [39]. The spatial distribution of nicotine was measured for each ROI, and statistically significant variations in the relative nicotine abundance were observed.

5.8 LEMS

The recent development of atmospheric pressure femtosecond laser mass spectrometry for ambient MS imaging applications [40,41] gave rise to **L**aser **E**lectrospray **M**ass **S**pectrometry (LEMS) upon coupling to ESI for postionization (Fig. 5.8). A Ti:Sapphire laser system operating at 10 Hz, producing a pulse laser light at 800 nm with 400 μJ per pulse and 70 fs pulse duration, is used to vaporize samples. No matrix material is needed for LEMS studies due to the very high peak power density of the laser system used. This allows a nonresonant desorption process through use of a highly focused near IR femtosecond laser pulse without employing a selective laser absorbing matrix described in the above sections. As with other ambient ionization techniques, LEMS offers simplified sample preparation and handling. The experiments are performed using a custom-made femtosecond laser ion source that contains a motorized *xy* translation stage that holds the sample at an approximately 5 mm distance from the sample cone of a time-of-flight mass spectrometer. Comparative high energy/low energy (at 1042 nm, less than 50 μJ per pulse, fiber-based systems) LEMS studies of proteins have concluded that the LEMS desorption mechanism likely involves a nonresonant, multiphoton process, as opposed to being thermally driven or photoacoustically induced [42]. Femtosecond laser

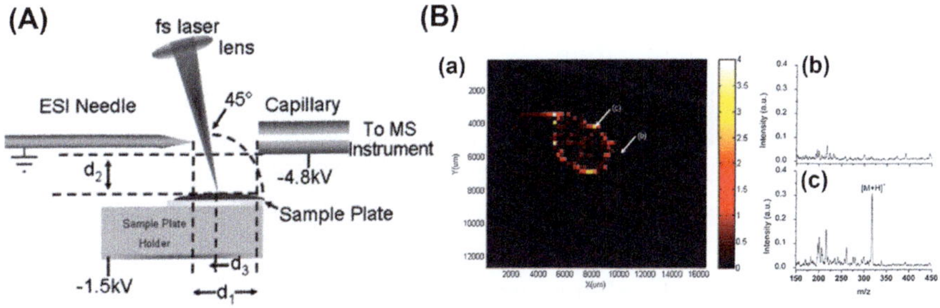

Figure 5.8 ***Laser Electrospray Mass Spectrometry (LEMS) setup using femtosecond laser.*** (A) The schematics of the atmospheric pressure femtosecond laser desorption ionization imaging mass spectrometer (LEMS). (B) The positive ion LEMS image of the distribution of oxycodone (*m/z 316*) from a dried spot forming a ring structure. *(A) Reprinted (adapted) with permission from J.J. Brady, E.J. Judge, R.J. Levis, Mass spectrometry of intact neutral macromolecules using intense non-resonant femtosecond laser vaporization with electrospray post-ionization, Rapid Commun. Mass Spectrom. 23 (2009) 3151–3157. Copyright (2009) John Wiley and Sons, Ltd. (B) Reprinted (adapted) with permission from E.J. Judge, J.J. Brady, D. Dalton, R.J. Levis, Analysis of pharmaceutical compounds from glass, fabric, steel, and wood surfaces at atmospheric pressure using spatially resolved, nonresonant femtosecond laser vaporization electrospray mass spectrometry, Anal. Chem. 82 (2010) 3231–3238. Copyright (2010) American Chemical Society.)*

pulses (at 70 fs at 800 nm with a pulse energy of 1.5 mJ) were focused onto the specimen with a typical lateral resolution (i.e., laser spot diameter) of approximately 300 μm [43]. In LEMS, proper focusing of the laser is necessary to ensure reproducibility and consistency of MS analysis results.

To date, LEMS has been applied to direct analysis of both plant and animal tissue. In the metabolic profile of petal and leaf samples, compounds such as anthocyanins, a variety of sugars, and other metabolites were detected. Further work on animal tissue (mouse brain) revealed phospholipids, especially phosphatidylcholine. These lipids were suppressed in brain sample that was prepared using the optimal cutting temperature (OCT) compounds in both fiber-based and open beam LEMS measurements [42,44]. This finding highlights the requirement for fresh or frozen tissue that further offers prospects of in situ or future in vivo use.

5.9 AP-fsLDI-MS

A close variant of LEMS, entitled **A**tmospheric **P**ressure **f**emto**s**econd **L**aser **D**esorption **I**onization (AP-fsLDI-MS) [45], uses a similar Ti:Sapphire (800 nm, 1 KHz, 10^{14} W/cm^2) femtosecond laser and no concurrent ESI

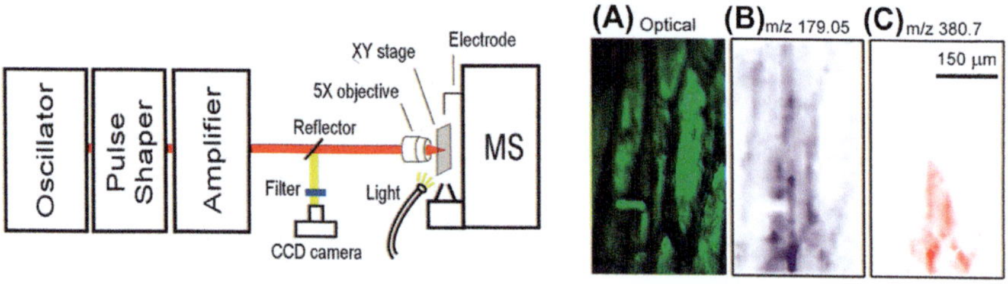

Figure 5.9 ***The schematics of Atmospheric Pressure femtosecond Laser Desorption Ionization setup and reported applications.*** High power femtosecond pulses create desorption and ionization in the absence of **E**lectro**S**pray **I**onization–mediated postionization. Imaging of onion epidermis with a lateral resolution of 10–15 μm in the negative ions mode, comparing optical, deprotonated glucose (*m/z* 179.06) and introduced triiodide (*m/z* 380.6) with expected concordance. *(Reprinted (adapted) with permission from Y. Coello, A.D. Jones, T.C. Gunaratne, M. Dantus, Atmospheric pressure femtosecond laser imaging mass spectrometry, Anal. Chem. 82 (2010) 2753–2758. Copyright (2010) American Chemical Society.)*

post ionization. In AP-fsLDI-MS, the specimen is positioned between the sample cone and an electrode that acts as ion repellent such that the potential difference can assist in directing the ionized plume toward the mass analyzer inlet (Fig. 5.9). Near IR femtosecond laser pulses are characterized by having very high peak power densities (10^{14} W/cm^2). This peak power allows direct nonresonant desorption and concurrent ionization of a large array of analytes. Due to this power density effect, these two closely related desorption techniques do not require addition of laser absorbing matrix materials to assist in desorption, and therefore greatly simplify the sample preparation workflow. AP-fsLDI-MS takes advantage of this high peak power to cause concurrent desorption ionization in close proximity to MS entrance where transfer of ions is aided by a strategically placed electrode. Here, it must also be emphasized that in the absence of an absorbing matrix, analytes contained in the sample may directly absorb the laser light. For example, while LEMS analysis of proteins using nonresonant femtosecond systems have been proposed to be gentle, desorbing neutrals and also preserving the folded structure [46], fragmentation of cellulose has been shown to result in residual glucose which has been detected during analysis using onion epidermis [45]. Analytical capabilities of AP-fsLDI-MS (Fig. 5.9) have also been reported in the literature for imaging applications using onion epidermis as well as deposited ink, where tight focusing of the femtosecond laser allowed superior lateral resolution on the order of 10 μm [45].

5.10 LA-FAPA-MS

Laser **A**blation **F**lowing **A**tmospheric **P**ressure **A**fterglow **M**ass **S**pectrometry uses "corona discharge" under ambient conditions to bring about ionization of LA plume [47]. Atmospheric pressure chemical ionization (APCI) generates reactive charged species that then will ionize gaseous analytes via ion—molecule reactions (IMRs) [47]. Compared to ESI, this postionization technique is more efficient in analyzing small nonpolar compounds, but may not necessarily offer the multiple charging required for the analysis of large molecules such as proteins. In essence, LA-FAPA-MS can be considered to be a hybrid laser and ambient ionization method that uses a corona discharge system, where laser ablated material are transferred by a nitrogen stream to the afterglow of a helium glow discharge ionization source under atmospheric pressure [48—51] (Figs. 5.10 and 5.11A for ionization and analysis setups). Since LA-FAPA requires both dedicated LA and ionization chambers that use nonatmospheric composition of gasses, the method may not be strictly ambient, but it is certainly closely related, as the enclosures only optimize the method, and do not prevent it from performing altogether under atmospheric conditions. In a typical LA-FAPA experiment, a Nd:YAG laser producing UV pulses at 266 nm, 20 Hz is focused on the sample surface (lateral resolutions of 10

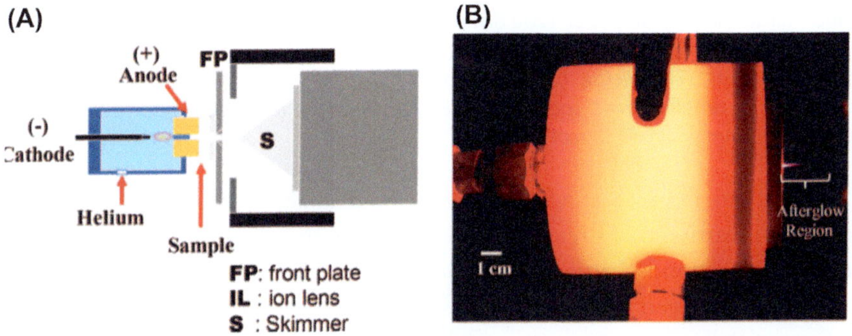

Figure 5.10 ***The schematic diagram of the Laser Ablation Flowing Atmospheric Pressure Afterglow Mass Spectrometry ionization source.*** (A) A glow discharge is formed between a tungsten pin cathode and a brass anode plate. A hole in the anode plate allows the excited discharge gas molecules to exit and to interact with a sample outside the chamber. (B) A photograph of the FAPA cell with a neon discharge system that illustrates the afterglow region. *(Reprinted (adapted) with permission from J.T. Shelley, S.J. Ray, G.M. Hieftje, Laser ablation coupled to a flowing atmospheric pressure afterglow for ambient mass spectral imaging, Anal. Chem. 80 (2008) 8308—8313. Copyright (2008) American Chemical Society.)*

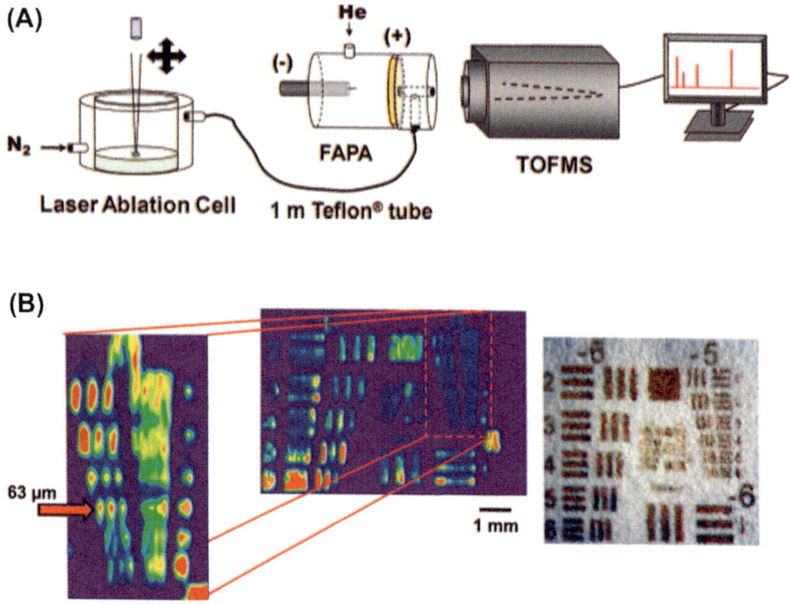

Figure 5.11 ***The schematic diagram of a typical Laser Ablation Flowing Atmospheric Pressure Afterglow (LA-FAPA) analysis system and reported application.*** (A) The analysis system where the sample is ablated with 266 nm laser. The resulting aerosols are transferred into a stream of N_2 through a 1 m Teflon tube to the afterglow region for ionization and subsequent detection, in this case by a time-of-flight mass analyzer. (B) The generation of an LA-FAPA-MS chemical image. The chemical image of resolution target printed on paper 1200 dot per inch (dpi) resolution indicating vertical and horizontal resolution of 178 and 63 μm. *(Reprinted (adapted) with permission from J.T. Shelley, S.J. Ray, G.M. Hieftje, Laser ablation coupled to a flowing atmospheric pressure afterglow for ambient mass spectral imaging, Anal. Chem. 80 (2008) 8308–8313. Copyright (2008) American Chemical Society.)*

and 300 μm being typical). The generated ablation specimen vapor or aerosols is carried in a stream of nitrogen through a Teflon tube where it is mixed in a second chamber with the afterglow from the corona discharge system. It has been shown that the spatial resolution as well as the adverse effects of thermolysis are improved by choice of laser as opposed to plasma for desorption. Ablation with pulsed laser also allows depth profiling using successive laser pulses, where a sampling depth of approximately 36 μm with each burst has been reported [49].

While LA-FAPA-MS analysis is largely restricted to small molecules (less than 1000 Da), a broad range of small molecule compounds can be detected with this method [49]. LA-FAPA-MS has been used to characterize medicinal tablets as well as to perform MSI analysis of turkey tissue wherein an

LOD of 5 fmol (using spiked caffeine) and a spatial resolution of 20 μm were achieved [49]. LA-FAPA-MS has also been used in the imaging mode for spatially resolved analysis of small molecules from solid samples. A solution of ink doped with caffeine was used to print the Indiana University logo (5.6 mm in width) which was then imaged using LA-FAPA-MS in a total time of less than 30 min, with a 63 μm (x-direction) and a 178 μm (y-direction) lateral resolution [49]. In the same study, the applicability of LA-FAPA-MS for imaging was also investigated (Fig. 5.11B). In the first example, lidocaine was spotted onto wet turkey tissue and imaged with concordance between MS and optical images. In the second example, celery veins into which caffeine had been introduced were imaged successfully using LA-FAPA-MS. The celery stick was immersed into a solution containing a blue dye and caffeine molecules and then sliced and subsequently imaged to identify the location of veins through detecting infused molecules [49].

5.11 LA-APCI-MS

Laser **A**blation **A**tmospheric **C**hemical **I**onization **M**ass **S**pectrometry (LA-APCI-MS) is yet another variant with published applications in imaging as well as in analysis of drug tablets, thin layer chromatography (TLC) plates, and biological tissue sections [52,53] (Fig. 5.12). The sample transfer component used in LA-APCI-MS is very similar to that described above for LA-FAPA-MS. In this method, specimens are placed in an ablation cell and then irradiated by a frequency quintupled, Q-switched Nd:YAG laser outputting UV radiation at 213 nm, where a nitrogen gas flow carries the ablation plume via a polyamide tubing to a corona discharge source for postionization (Fig. 5.12A). In the first iteration, the neutral plume was split by a T-connector for the detection by both ICP-MS and APCI-MS, enabling simultaneous elemental and molecular analysis with mass spectrometry [53]. Subsequently, a commercially available laser capture microdissection device was coupled to the APCI-MS, enabling additional multimodal imaging capability [54]. Here, imprinted dyes, inks, as well as biological tissues were analyzed in transmission configuration (i.e., geometry), where a Venturi pump equipped with a sampling capillary was used to deliver the ablated molecules (largely neutral) to the APCI source for subsequent ionization. In this work using LA-APCI-MS for the MSI analysis of lipid molecules present in the mouse brain, a spatial resolution of 13 μm has been reported [54].

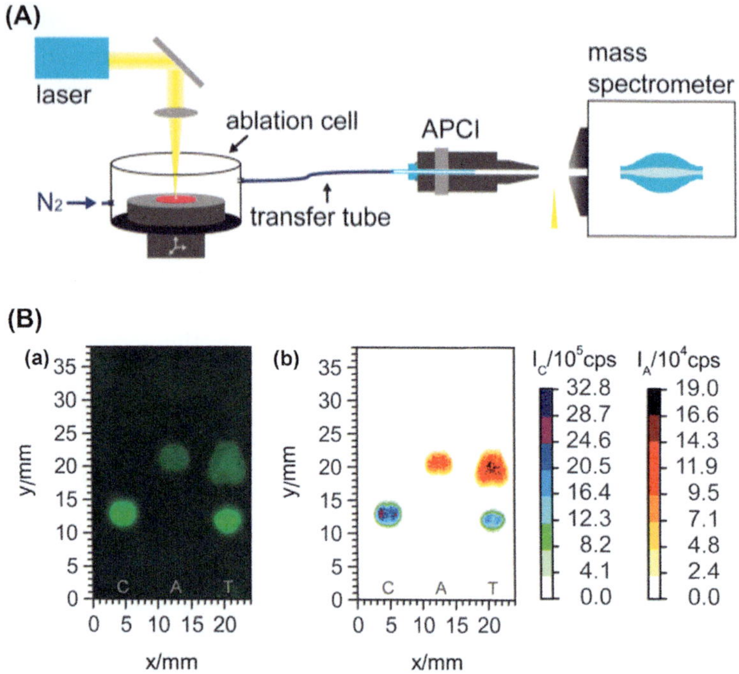

Figure 5.12 ***The schematics of the Laser Ablation Flowing Atmospheric Pressure Afterglow Mass Spectrometry (LA-APCI-MS) setup and reported applications.*** (A) The LA-APCI-MS system. Here, laser ablation inside a chamber creates an ablation plume that is transported to an Atmospheric Pressure Chemical Ionization (APCI) source for ionization prior to detection by mass analyzer. (B) The LA-APCI-MS imaging analysis of a thin layer chromatography plate at 200 μm spatial resolution where (a) is an optical image (fluorescence excitation at 470 nm, and emission at 535 nm), and (b) denotes overlay of the ion images of acetaminophen molecule (*m/z* 152.07, "A") and of caffeine (*m/z* 195.08, "C") obtained by LA-APCI-MS is shown in "T". *(A) Reprinted (adapted) with permission from C. Herdering, O. Reifschneider, C.A. Wehe, M. Sperling, U. Karst, Ambient molecular imaging by laser ablation atmospheric pressure chemical ionization mass spectrometry, Rapid Commun. Mass Spectrom. 27 (2013) 2595–2600. Copyright (2013) John Wiley and Sons, Ltd. (B) Reprinted (adapted) with permission from C. Herdering, O. Reifschneider, C.A. Wehe, M. Sperling, U. Karst, Ambient molecular imaging by laser ablation atmospheric pressure chemical ionization mass spectrometry, Rapid Commun. Mass Spectrom. 27 (2013) 2595–2600. Copyright (2013) John Wiley and Sons, Ltd.)*

5.12 PAMLDI-MS

Plasma **A**ssisted **M**ultiwavelength **L**aser **D**esorption **I**onization **M**ass **S**pectrometry (PAMLDI-MS) builds on the success of Direct Analysis in Real Time (DART). DART is a well-known mass spectrometry analysis method that can be used to create reactive species in addition to its direct desorption/

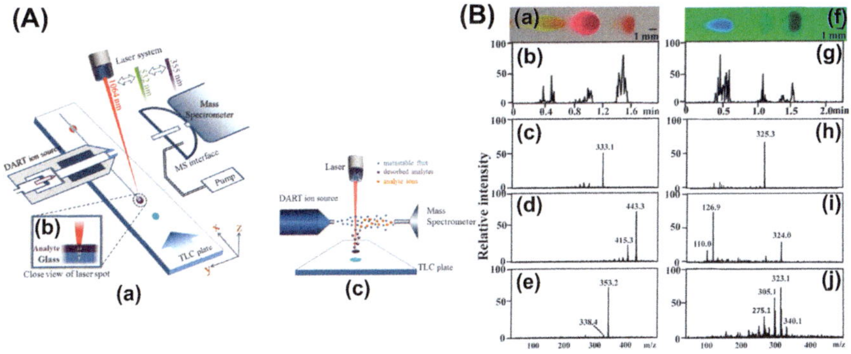

Figure 5.13 ***Plasma Assisted Multiwavelength Laser Desorption Ionization Mass Spectrometry (PAMLDI-MS) schematics and example application.*** (A) The schematic illustration of a typical PAMLDI-MS setup utilizing thin layer chromatography (TLC) plate as a substrate. The inset displays a closeup view of the relative position of the laser ablation point and the TLC surface. The desorption and ionization processes take place in PAMLDI-MS where reactive species created by DART serve to postionize laser ablation plume. (B) Analysis of TLC resolved samples by PAMLDI-MS as illustrated by (a) photograph of three separated dye standards resolved by chromatography on a silica gel-coated TLC plate where (b) shows the extracted ion chromatograms for fluorescein, rhodamine B, and Sudan III dyes are seen from left to right. The positive-ion TLC/PAMLDI mass spectra of (c) fluorescein, (d) rhodamine B, and (e) Sudan III. (f) The photograph of three drug standards separated on a silica gel-coated TLC plate. (g) Extracted ion chromatograms of quinine, gliclazide, and chloramphenicol are shown from left to right. The positive TLC/PAMLDI mass spectra of (h) quinine, (i) gliclazide, and (j) chloramphenicol. *(Reprinted (adapted) with permission from J. Zhang, Z. Zhou, J. Yang, W. Zhang, Y. Bai, H. Liu, Thin layer chromatography/plasma assisted multiwavelength laser desorption ionization mass spectrometry for facile separation and selective identification of low molecular weight compounds, Anal. Chem. 84 (2012) 1496–1503. Copyright (2012) American Chemical Society.)*

ionization capabilities (Fig. 5.13A). The source can be utilized in a two-step ionization scheme upon coupling with a laser. The tandem of laser sampling and DART is reported as PAMLDI-MS or Surface Assisted Laser Desorption Ionization Mass Spectrometry (SALDI-MS). In this report, the platform utilized a modified pulsed Nd:YAG laser producing 10 ns pulses of laser light at wavelengths of 355, 532, and 1064 nm operating at a repetition rate of 10 Hz [55,56]. The laser beam was set to irradiate the sample at an incident angle of 45° for the analysis of solid samples. This irradiation generates a plume of neutral analytes that are subsequently sent for the ionization in a stream of reactive species generated by a DART source for ion molecule interactions and ionization. The multiwavelength aspect allows selective boosting of analyte signal by absorption matching to UV, IR, and visible regions of the electromagnetic spectrum.

PAMLDI-MS has been utilized in the analysis of imprinted dyes and plant extracts separated on TLC plates (Fig. 5.13B). The method offers better sensitivity upon utilization of temperature modulation at the DART source [56]. In another study, graphene and graphite were used as matrix material to further assist laser desorption, increasing ion intensities up to 600 folds [57]. Plant tissues and natural products such as Chinese tea extract have also been analyzed by PALDI-MS in the imaging mode, producing molecular images of dyes and metabolites with a pixel resolution of 60 μm × 60 μm.

5.13 LIAD-ESI-MS

Laser **I**nduced **A**coustic **D**esorption **E**lectro**S**pray **I**onization **M**ass **S**pectrometry (LIAD-ESI-MS) uses a pulsed laser irradiating the backside of a thin metallic foil present within a sapphire substrate, creating high-amplitude shockwaves that propagate through the foil to desorb analytes deposited on the opposite side of said foil. LIAD-ESI-MS was initially performed in a vacuum chamber, generating a limited number of ions alongside a significant population of neutral molecules. In 2009, LIAD-ESI reported the first marriage between LIAD and ESI source under ambient conditions [58,59]. While the LIAD-ESI setup is similar to that of ELDI described above, there are fundamental differences in the position of the pulsed laser beam and how it has been optimized to induce acoustic desorption as opposed to direct desorption (Fig. 5.14A). During the analysis, a pulsed Nd:YAG laser (1064 nm, 9 ns, 1–2 Hz, 1–30 mJ range) is used to irradiate the backside of metallic Al and Ti foils (10–25 μm in thickness) to desorb solid as well as liquid samples deposited on the other side of said foil. The plume of desorbed analyte is then intercepted with an orthogonal ESI spray generating ESI-like ions in a process similar to what is taking place in ELDI-MS. During the course of optimization, it was discovered that the optimal laser energy is around 10^9 W/cm^2. It is worth noting that metal substrates with a high thermal expansion coefficient are ideal for LIAD, indicating there could be a thermal component associated with efficacy or desorption mechanism of LIAD-ESI-MS. Overall, LIAD-ESI-MS has been used to characterize proteins, peptides, amino acids, and dyes in solid samples without producing significant fragmentation of

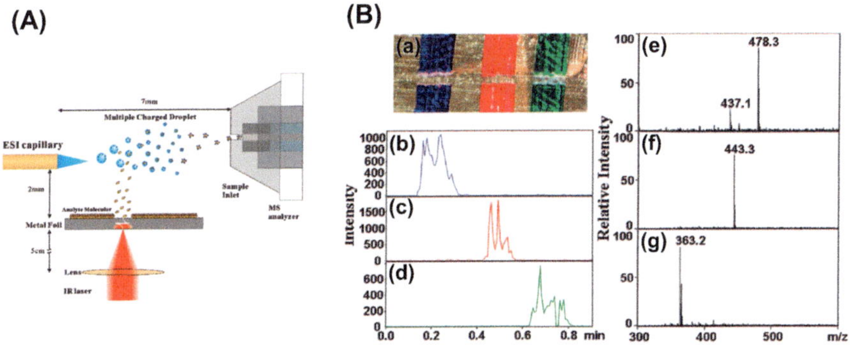

Figure 5.14 ***The schematic representation of the desorption and ionization during Laser Induced Acoustic Desorption ElectroSpray Ionization Mass Spectrometry (LIAD-ESI-MS) analysis and example application.*** (A) Here, the laser beam, metal support foil, and ESI ionization spray are shown. Propagation of shockwave through foil allows acoustic desorption. (B) Analysis of deposited ink on metallic support with LIAD-ESI-MS where (a) is the photograph of the support aluminum foil displaying desorption region of dry inks after irradiating the rear side of the foil with the infrared laser used in LIAD-ESI-MS, (b–d) show the extracted ion chromatograms of the ink ions: blue (*m/z* 478.3), red (*m/z* 443.3), and green (*m/z* 363.2), and (e–g) illustrate the positive-ion LIAD-ESI mass spectra of the inks used. *(Reprinted (adapted) with permission from S.C. Cheng, T.L. Cheng, H.C. Chang, J. Shiea, Using laser-induced acoustic desorption/electrospray ionization mass spectrometry to characterize small organic and large biological compounds in the solid state and in solution under ambient conditions, Anal. Chem. 81 (2009) 868–874. Copyright (2009) American Chemical Society.)*

desorbed analytes (Fig. 5.14B) [59]. In another study, LIAD-ESI-MS was used to characterize chemical compounds on TLC plates [58]. Since aluminum-based TLC plates are thicker (100 μm for an aluminum sheet and 200 μm for a C_{18} or silica gel stationary phase) than the typical metal foils used in conventional LIAD-ESI-MS, a glass slide with a viscous solution deposited onto it could be used to enhance desorption. Using this technique, drugs, dyes, and plant extracts with LOD on the order of nanograms have been reported [58].

5.14 LIAD-APCI-MS

Following its combination with ESI described above, LIAD has also been coupled to other API sources including APCI and Atmospheric Pressure Photoionization (APPI). The rationale behind this development was to increase the analysis domain of the technique to encompass compounds

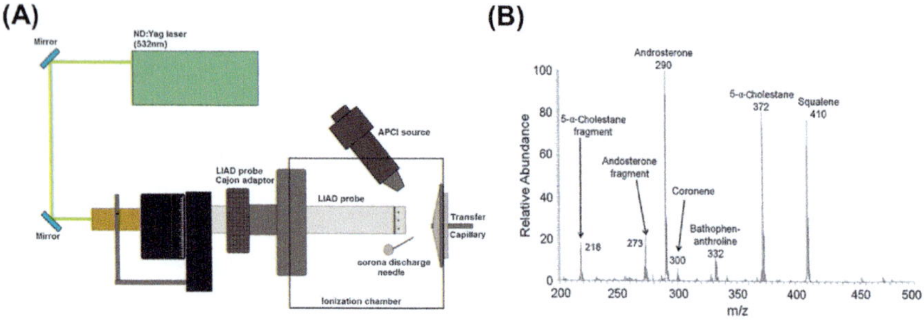

Figure 5.15 ***The Laser Induced Acoustic Desorption Atmospheric Pressure Chemical Ionization Mass Spectrometry (LIAD-APCI-MS) setup and reported application.*** (A) The setup where the glass window in the front of the source was removed such that the LIAD probe could be inserted into the APCI source. About 5″ of the probe (from the front side of a Cajon fitting adaptor) was inserted into the source. This placed the probe tip 5/8″ away from the Corona discharge source. The LIAD laser was directed by using silver mirrors to the back side of the foil. The outer cylinder of the LIAD probe, containing the sample foil, could be rotated for multiple deposition of analytes. (B) Analysis with LIAD-APCI-MS. A mixture of 5-α-cholestane (molecular weight MW 372), coronene (MW 300), squalene (MW 410), androsterone (MW 290), and bathophenanthroline (MW 332) (all in equimolar ratios) were subjected to LIAD-APCI-MS producing expected mass spectra consistent with the identity of each molecule. *(Reprinted (adapted) with permission from J. Gao, D.J. Borton, B.C. Owen, Z. Jin, M. Hurt, L.M. Amundson, J.T. Madden, K. Qian, H.I. Kenttämaa, Laser-induced acoustic desorption/atmospheric pressure chemical ionization mass spectrometry, J. Am. Soc. Mass Spectrom. 22 (2011) 531–538. Copyright (2011) American Society for Mass Spectrometry.)*

with low polarity that could be less amenable to ionization by ESI [60,61]. Alongside the necessary optics and titanium foil (12.5 μm in thickness) which was mounted on a glass support, an LIAD probe was inserted into the ionization chamber of a commercial APCI source (Fig. 5.15A). The APCI probe was operated at 375°C to vaporize gaseous reagents (e.g., benzene, MeOH/H_2O, and carbon disulfide CS_2 dopant) that were ionized by a corona needle. A Nd:YAG laser (532 nm, 3 ns pulse duration, 10 Hz, providing 20 mJ of energy per pulse) was focused on the backside of the Ti foil to desorb the analytes deposited on the opposite side for subsequent ionization by the reagent ions. Using this method, aromatic and aliphatic compounds were analyzed from which protonated and radical cation analyte ions were detected (Fig. 5.15B) [60]. Here, an estimate for LOD on the order of 2.3 ng for bathophenanthroline was reported [60].

5.15 LIAD-APPI-MS

This technique is a closely related variant of LIAD-MS technology described above that now uses **A**tmospheric **P**ressure **P**hoto-**I**onization as opposed to APCI for efficient analysis of low-polarity compounds such as polycyclic aromatic hydrocarbons (PAHs), cholesterols, and α-tocopherols [62] (Fig. 5.16A). A typical LIAD-APPI-MS setup comprises a pulsed IR laser, a cathode discharge plasma source, and a similar Ti foil (12.7 μm in thickness) as described above for desorption in LIAD-APCI-MS. The IR laser (1064 nm, 10 Hz, power density of ~10^8 W/cm^2) impacts the rear of the Ti foil positioned 5 mm from the MS inlet, vaporizing the samples on the opposite side. The ejected plume travels toward the MS inlet and is

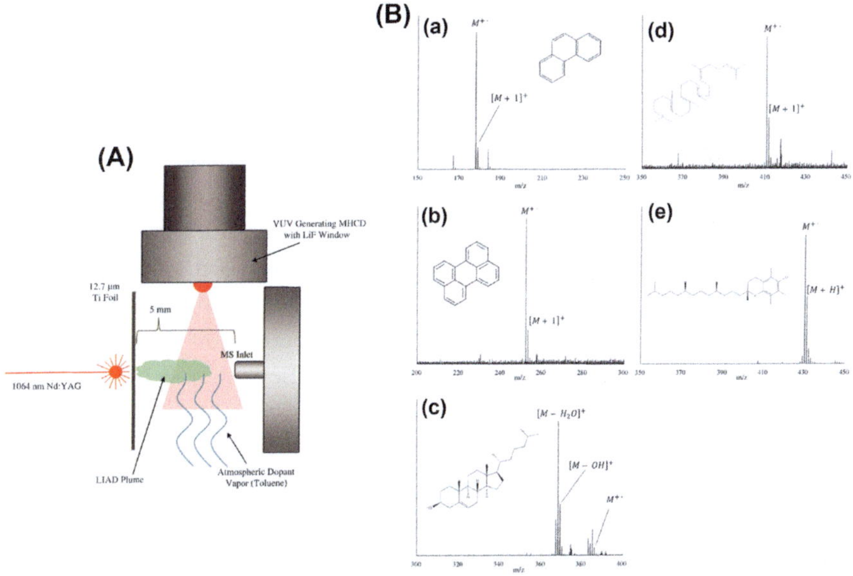

Figure 5.16 ***Schematic of the Laser Induced Acoustic Desorption Atmospheric Pressure Photoionization Mass Spectrometry (LIAD-APPI-MS) ion source and example application.*** (A) The source schematics where laser ablation creates the plume in a similar manner described for LIAD-APCI above. Here, however, the desorbed molecules are transferred to a region containing reactive photoions for ionization and subsequent transfer to the mass spectrometer. (B) The positive ion mode spectra of low polarity species deposited on titanium foil obtained with LIAD-APPI-MS where phenanthrene (a), perylene (b), cholesterol (c), squalene (d), and α-tocopherol (e) spectra are shown. *(Reprinted (adapted) with permission from K. Benham, R. Hodyss, F.M. Fernández, T.M. Orlando, Laser-induced acoustic desorption atmospheric pressure photoionization via VUV-generating microplasmas, J. Am. Soc. Mass Spectrom. 27 (2016) 1805–1812. Copyright (2016) American Society for Mass Spectrometry.)*

irradiated by an orthogonal Lyman-α light (10.2 eV) for photoionization (via a microhollow cathode discharge (MHCD) microplasma photon source in conjunction with the addition of a gas-phase molecular dopant). With introduction of the toluene vapor as a dopant for ionization, efficient ionization is seen. Under these conditions, radical ions (Mc^{+}) of α-tocopherol were detected at an LOD level in the pmol range (Fig. 5.16B) [62].

5.16 PIR-LAESI-MS

Recently, picosecond IR laser desorption via impulsive vibrational excitation (DIVE) has provided a promising mechanism for the development of IR laser-based MS ion sources [63–65] described as **P**icosecond **I**nfra**R**ed **L**aser **A**blation **E**lectro**S**pray **I**onization **M**ass **S**pectrometry (Fig. 5.17). During the DIVE process, the photon energy of the IR laser pulse is selectively deposited into the O–H vibrational stretch mode of water molecules present in biological tissues. The O–H stretch mode of water is directly coupled through the water hydrogen bond network to the O–O translation motion, the motions that drive molecular ablation [66]. The deposited energy is thus transferred into translational motion on time scales faster than the thermal or acoustic energy losses. It is worth noting that the IR LA at a similar wavelength but driven through longer pulse duration has been implemented using Er:YAG lasers (pulse duration 100 ns–100 μs) and with Nd:YAG pumped OPO systems offering pulse durations on the order of ~10 ns. These developments constituted the basis for coupling these ablative systems with secondary postionization sources, such as electrospray in Laser Ablation Electrospray Ionization (LAESI), a very useful tool for direct MS analysis of biological samples [5,67]. Long pulse durations, especially for microsecond systems, result in an energy deposition process that may transfer some of the deposited energy to the analyte molecules or penetrate the surrounding tissue, resulting in tissue damage or analyte fragmentation. In PIR-LAESI-MS, the short pulse duration of PIRL allows rapid energy deposition into water molecules to act as a propellant to drive the analyte molecules "dissolved in water" or present in biological tissues to the gas phase efficiently, intact and without significant damage, in a very similar manner to other nanosecond IR sources reviewed so far. These undamaged gas phase molecules are then available for subsequent ionization and MS analysis. To date, all reported cases of PIRL-MS have taken place under atmospheric conditions where efficient extraction of molecules from

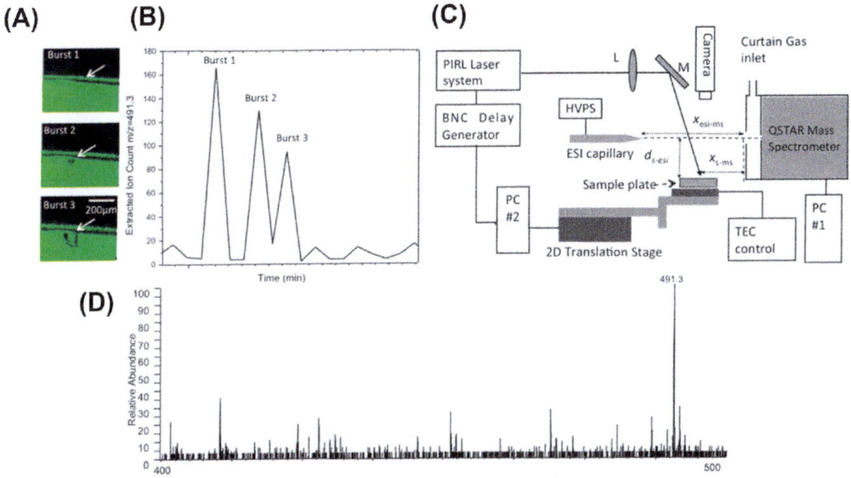

Figure 5.17 ***Characterization of the Picosecond InfraRed Laser (PIRL) ablation crater with optical imaging and Picosecond InfraRed Laser Ablation ElectroSpray Ionization (PIR-LAESI) mass spectrometry imaging performed on 2% agar.*** (A) The ablation crater from PIRL interrogation on 2% agar revealed through dark-field imaging allows characterization of tissue removal per laser burst. Three bursts of 10 PIRL pulses (with fluence of 1.9 J/cm^2 with the energy of 150 μJ/pulse) were deposited onto a 2% agar gel containing a reporter Rhodamine 640 dye. The crater created through the initial burst (i.e., first 10 pulses irradiating the sample) was not visible in microscopy but it produced MS signal (see panel B). (B) Abundance of Rhodamine 640 (*m/z* 491.3) in a 100 μm thick agar layer (see part D for its PIRL-MS spectrum) as a function of PIRL pulse bursts (10 pulses per burst) using a laboratory custom designed PIR-LAESI system further illustrated in panel C, where the LAESI source geometry is also shown. The lateral resolution (i.e., the laser focus spot) was 100 μm. In part B the 2% agar substrate contained 100 μM Rhodamine 640. This design allowed the authors to monitor the abundance of the postionized Rhodamine *m/z* 491.3 to confirm depth profiling shown in part A. The vertical resolution for MS analysis was suggested to be 30 μm (spectrum D). *(Reprinted (adapted) with permission from J. Zou, F. Talbot, A. Tata, L. Ermini, K. Franjic, M. Ventura, J.Z. Zheng, H. Ginsberg, M. Post, D.R. Ifa, D. Jaffray, R.J.D. Miller, A. Zarrine-Afsar, Ambient mass spectrometry imaging with picosecond infrared laser ablation electrospray ionization (PIR-LAESI), Anal. Chem. 87 (2015) 12071–12079. Copyright (2015) American Chemical Society.)*

water-rich biological materials has been demonstrated [68–70]. PIRL-MS is not restricted to aqueous solutions: it is also readily applicable to biological tissue characterizations where water content can be as high. It is worth noting that ablation with PIRL has been shown to be very effective in bone and dentin, two tissue types with low water content as well, albeit with no MS collection and characterization [69].

A tandem of PIRL ablation coupled to electrospray ionization (PIR-LAESI) mass spectrometry was demonstrated and characterized for tissue molecular imaging, with an LOD in the range of 100 nM for reserpine or better than 5 nM for verapamil in aqueous solution. PIRL ablation crater size has been characterized using agar films containing rhodamine reporter dyes. PIR-LAESI was shown to offer a 20–30 μm vertical resolution (with ~3 μm removal per pulse) and consequent lateral resolution of ~100 μm. Closer to 25 fmol of rhodamine was detected in agar ablation experiments. PIR-LAESI has also been used to map the distribution of endogenous methoxykaempferol glucoronide in zebra plant (*A. squarrosa*) leaves subjected to LAESI-MS, producing a localization map that closely matches the optical white field image of the leaves as well as to the LAESI-MS image generated using a parallel experiment with a nanosecond source (Fig. 5.18) [63].

PIR-LAESI was further used to image the distribution, inside mouse kidneys, of gadoteridol, an exogenous magnetic resonance contrast agent intravenously injected into the tail vein of mice (Fig. 5.19). It is also further suggested that PIR-LAESI is capable of desorption ionization of proteins as well as phospholipids [63]. This study concluded that PIR-LAESI was a suitable ion source for potential future ambient mass spectrometry applications. As such, a PIRL probe combined with secondary ionization such as ESI and mass spectrometry has the potential to provide molecular feedback to guide laser surgery. Since this debut, additional applications of PIRL-DIVE in other investigations suggests that the technique produces minimal tissue damage outside the ablation zone [63,71].

5.17 PIRL-MS

Capitalizing on PIRL's efficient rapid extraction capabilities and rapid cooling of the plume due to expansion [72], the possibility of coupling PIRL ablation to soft ionization methods such as thermal desolvation for the analysis of precharged lipid molecules that carry a negative charge forming a new platform entitled **P**icosecond **I**nfra**R**ed **L**aser **M**ass **S**pectrometry (PIRL-MS) has been explored [64,65]. PIRL (2.96 μm, 1 kHz with pulse duration of 300 ps, outputting ~0.15 J/cm^2 of energy) was directly attached to the inlet of time of flight (TOF) instrument utilizing a metallic capillary extension (Fig. 5.20) [64,65].

After demonstration of the efficient classification of mouse organs with this method based on extracted and desolvated lipid molecules (MS analysis

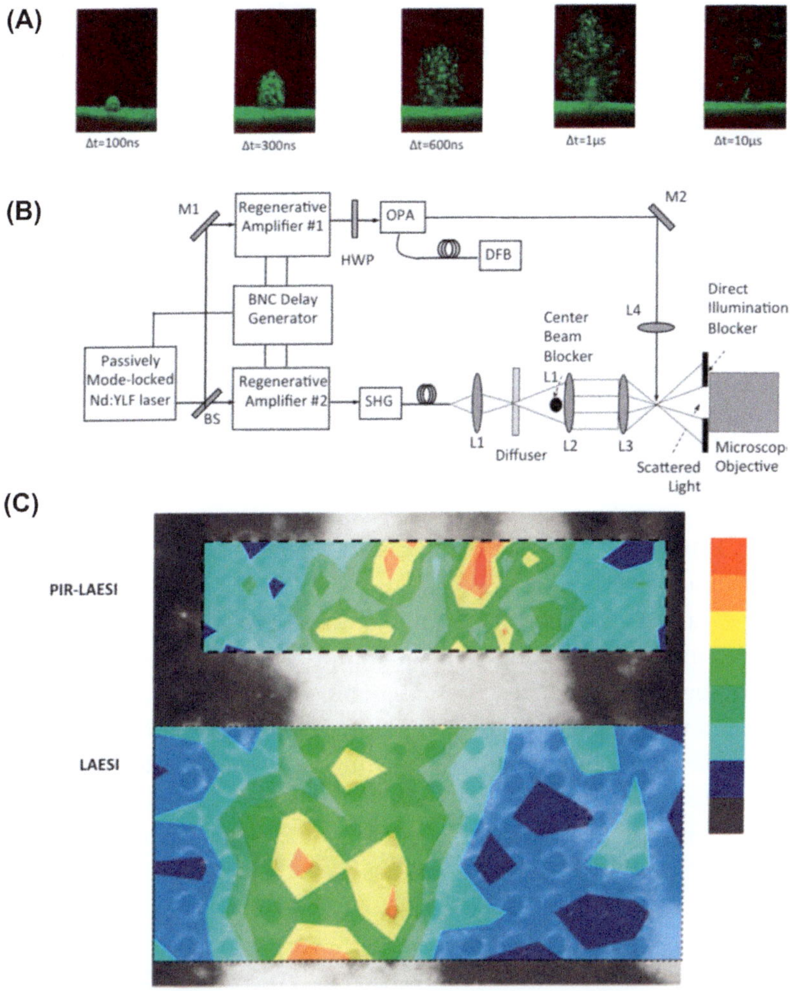

Figure 5.18 ***Summary of the characterization of Picosecond InfraRed Laser (PIRL) ablation plume with dark-field imaging and with Picosecond InfraRed Laser Ablation ElectroSpray Ionization (PIR-LAESI) mass spectrometry imaging using a zebra plant leaf as a test bed.*** (A) Dark-field imaging of the ablated leaf (using single PIRL pulse) at time points after the ablation pulse. (B) The schematics of the dark field imaging platform used to reveal plume propagation. In this diagram, BS is a 50% beam splitter for 1053 nm laser, M1 is a flat mirror with a high reflective coating at 1053 nm, HWP is a half wave plate, M2 is a flat mirror with high reflective coating at 2910 nm, L1–3 are BK7 are lenses with focal lengths of L1 (f = 40 mm), L2 (f = 25 mm), and L3 (f = 40 mm), L4 is a CaF_2 lens with f = 100 mm, OPA indicates an optical parametric amplifier, SHG is second harmonic generator, DFB is a distributed feedback diode laser. (C) The molecular map of methoxykaempferol glucoronide (*m/z* 493.1) in zebra plant leaf is shown using heat map plots, and compared to an LAESI-MS setup that uses an

in the negative ion mode [64]), authors explored the possibility of coupling PIRL-MS to other imaging modalities such as wide-field tissue polarimetry for efficient guidance to regions of cancer [64]. Polarimetric imaging reveals tissue heterogeneities to which authors could guide the PIRL desorption source (Fig. 5.21). One key attribute is that the PIRL-MS plume was transported to MS through a 2 m long tygon tube, thus providing a handheld MS point sampling device capable of analyzing thick tissue. With this handheld MS point sampling device developed, polarimetry was investigated on thick tissue sections for possible in situ MS guidance [64].

The point sampling with PIRL-MS combined with reflection mode polarimetry was used to identify areas of necrosis in metastatic breast tumors and verify the presence of necrotic cancer marker ion *m/z* 572.48 in an efficient manner. Polarimetry guided the PIRL probe to areas of tissue heterogeneity that from MS sampling were determined to be necrotic. Of special note was the presence of many common fatty acids and phospholipids in PIRL-MS spectra of breast cancer reported previously using other MS technologies [73–75].

In yet a second debut application, medulloblastoma (MB), the most prevalent malignant childhood brain tumor that consists of at least four distinct subgroups which each possess a unique survival rate, the chance for relapse and response to treatment were subjected to PIRL-MS analysis. To rapidly determine MB subgroups in a manner that would be actionable during surgery (allowing personalized treatment to begin as soon as the cancer is encountered intrasurgically), mouse xenograft tumors of 2 MB subgroups of SHH and Group 3 were subjected to lipid profiling using the

OPO laser to produce nanosecond IR pulses [63]. The panel on top uses a 100 μm PIRL laser spot size, with a step size (i.e., center to center distance between the two consecutive laser craters) of 200 μm. The molecular map of the same molecule with a reproduced LAESI-MS setup using a nanosecond optical parametric oscillator (OPO) laser is shown in the bottom that used OPO output IR the fluence of 2 J/cm^2 (300 μm laser spot size, with a step size of 400 μm). Each pixel was formed from output of 20 laser pulses, with a repetition rate of 10 Hz. With PIR-LAESI-MS authors were able to achieve a lateral resolution that was ~3 times higher compared to LAESI with an OPO source at a matching laser fluence values. *(Reprinted (adapted) with permission from J. Zou, F. Talbot, A. Tata, L. Ermini, K. Franjic, M. Ventura, J.Z. Zheng, H. Ginsberg, M. Post, D.R. Ifa, D. Jaffray, R.J.D. Miller, A. Zarrine-Afsar, Ambient mass spectrometry imaging with picosecond infrared laser ablation electrospray ionization (PIR-LAESI), Anal. Chem. 87 (2015) 12071–12079. Copyright (2015) American Chemical Society.)*

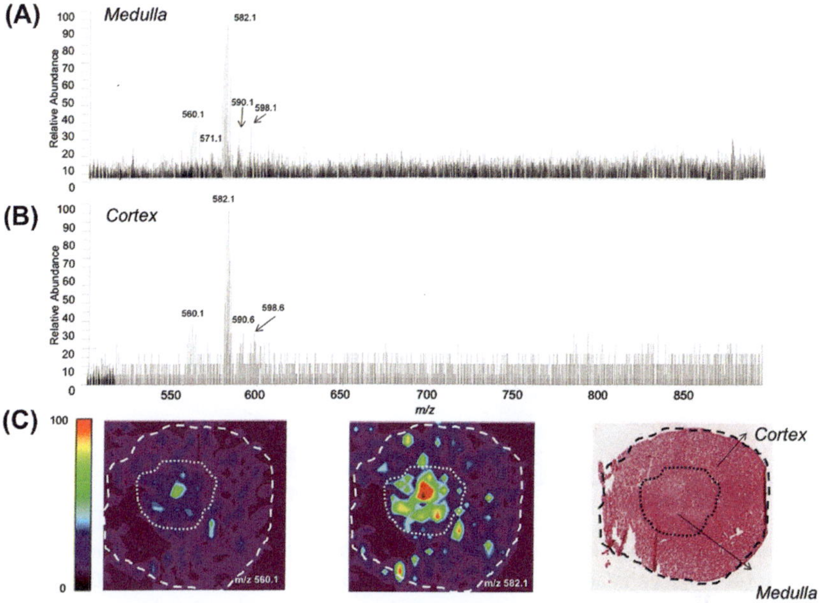

Figure 5.19 ***Picosecond InfraRed Laser Ablation ElectroSpray Ionization Mass Spectrometry (PIR-LAESI MS) molecular imaging of the contrast agent gadoteridol in mouse kidney recorded in positive ion mode.*** (A) PIR-LAESI MS averaged spectrum from medulla region showing the ion of [gadoteridol + H]$^+$ with *m/z* of 560.1, the ion of [2gadoteridol + Na + Na]$^{2+}$ with *m/z* value of 571.1, the ion of [gadoteridol + Na]$^+$ with *m/z* 582.1, the ion of [2gadoteridol + Na + K]$^{2+}$ with *m/z* if 590.1, and the [gadoteridol + K]$^+$ adduct with *m/z* of 598.1. (B) The PIR-LAESI averaged spectrum of the cortex kidney region where a lower total gadoteridol ion count compared to the medulla region was observed. (C) shows molecular images. *(Reprinted (adapted) with permission from J. Zou, F. Talbot, A. Tata, L. Ermini, K. Franjic, M. Ventura, J.Z. Zheng, H. Ginsberg, M. Post, D.R. Ifa, D. Jaffray, R.J.D. Miller, A. Zarrine-Afsar, Ambient mass spectrometry imaging with picosecond infrared laser ablation electrospray ionization (PIR-LAESI), Anal. Chem. 87 (2015) 12071–12079. Copyright (2015) American Chemical Society.)*

handheld PIRL-MS probe developed by Zarrine Afsar group [65]. The PIRL-MS handheld probe provides real-time MS profiles of tissue based on laser desorbed lipids and small molecule content in only 5–10 s of sampling (Fig. 5.22). PIRL-MS analysis of ex vivo MB tumors offered a 98% success rate in subgroup affiliation determination, observed with 194 PIRL-MS datasets from 19 independent tumors and 10 repetitions each. This study utilized six different established MB cell lines to capture some of the intrinsic biological heterogeneity of said cancer type. The rapid classification of MB with PIRL-MS utilized a variety of fatty acid chains,

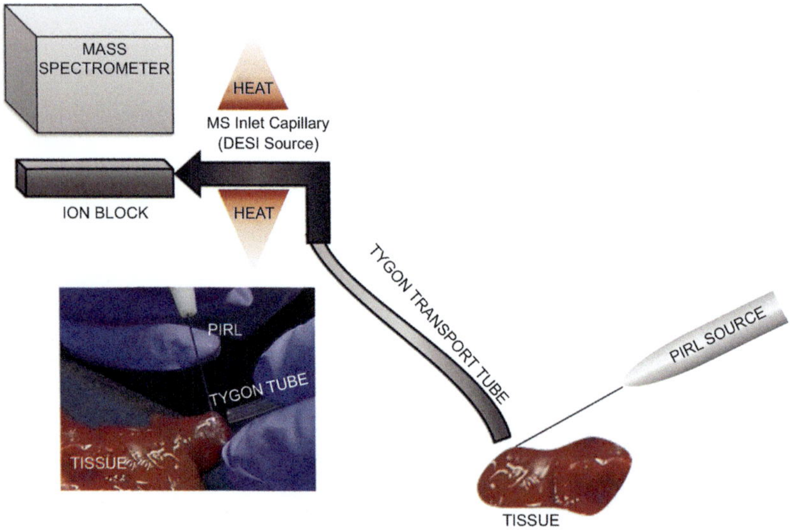

Figure 5.20 ***Picosecond InfraRed Laser Mass Spectrometry setup.*** An extension tube to MS inlet (or collection capillary such as, but not limited to the one used in **D**esorption **E**lectrospray **I**onization **M**ass **S**pectrometry), is extended with a 2 m long tygon tube. Heat is provided to facilitate desolvation and soft ionization of the laser ablated material prior to direct injection into the ion block of a time-of-flight instrument. *(Reprinted (adapted) with permission from J.M. Woolman, A. Gribble, E. Bluemke, J. Zou, M. Ventura, N. Bernards, M. Wu, H.J. Ginsberg, S. Das, A. Vitkin, A. Zarrine-Afsar, Optimized mass spectrometry analysis workflow with polarimetric guidance for ex vivo and in situ sampling of biological tissues, Sci. Rep. 7 (2017) 468. Copyright (2017) The author; Creative Commons Attribution 4.0 International License.)*

glycerophosphates, glycerophosphoglycerols, and glycerophosphocholines that were rapidly extracted from the MB tumor tissue [65].

In conclusion, it has been shown that 5–10 s of sampling with a handheld PIRL-MS probe is sufficient to distinguish xenograft tumors of the Group 3 MB from the SHH subgroup. A picosecond IR laser is capable of cutting biological tissues without considerable thermal damage [68,70]. As such, PIRL may become a stand-alone surgical scalpel with the added attribute of minimal postoperative scar tissue formation that can also provide information regarding tissue types being processed via PIRL-MS.

5.18 SpiderMass

In 2016, a nanosecond IR laser-based MS platform allowing real-time analysis of biological tissues with potential applicability in vivo and

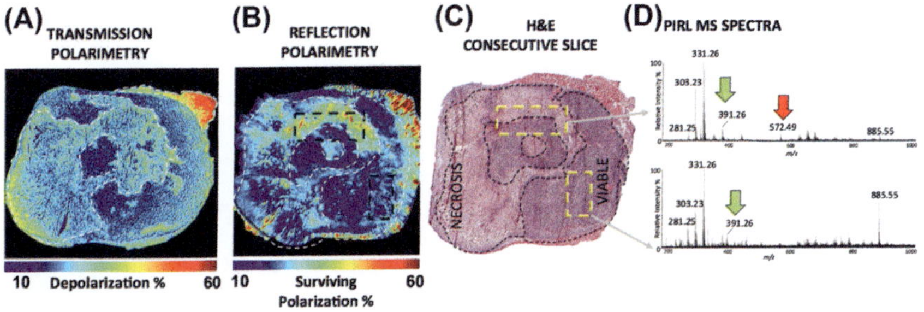

Figure 5.21 ***Guidance of Picosecond InfraRed Laser Mass Spectrometry (PIRL-MS) probe to areas of tissue heterogeneity using polarimetric imaging of a necrotic tissue slice in the reflection mode.*** The handheld laser MS sampling probe was guided to areas of polarimetric heterogeneity typical of necrosis using a 50 μm tissue slice. (A) Transmission polarimetry image from 10 μm tissue slice suggests concordance between reflection mode polarimetry and transmission mode polarimetry to guide PIRL-MS. (B) Polarimetric imaging of the 50 μm tissue slice in the reflection mode. (C) Hematoxylin and eosin (H&E) image of a 10 μm slice consecutive to the 50 μm section used in polarimetric imaging. (D) PIRL-MS analysis of said 50 μm tissue slice based on reflection mode polarimetry. The laser ablation sites are indicated on the H&E image of panel C. After 5–10 s of PIRL ablation MS spectra were generated. The viable cancer maker ion of *m/z* 391.25 was present in areas of low depolarization (for transmission) and low "surviving" polarization (for reflection mode measurements). The laser ablated region with necrotic and viable cancer as indicated from polarimetry feedback resulted in the detection of necrotic (*m/z* 572.48) and viable (*m/z* 391.25) biomarker ions. The necrotic (red) and viable (green) regions are highlighted in overlays. All relevant biomarker ions for breast tissue including *m/z* 281.25 [FA(18:1)-H]− (oleic acid), 303.23 [FA(20:4)-H]− (arachidonic acid), *m/z* 331.26 [FA(22:4)-H]− (adrenic acid) were detectable with PIRL-MS. *(Reprinted (adapted) with permission from J.M. Woolman, A. Gribble, E. Bluemke, J. Zou, M. Ventura, N. Bernards, M. Wu, H.J. Ginsberg, S. Das, A. Vitkin, A. Zarrine-Afsar, Optimized mass spectrometry analysis workflow with polarimetric guidance for ex vivo and in situ sampling of biological tissues, Sci. Rep. 7 (2017) 468. Copyright (2017) The author; Creative Commons Attribution 4.0 International License.)*

minimal damage was reported [76]. Initial applications ranged from apple peel analysis for food safety and human skin analysis for real-time drug metabolism and pharmacokinetics (DMPK) or cosmetic applications. In a similar manner to PIRL-MS setup, the remote desorption and real-time transportation of the gas-phase material fulfills the requirements for applications to human body in surgery [76]. The SpiderMass instrument is comprised of three different parts: (1) a microsampling probe, (2) a transfer line, and (3) an MS analyzer. Fig. 5.23 provides a representation alongside a workflow for instrument use [76].

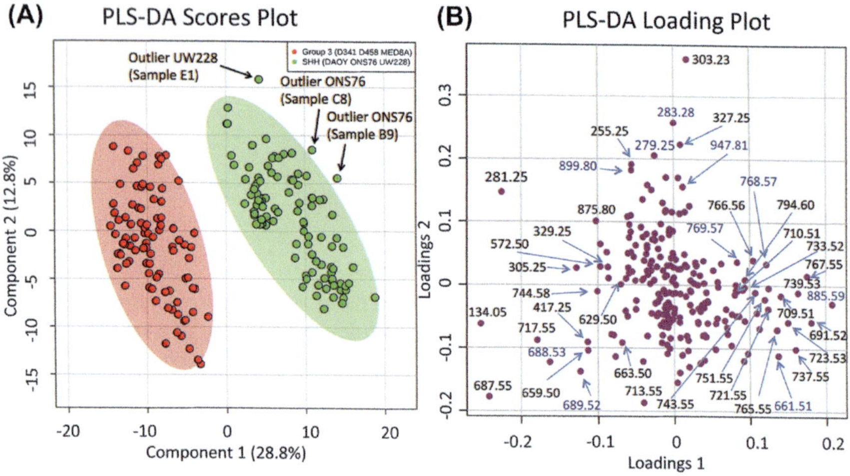

Figure 5.22 ***Statistical discrimination of the Sonic Hedgehog (SHH) and Group 3 MB from 10 s Picosecond InfraRed Laser Mass Spectrometry (PIRL-MS) analysis.*** MS spectra were collected in the negative ion mode in less than 10 s using 19 tumors comprised of mouse xenografts from six established medulloblastoma (MB) cell lines of D341, D458 and MED8A (for Group 3 MB) and DAOY, ONS76, and UW228 (for SHH MB) subjected multivariate analysis using partial least squares discriminant data analysis (PLS-DA) through the MetaboAnalyst (10 repetition laser sampling from each tumor). (A) PLS-DA scores plot using 194 data points, each being derived from a single 5–10 s PIRL-MS spectrum, show a clear statistical separation. The shaded ovals show a 95% confidence interval. No misclassification within 95% interval is seen. Three outliers were detected outside the 95% confidence interval. (B) The PLS-DA loading plot illustrating single *m/z* values in the rank order they contribute to the statistical discrimination between PIRL-MS profiles of SHH and Group 3 MB. The *m/z* values that are located to the outside regions of loading plot center contribute most strongly. *(Reprinted (adapted) with permission from M. Woolman, I. Ferry, C.M. Kuzan-Fischer, M. Wu, J. Zou, T. Kiyota, S. Isik, D. Dara, A. Aman, S. Das, M.D. Taylor, J.T. Rutka, H.J. Ginsberg, A. Zarrine-Afsar, Rapid determination of medulloblastoma subgroup affiliation with mass spectrometry using a handheld picosecond infrared laser desorption probe, Chem. Sci. 8 (2017) 6508–6519. Copyright (2017) The author; Creative Commons Attribution 4.0 International License.)*

The SpiderMass sampling probe was designed such that it can be performed remotely from the MS analyzer itself. The low invasiveness, biocompatibility, and remote (from mass analyzer position) operations of the sampling probe allow for intrasurgical utility. Thus, the microsampling probe is based on LA in the IR range in a manner very similar to PIRL-MS, albeit using a different pulse duration. In comparison to PIRL-MS, the SpiderMass is equally applicable to water-rich biological tissue investigations. The

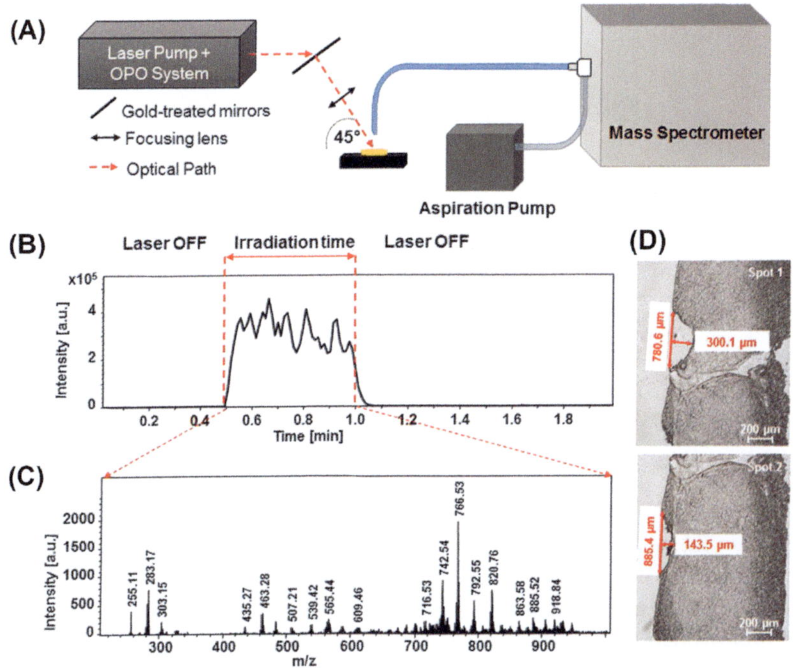

Figure 5.23 ***SpiderMass.*** (A) A schematic representation of the SpiderMass setup. (B) Time evolution of the total ion current over 30 s of data collection, and (C) average **M**ass **S**pectrometry spectrum in the negative mode thereof, integrated over said sampling period. (D) Optical microscope evaluation of tissues sections after SpiderMass sampling using bovine liver indicating the maximum ablation. *(Reprinted (adapted) with permission from B. Fatou, P. Saudemont, E. Leblanc, D. Vinatier, V. Mesdag, M. Wisztorski, C. Focsa, M. Salzet, M. Ziskind, I. Fournier, In vivo real-time mass spectrometry for guided surgery application, Sci. Rep. 6 (2016) 25919. Copyright (2016) The author; Creative Commons Attribution 4.0 International License.)*

SpiderMass instrument uses an OPO system pumped by an Nd:YAG laser in order to fine-tune the laser parameters at 2.96 μm to match the intense absorption bands of water. In addition, SpiderMass has been able to retrieve molecular classification information using lipid species in real time.

Fig. 5.24 presents the MS spectra obtained in real time from two unprocessed surgical tissue pieces from biopsies of ovarian cancer (recorded with 5 mJ/pulse of energy). One of these two pieces was from the primary tumor site (high-grade serous ovarian carcinoma), and the other normal tissue taken from bilateral control ovary (Fig. 5.24). The MS spectra were averaged over 30 s of rastering the laser beam over tissue surface, leading to

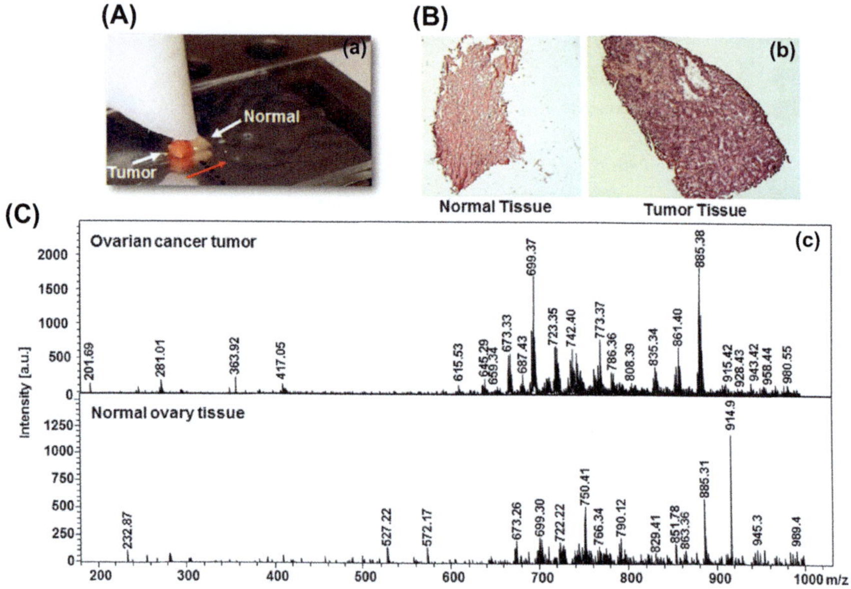

Figure 5.24 ***Signal acquisition from two cancer and control biopsies of a single patient with ovarian cancer.*** The red arrow indicates the position of laser beam raster over cancer. (A) The microscope image of the biopsies prior to sampling. (B) Hematoxylin and eosin staining of a tissue after the sampling with SpiderMass. (C) The average **M**ass **S**pectrometry spectrum averaged from the entire irradiation period for each of the two samples analyzed. *(Reprinted (adapted) with permission from B. Fatou, P. Saudemont, E. Leblanc, D. Vinatier, V. Mesdag, M. Wisztorski, C. Focsa, M. Salzet, M. Ziskind, I. Fournier, In vivo real-time mass spectrometry for guided surgery application, Sci. Rep. 6 (2016) 25919. Copyright (2016) The author; Creative Commons Attribution 4.0 International License.)*

different spectral profiles for these two samples. While abundant lipid species such as *m/z* 885.4 (PI 18:0/20:4) were observed from both normal and tumor tissue, many lipid signals specific to tumor or normal tissue were also observed. This observation indicates that, without causing significant damage to biopsy tissue, molecular profiles specific to tissue type can be obtained to differentiate phenotypes of tissues. The SpiderMass technique is of particular interest in dermatology applications since the laser-induced damage from this technique, similar to PIRL, is low.

SpiderMass for real-time in vivo analysis was evaluated using skin tissue and 10 human volunteers that consented to sampling of their fingertips. In these experiments, a finger of the right hand (Fig. 5.25) was sampled for a

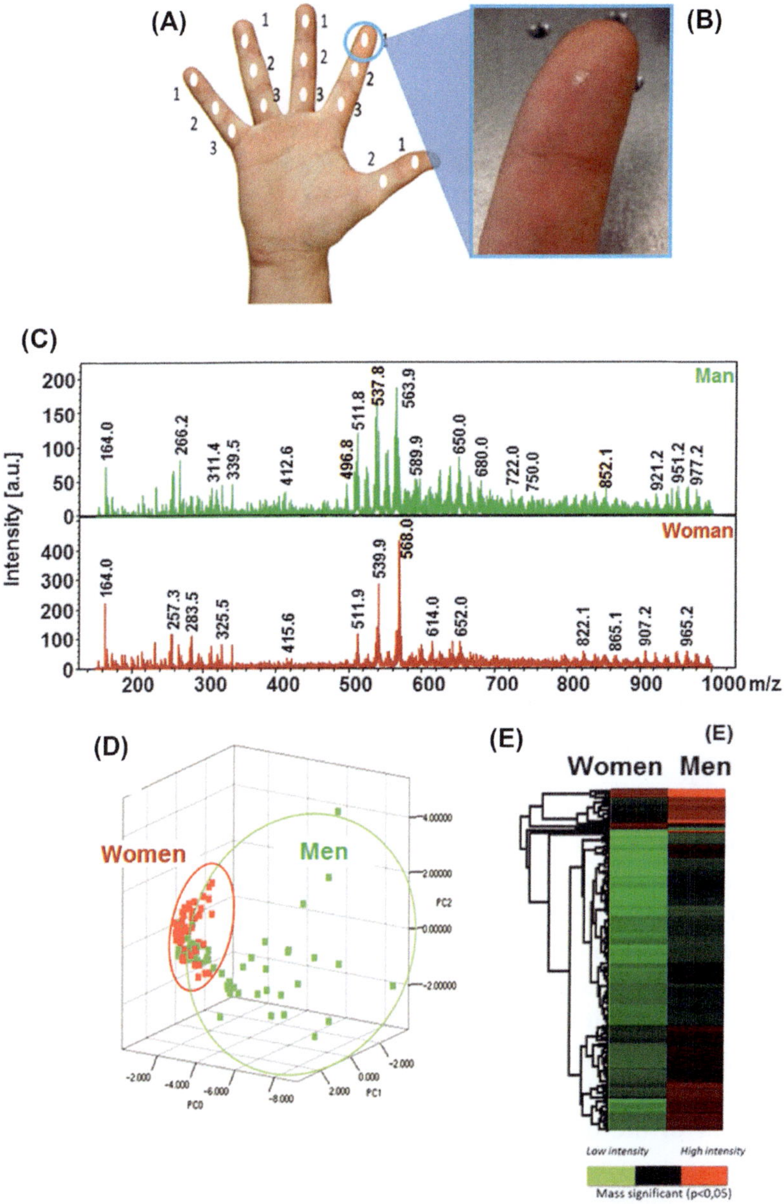

Figure 5.25 ***In vivo analysis of human finger skin using SpiderMass.*** (A) The schematic mapping of the various irradiation spots on the different phalanxes belonging to men or women. (B) Zoomed-in photo of a finger after laser sampling. The white spots arising as a consequence of dehydration mark the irradiation zone. (C) The averaged **M**ass **S**pectrometry spectrum of corresponding phalanx from men versus women. (D) Principal component analysis suggests discrimination according to gender. (E) Hierarchical clustering using significant masses ($P < .05$) according to the gender. *(Reprinted (adapted) with permission from B. Fatou, P. Saudemont, E. Leblanc, D. Vinatier, V. Mesdag, M. Wisztorski, C. Focsa, M. Salzet, M. Ziskind, I. Fournier, In vivo real-time mass spectrometry for guided surgery application, Sci. Rep. 6 (2016) 25919. Copyright (2016) The author; Creative Commons Attribution 4.0 International License.)*

period of 10 s. This figure also shows finger skin after 10 s sampling with SpiderMass laser. A white trace was observable that corresponded to tissue dehydration upon laser sampling. This trace disappeared after a few minutes.

SpiderMass can also be used to detect exogenous compounds in the skin such as the photo-protector agent Melascreen. In this case, SpiderMass spectra indicate the impact of the drug on the skin lipidome, alongside pollutants that can also be detected on skin directly in vivo and in real time. Of particular note, the polyaromatic hydrocarbon at m/z 178.08 Da, anthracene or phenanthrene molecule that contains the aromatic cycles sharing delocalized electrons with a lower threshold for ionization are shown to produce cation species. In negative ion mode, there may be plasma created at the laser sampling site to negatively charge molecules (this has yet to be verified). In the majority of situations, the laser energy may not be high enough to directly ionize tissue molecules, and SpiderMass then relies on a water-assisted laser desorption ionization. Taken together, these results show that SpiderMass is a suitable instrument for in vivo and real-time DMPK and ADME analyses with importance in pharmacological, environmental, and biological tissue studies [77].

In summary, this chapter provided a review of major ambient laser-based mass spectrometry technologies. Table 5.1 summarizes each technique in terms of desorption ionization methods and simplified schematics. As illustrated in this table, a variety of laser desorption and ionization technologies have been combined together to form the technologies reviewed above. As alluded to earlier, subtle changes in laser pulse duration, ionization methods, or desorption geometry have been taken to indicate novelty. This chapter provides an unbiased overview of these technological implementations without taking into consideration the extent of the desorption/ionization method variations that may be required to claim a new technology. In this realm, comparative studies to critically examine the analytical competitive advantages of each "so-called" ion source in terms of ion yield, LOD, dynamic analyte range, complexity and depth of molecular information, spatial resolution among others must be undertaken.

Table 5.1 Summary of the common ambient laser-based mass spectrometry methods.

Method	Abbreviation	Desorption source(s)	Ionization method	References	Schematics
ELDI-MS	**E**lectrospray-assisted **L**aser **D**esorption/ **I**onization **M**ass **S**pectrometry	337 nm (4 ns pulse duration, 20 μJ/pulse energy) nitrogen laser operating at 10 Hz	**E**lectro**S**pray **I**onization (**ESI**) of the laser plume	[1–4]	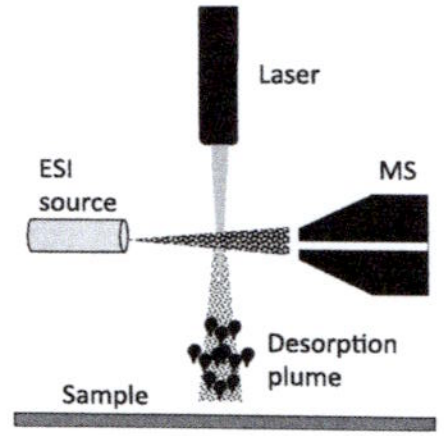
LAESI-MS	**L**aser **A**ssisted **D**esorption **E**lectrospray **I**onization **M**ass **S**pectrometry	2.94 μm (pulse duration <100 ns, 3.5 mJ/ pulse energy) Er:YAG laser operating at 5–10 Hz; OPO generating 2.94 μm (pulse duration 4 ns) from Nd:YAG laser operating at 10 Hz, 1064 nm	**ESI** of the laser plume	[5,6,15–30]	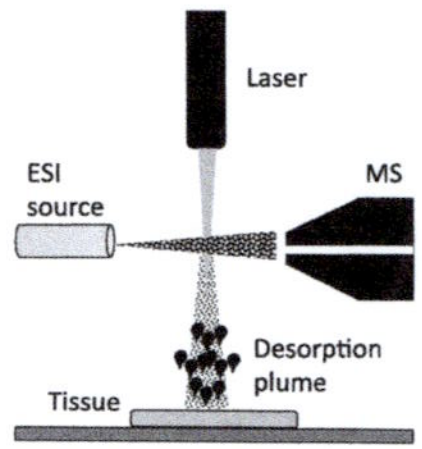
IR-MALDESI-MS	Infra**R**ed-**M**atrix **A**ssisted **L**aser **D**esorption Electro**S**pray **I**onization **M**ass **S**pectrometry	2.94 μm (pulse duration 100 ns, 2.8 mJ/pulse energy) Er:YAG laser operating at 10 Hz	**ESI** of the laser plume	[8–14,17, 31–36]	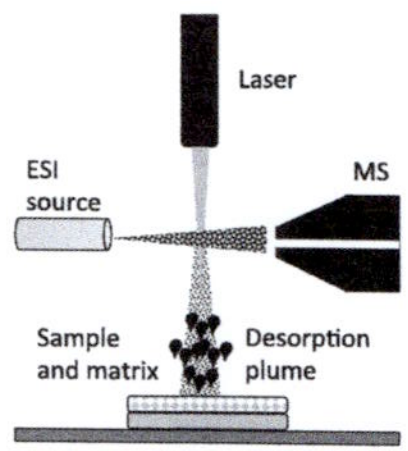

Continued

Table 5.1 Summary of the common ambient laser-based mass spectrometry methods.—cont'd

Method	Abbreviation	Desorption source(s)	Ionization method	References	Schematics
IR-LADESI-MS	**I**nfra**R**ed **L**aser-**A**ssisted **D**esorption **E**lectrospray **I**onization **M**ass **S**pectrometry	2.94 μm (pulse duration 100 ns) Er:YAG laser operating at 10 Hz	**ESI** of the laser plume	[7]	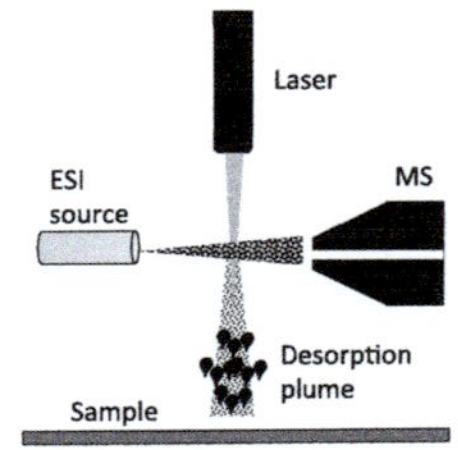
LDSPI-MS	**L**aser **D**esorption **S**pray **I**onization **M**ass **S**pectrometry	1064 nm (pulse duration 10 ns) Nd:YAG laser operating at 20 Hz producing 11 mJ energy	**ESI** of the laser plume	[37,38]	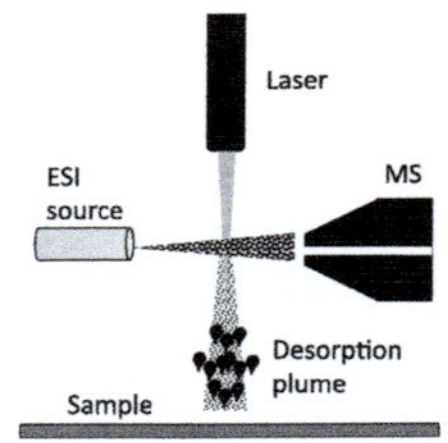
AIRLAB-MS	**A**mbient **I**nfrared **L**aser **Ab**lation **M**ass **S**pectrometry	2.94 μm laser light producing power density of 5.3 J/cm^2 operating at 20 Hz	Liquid transport of plume to **ESI**	[39]	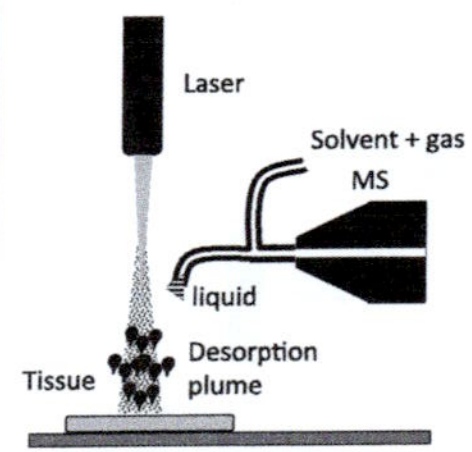

LEMS	**L**aser **E**lectrospray **M**ass **S**pectrometry	800 nm (pulse duration 70 fs, 400 μJ/pulse energy) Ti:Sapphire laser operating at 10 Hz; Fiber based 1042 nm, <50 μJ/pulse energy	**ESI** of the laser plume	[40–44,46]	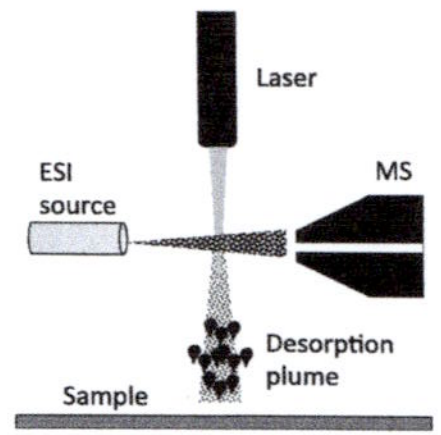
AP-fsLDI-MS	**A**tmospheric **P**ressure **f**emto**s**econd **L**aser **D**esorption **I**onization	800 nm (10^{14} W/cm^2) Ti:Sapphire laser operating at 1 kHz	Concurrent **L**aser **D**esorption **I**onization	[45]	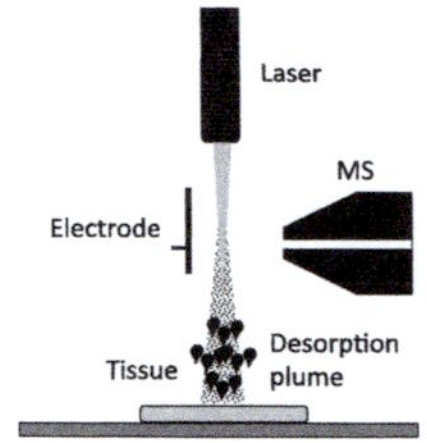
LA-FAPA-MS	**L**aser **A**blation **F**lowing **A**tmospheric **P**ressure **A**fterglow **M**ass **S**pectrometry	266 nm Nd:YAG laser operating at 20 Hz	**A**tmospheric **P**ressure **C**hemical **I**onization (**APCI**)	[47–51]	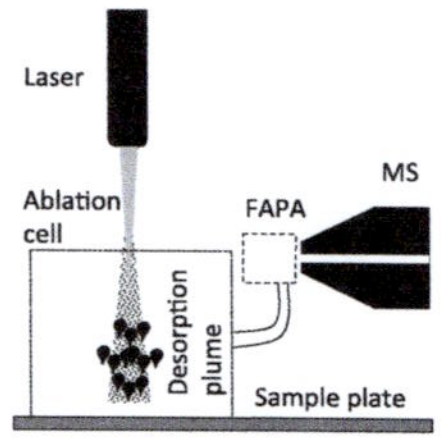

Continued

Table 5.1 Summary of the common ambient laser-based mass spectrometry methods.—cont'd

Method	Abbreviation	Desorption source(s)	Ionization method	References	Schematics
LA-APCI-MS	**L**aser **A**blation **A**tmospheric **C**hemical **I**onization **M**ass **S**pectrometry	213 nm Nd:YAG laser	**A**tmospheric **P**ressure **C**hemical **I**onization (**APCI**)	[52–54]	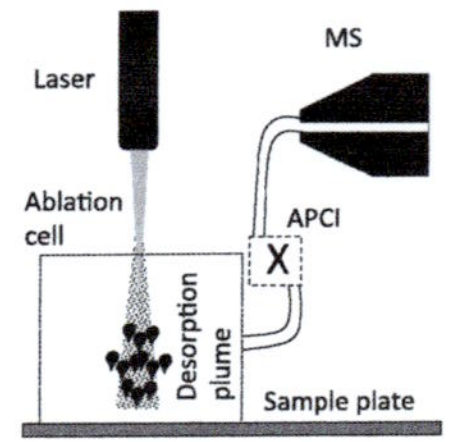
PAMLDI-MS	**P**lasma **A**ssisted **M**ultiwavelength **L**aser **D**esorption **I**onization **M**ass **S**pectrometry	355, 532, 1064 nm Nd:YAG laser (pulse duration 10 ns) operating at 10 Hz	Reactive species in DART plasma gas	[55–57]	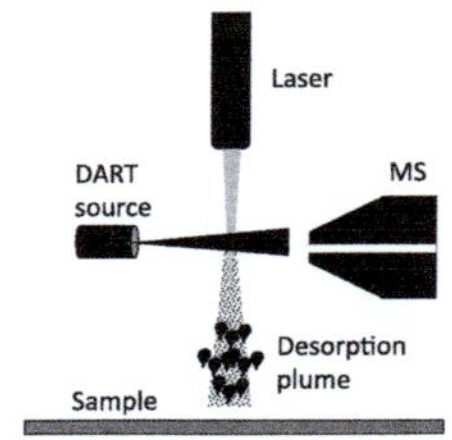
LIAD-ESI-MS	**L**aser **I**nduced **A**coustic **D**esorption **E**lectro**S**pray **I**onization **M**ass **S**pectrometry	1064 nm (pulse duration 9 ns) Nd:YAG laser operating at 1–2 Hz producing 1–30 mJ energy	**ESI** of the acoustically desorbed plume	[58,59]	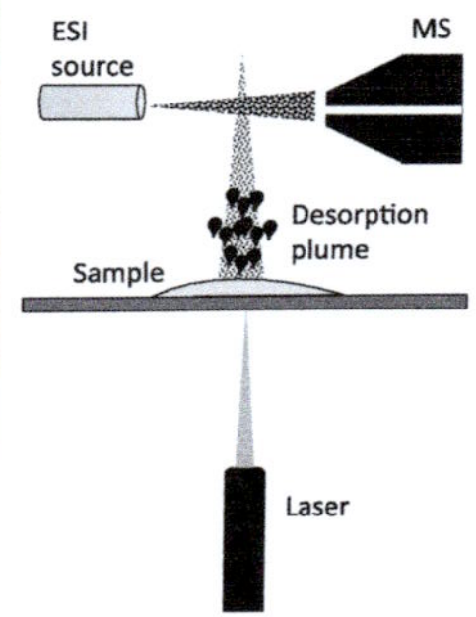

LIAD-APCI-MS	**L**aser **I**nduced **A**coustic **D**esorption **A**tmospheric **P**ressure **C**hemical **I**onization **M**ass **S**pectrometry	532 nm (pulse duration 3 ns, 20 mJ/pulse energy) From Nd:YAG laser operating at 10 Hz	**A**tmospheric **P**ressure **C**hemical **I**onization (**APCI**) of the acoustically desorbed plume	[60,61]	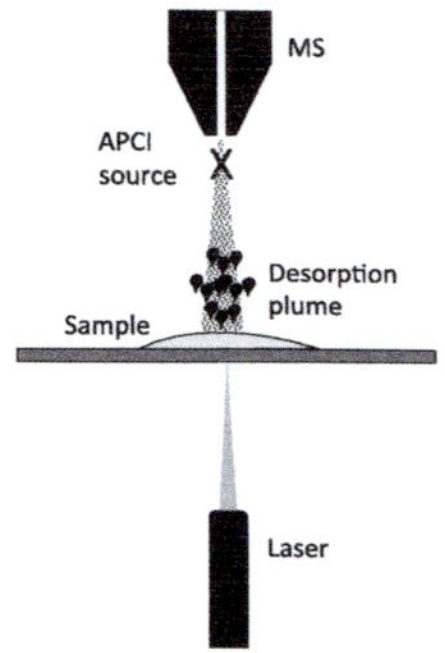
LIAD-APPI-MS	**L**aser **I**nduced **A**coustic **D**esorption **A**tmospheric **P**ressure **P**hoto-**I**onization **M**ass **S**pectrometry	1064 nm (10^8 W/cm^2) operating at 10 Hz	**A**tmospheric **P**ressure **P**hoto **I**onization (**APPI**)	[62]	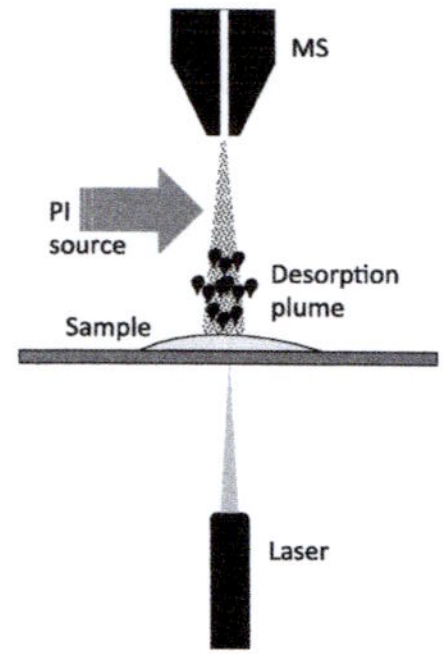

Continued

Table 5.1 Summary of the common ambient laser-based mass spectrometry methods.—cont'd

Method	Abbreviation	Desorption source(s)	Ionization method	References	Schematics
PIR-LAESI-MS	**P**icosecond **I**nfra**R**ed **L**aser **A**blation **E**lectro**S**pray **I**onization **M**ass **S**pectrometry	2.88 μm (pulse duration 80 ps, 150 μJ/pulse energy) operating at 100 Hz	**ESI** of the laser plume	[65]	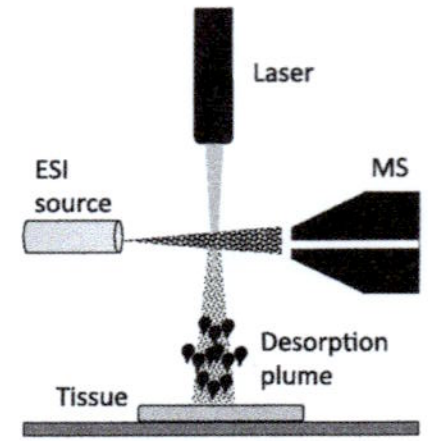
PIRL-MS	**P**icosecond **I**nfra**R**ed **L**aser **M**ass **S**pectrometry	3000 ± 100 nm (pulse duration ~300 ps) operating at 1 kHz, fiber coupled outputting ~0.15 J/cm^2	No external ionization source used Desolvation, inlet ionization (?)	[63,64]	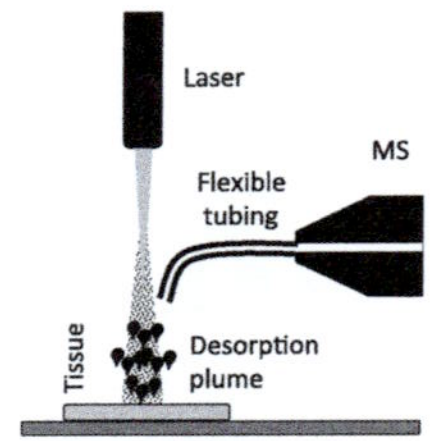
SpiderMass		2.9–3.0 μm (OPO mediated) generating ~10 mJ/pulse From 4 ns Nd:YAG 1064 nm operating at 10 Hz	No external ionization source used	[74,75]	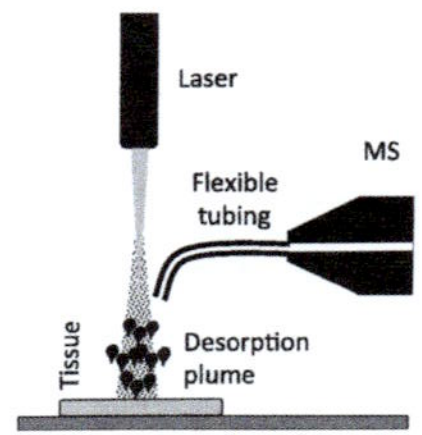

In this table we have also included simplified schematics of each technology. The desorption laser parameters cited are taken from each reported reference and do not conform to the unified prescribed conventions in laser science and may be incomplete. We thus encourage the readers to consult each reference and to make additional inquiries if desired.

References

[1] J. Shiea, M.Z. Huang, H.J. Hsu, C.Y. Lee, C.H. Yuan, I. Beech, J. Sunner, Electrospray-assisted laser desorption/ionization mass spectrometry for direct ambient analysis of solids, Rapid Commun. Mass Spectrom. 19 (2005) 3701–3704.

[2] M.Z. Huang, H.J. Hsu, J.Y. Lee, J. Jeng, J. Shiea, Direct protein detection from biological media through electrospray-assisted laser desorption ionization/mass spectrometry, J. Proteome Res. 5 (2006) 1107–1116.

[3] J. Shiea, C.H. Yuan, M.Z. Huang, S.C. Cheng, Y.L. Ma, W.L. Tseng, H.C. Chang, W.C. Hung, Detection of native protein ions in aqueous solution under ambient conditions by electrospray laser desorption/ionization mass spectrometry, Anal. Chem. 80 (2008) 4845–4852.

[4] S.C. Cheng, Y.S. Lin, M.Z. Huang, J. Shiea, Applications of electrospray laser desorption ionization mass spectrometry for document examination, Rapid Commun. Mass Spectrom. 24 (2010) 203–208.

[5] P. Nemes, A. Vertes, Laser ablation electrospray ionization for atmospheric pressure, in vivo, and imaging mass spectrometry, Anal. Chem. 79 (2007) 8098–8106.

[6] P. Nemes, A.A. Barton, Y. Li, A. Vertes, Ambient molecular imaging and depth profiling of live tissue by infrared laser ablation electrospray ionization mass spectrometry, Anal. Chem. 80 (2008) 4575–4582.

[7] Y.H. Rezenom, J. Dong, K.K. Murray, Infrared laser-assisted desorption electrospray ionization mass spectrometry, Analyst 133 (2008) 226–232.

[8] J.S. Sampson, K.K. Murray, D.C. Muddiman, Intact and top-down characterization of biomolecules and direct analysis using infrared matrix-assisted laser desorption electrospray ionization coupled to FT-ICR mass spectrometry, J. Am. Soc. Mass Spectrom. 20 (2009) 667–673.

[9] F. Huang, K.K. Murray, Continuous flow infrared matrix-assisted laser desorption electrospray ionization mass spectrometry, Rapid Commun. Mass Spectrom. 24 (2010) 2799–2804.

[10] J.A. Barry, D.C. Muddiman, Global optimization of the infrared matrix-assisted laser desorption electrospray ionization (IR MALDESI) source for mass spectrometry using statistical design of experiments, Rapid Commun. Mass Spectrom. 25 (2011) 3527–3536.

[11] G. Robichaud, J.A. Barry, K.P. Garrard, D.C. Muddiman, Infrared matrix-assisted laser desorption electrospray ionization (IR-MALDESI) imaging source coupled to a FT-ICR mass spectrometer, J. Am. Soc. Mass Spectrom. 24 (2013) 92–100.

[12] J.A. Barry, G. Robichaud, M.T. Bokhart, C. Thompson, C. Sykes, A.D. Kashuba, D.C. Muddiman, Mapping antiretroviral drugs in tissue by IR-MALDESI MSI coupled to the Q Exactive and comparison with LC-MS/MS SRM assay, J. Am. Soc. Mass Spectrom. 25 (2014) 2038–2047.

[13] F. Meier, K.P. Garrard, D.C. Muddiman, Silver dopants for targeted and untargeted direct analysis of unsaturated lipids via infrared matrix-assisted laser desorption electrospray ionization (IR-MALDESI), Rapid Commun. Mass Spectrom. 28 (2014) 2461–2470.

[14] G. Robichaud, J.A. Barry, D.C. Muddiman, IR-MALDESI mass spectrometry imaging of biological tissue sections using ice as a matrix, J. Am. Soc. Mass Spectrom. 25 (2014) 319–328.

[15] J.A. Barry, M.R. Groseclose, G. Robichaud, S. Castellino, D.C. Muddiman, Assessing drug and metabolite detection in liver tissue by UV-MALDI and IR-MALDESI mass spectrometry imaging coupled to FT-ICR MS, Int. J. Mass Spectrom. 377 (2015) 448–455.

[16] M.T. Bokhart, E. Rosen, C. Thompson, C. Sykes, A.D. Kashuba, D.C. Muddiman, Quantitative mass spectrometry imaging of emtricitabine in cervical tissue model using infrared matrix-assisted laser desorption electrospray ionization, Anal. Bioanal. Chem. 407 (2015) 2073–2084.

[17] K.H. Cochran, J.A. Barry, G. Robichaud, D.C. Muddiman, Analysis of trace fibers by IR-MALDESI imaging coupled with high resolving power MS, Anal. Bioanal. Chem. 407 (2015) 813–820.

[18] B. Shrestha, A. Vertes, In situ metabolic profiling of single cells by laser ablation electrospray ionization mass spectrometry, Anal. Chem. 81 (2009) 8265–8271.

[19] P. Nemes, A.A. Barton, A. Vertes, Three-dimensional imaging of metabolites in tissues under ambient conditions by laser ablation electrospray ionization mass spectrometry, Anal. Chem. 81 (2009) 6668–6675.

[20] G. Parsiegla, B. Shrestha, F. Carrière, A. Vertes, Direct analysis of phycobilisomal antenna proteins and metabolites in small cyanobacterial populations by laser ablation electrospray ionization mass spectrometry, Anal. Chem. 84 (2012) 34–38.

[21] B. Shrestha, P. Nemes, J. Nazarian, Y. Hathout, E.P. Hoffman, A. Vertes, Direct analysis of lipids and small metabolites in mouse brain tissue by AP IR-MALDI and reactive LAESI mass spectrometry, Analyst 135 (2010) 751–758.

[22] P. Nemes, A.S. Woods, A. Vertes, Simultaneous imaging of small metabolites and lipids in rat brain tissues at atmospheric pressure by laser ablation electrospray ionization mass spectrometry, Anal. Chem. 82 (2010) 982–988.

[23] H. Li, P. Balan, A. Vertes, Molecular imaging of growth, metabolism, and antibiotic inhibition in bacterial colonies by laser ablation electrospray ionization mass spectrometry, Angew. Chem. Int. Ed. Engl. 55 (2016) 15035–15039.

[24] J.A. Stolee, B. Shrestha, G. Mengistu, A. Vertes, Observation of subcellular metabolite gradients in single cells by laser ablation electrospray ionization mass spectrometry, Angew. Chem. Int. Ed. Engl. 51 (2012) 10386–10389.

[25] J.A. Stolee, A. Vertes, Toward single-cell analysis by plume collimation in laser ablation electrospray ionization mass spectrometry, Anal. Chem. 85 (2013) 3592–3598.

[26] B. Shrestha, J.M. Patt, A. Vertes, In situ cell-by-cell imaging and analysis of small cell populations by mass spectrometry, Anal. Chem. 83 (2011) 2947–2955.

[27] A. Vaikkinen, B. Shrestha, J. Nazarian, R. Kostiainen, A. Vertes, T.J. Kauppila, Simultaneous detection of nonpolar and polar compounds by heat-assisted laser ablation electrospray ionization mass spectrometry, Anal. Chem. 85 (2013) 177–184.

[28] B. Shrestha, P. Sripadi, B.R. Reschke, H.D. Henderson, M.J. Powell, S.A. Moody, A. Vertes, Subcellular metabolite and lipid analysis of *Xenopus laevis* eggs by LAESI mass spectrometry, PLoS One 9 (2014) e115173.

[29] H. Li, E. Ha, R.P. Donaldson, A.M. Jeremic, A. Vertes, Rapid assessment of human amylin aggregation and its inhibition by copper(II) ions by laser ablation electrospray ionization mass spectrometry with ion mobility separation, Anal. Chem. 87 (2015) 9829–9837.

[30] B. Shrestha, A. Vertes, High-throughput cell and tissue analysis with enhanced molecular coverage by laser ablation electrospray ionization mass spectrometry using ion mobility separation, Anal. Chem. 86 (2014) 4308–4315.

[31] K.H. Cochran, J.A. Barry, D.C. Muddiman, D. Hinks, Direct analysis of textile fabrics and dyes using infrared matrix-assisted laser desorption electrospray ionization mass spectrometry, Anal. Chem. 85 (2013) 831–836.

[32] M.T. Bokhart, D.C. Muddiman, Infrared matrix-assisted laser desorption electrospray ionization mass spectrometry imaging analysis of biospecimens, Analyst 141 (2016) 5236–5245.

[33] E.P. Rosen, M.T. Bokhart, M. Nazari, D.C. Muddiman, Influence of C-trap ion accumulation time on the detectability of analytes in IR-MALDESI MSI, Anal. Chem. 87 (2015) 10483–10490.
[34] E.P. Rosen, M.T. Bokhart, H.T. Ghashghaei, D.C. Muddiman, Influence of desorption conditions on analyte sensitivity and internal energy in discrete tissue or whole body imaging by IR-MALDESI, J. Am. Soc. Mass Spectrom. 26 (2015) 899–910.
[35] M. Nazari, D.C. Muddiman, Polarity switching mass spectrometry imaging of healthy and cancerous hen ovarian tissue sections by infrared matrix-assisted laser desorption electrospray ionization (IR-MALDESI), Analyst 141 (2016) 595–605.
[36] M. Ekelöf, E.K. McMurtrie, M. Nazari, S.D. Johanningsmeier, D.C. Muddiman, Direct analysis of triterpenes from high-salt fermented cucumbers using infrared matrix-assisted laser desorption electrospray ionization (IR-MALDESI), J. Am. Soc. Mass Spectrom. 28 (2017) 370–375.
[37] J. Liu, B. Qiu, H. Luo, Fingerprinting of yogurt products by laser desorption spray post-ionization mass spectrometry, Rapid Commun. Mass Spectrom. 24 (2010) 1365–1370.
[38] J. Liu, C. Zhang, J. Sun, H. Luo, Unexpected complexation reaction during analysis of proteins using laser desorption spray post-ionization mass spectrometry, Analyst 137 (2012) 1764–1767.
[39] J.T. O'Brien, E.R. Williams, H.Y. Holman, Ambient infrared laser ablation mass spectrometry (AIRLAB-MS) of live plant tissue with plume capture by continuous flow solvent probe, Anal. Chem. 87 (2015) 2631–2638.
[40] E.J. Judge, J.J. Brady, D. Dalton, R.J. Levis, Analysis of pharmaceutical compounds from glass, fabric, steel, and wood surfaces at atmospheric pressure using spatially resolved, nonresonant femtosecond laser vaporization electrospray mass spectrometry, Anal. Chem. 82 (2010) 3231–3238.
[41] E.J. Judge, J.J. Brady, R.J. Levis, Mass analysis of biological macromolecules at atmospheric pressure using nonresonant femtosecond laser vaporization and electrospray ionization, Anal. Chem. 82 (2010) 10203–10207.
[42] F. Shi, P.M. Flanigan, J.J. Archer, R.J. Levis, Direct analysis of intact biological macromolecules by low-energy, fiber-based femtosecond laser vaporization at 1042 nm wavelength with nanospray postionization mass spectrometry, Anal. Chem. 87 (2015) 3187–3194.
[43] J.J. Brady, E.J. Judge, R.J. Levis, Mass spectrometry of intact neutral macromolecules using intense non-resonant femtosecond laser vaporization with electrospray post-ionization, Rapid Commun. Mass Spectrom. 23 (2009) 3151–3157.
[44] F. Shi, P.M. Flanigan, J.J. Archer, R.J. Levis, Ambient molecular analysis of biological tissue using low-energy, femtosecond laser vaporization and nanospray postionization mass spectrometry, J. Am. Soc. Mass Spectrom. 27 (2016) 542–551.
[45] Y. Coello, A.D. Jones, T.C. Gunaratne, M. Dantus, Atmospheric pressure femtosecond laser imaging mass spectrometry, Anal. Chem. 82 (2010) 2753–2758.
[46] J.J. Brady, E.J. Judge, R.J. Levis, Nonresonant femtosecond laser vaporization of aqueous protein preserves folded structure, Proc. Natl. Acad. Sci. U.S.A. 108 (2011) 12217–12222.
[47] M. Bruggemann, E. Karu, T. Hoffmann, Critical assessment of ionization patterns and applications of ambient desorption/ionization mass spectrometry using FAPA-MS, J. Mass Spectrom. 51 (2016) 141–149.
[48] J.T. Shelley, G.C. Chan, G.M. Hieftje, Understanding the flowing atmospheric-pressure afterglow (FAPA) ambient ionization source through optical means, J. Am. Soc. Mass Spectrom. 23 (2012) 407–417.

[49] J.T. Shelley, S.J. Ray, G.M. Hieftje, Laser ablation coupled to a flowing atmospheric pressure afterglow for ambient mass spectral imaging, Anal. Chem. 80 (2008) 8308–8313.
[50] J.N. Schaper, K.P. Pfeuffer, J.T. Shelley, N.H. Bings, G.M. Hieftje, Drop-on-demand sample introduction system coupled with the flowing atmospheric-pressure afterglow for direct molecular analysis of complex liquid microvolume samples, Anal. Chem. 84 (2012) 9246–9252.
[51] K.P. Pfeuffer, S.J. Ray, G.M. Hieftje, Measurement and visualization of mass transport for the flowing atmospheric pressure afterglow (FAPA) ambient mass-spectrometry source, J. Am. Soc. Mass Spectrom. 25 (2014) 800–808.
[52] C. Herdering, O. Reifschneider, C.A. Wehe, M. Sperling, U. Karst, Ambient molecular imaging by laser ablation atmospheric pressure chemical ionization mass spectrometry, Rapid Commun. Mass Spectrom. 27 (2013) 2595–2600.
[53] C. Herdering, C.A. Wehe, O. Reifschneider, I. Raj, G. Ciarimboli, K. Diebold, C. Becker, M. Sperling, U. Karst, Laser ablation based bioimaging with simultaneous elemental and molecular mass spectrometry: towards spatially resolved speciation analysis, Rapid Commun. Mass Spectrom. 27 (2013) 2588–2594.
[54] M. Lorenz, O.S. Ovchinnikova, V. Kertesz, G.J. Van Berkel, Laser microdissection and atmospheric pressure chemical ionization mass spectrometry coupled for multimodal imaging, Rapid Commun. Mass Spectrom. 27 (2013) 1429–1436.
[55] J. Zhang, Z. Li, C. Zhang, B. Feng, Z. Zhou, Y. Bai, H. Liu, Graphite-coated paper as substrate for high sensitivity analysis in ambient surface-assisted laser desorption/ionization mass spectrometry, Anal. Chem. 84 (2012) 3296–3301.
[56] J. Zhang, Z. Zhou, J. Yang, W. Zhang, Y. Bai, H. Liu, Thin layer chromatography/plasma assisted multiwavelength laser desorption ionization mass spectrometry for facile separation and selective identification of low molecular weight compounds, Anal. Chem. 84 (2012) 1496–1503.
[57] C. Chang, X. Li, Y. Bai, G. Xu, B. Feng, Y. Liao, H. Liu, Graphene matrix for signal enhancement in ambient plasma assisted laser desorption ionization mass spectrometry, Talanta 114 (2013) 54–59.
[58] S.C. Cheng, M.Z. Huang, J. Shiea, Thin-layer chromatography/laser-induced acoustic desorption/electrospray ionization mass spectrometry, Anal. Chem. 81 (2009) 9274–9281.
[59] S.C. Cheng, T.L. Cheng, H.C. Chang, J. Shiea, Using laser-induced acoustic desorption/electrospray ionization mass spectrometry to characterize small organic and large biological compounds in the solid state and in solution under ambient conditions, Anal. Chem. 81 (2009) 868–874.
[60] J. Gao, D.J. Borton, B.C. Owen, Z. Jin, M. Hurt, L.M. Amundson, J.T. Madden, K. Qian, H.I. Kenttämaa, Laser-induced acoustic desorption/atmospheric pressure chemical ionization mass spectrometry, J. Am. Soc. Mass Spectrom. 22 (2011) 531–538.
[61] D.J. Borton, L.M. Amundson, M.R. Hurt, A. Dow, J.T. Madden, G.J. Simpson, H.I. Kenttämaa, Development of a high-throughput laser-induced acoustic desorption probe and raster sampling for laser-induced acoustic desorption/atmospheric pressure chemical ionization, Anal. Chem. 85 (2013) 5720–5726.
[62] K. Benham, R. Hodyss, F.M. Fernández, T.M. Orlando, Laser-induced acoustic desorption atmospheric pressure photoionization via VUV-generating microplasmas, J. Am. Soc. Mass Spectrom. 27 (2016) 1805–1812.
[63] J. Zou, F. Talbot, A. Tata, L. Ermini, K. Franjic, M. Ventura, J.Z. Zheng, H. Ginsberg, M. Post, D.R. Ifa, D. Jaffray, R.J.D. Miller, A. Zarrine-Afsar, Ambient mass spectrometry imaging with picosecond infrared laser ablation electrospray ionization (PIR-LAESI), Anal. Chem. 87 (2015) 12071–12079.

[64] M. Woolman, A. Gribble, E. Bluemke, J. Zou, M. Ventura, N. Bernards, M. Wu, H.J. Ginsberg, S. Das, A. Vitkin, A. Zarrine-Afsar, Optimized mass spectrometry analysis workflow with polarimetric guidance for ex vivo and in situ sampling of biological tissues, Sci. Rep. 7 (2017).
[65] M. Woolman, I. Ferry, C.M. Kuzan-Fischer, M.G. Wu, J. Zou, T. Kiyota, S. Isik, D. Dara, A. Aman, S. Das, M.D. Taylor, J.T. Rutka, H.J. Ginsberg, A. Zarrine-Afsar, Rapid determination of medulloblastoma subgroup affiliation with mass spectrometry using a handheld picosecond infrared laser desorption probe, Chem. Sci. 8 (2017) 6508–6519.
[66] M.L. Cowan, B.D. Bruner, N. Huse, J.R. Dwyer, B. Chugh, E.T. Nibbering, T. Elsaesser, R.J. Miller, Ultrafast memory loss and energy redistribution in the hydrogen bond network of liquid H_2O, Nature 434 (2005) 199–202.
[67] P. Nemes, A. Vertes, Atmospheric-pressure molecular imaging of biological tissues and biofilms by LAESI mass spectrometry, J. Vis. Exp. 43 (2010) e2907.
[68] S. Amini-Nik, D. Kraemer, M.L. Cowan, K. Gunaratne, P. Nadesan, B.A. Alman, R.J. Miller, Ultrafast mid-IR laser scalpel: protein signals of the fundamental limits to minimally invasive surgery, PLoS One 5 (9) (2010) e13053.
[69] K. Franjic, M.L. Cowan, D. Kraemer, R.J. Miller, Laser selective cutting of biological tissues by impulsive heat deposition through ultrafast vibrational excitations, Opt. Express 17 (2009) 22937–22959.
[70] N. Jowett, W. Wollmer, A.M. Mlynarek, P. Wiseman, B. Segal, K. Franjic, P. Krotz, A. Bottcher, R. Knecht, R.J. Miller, Heat generation during ablation of porcine skin with erbium:YAG laser vs a novel picosecond infrared laser, JAMA Otolaryngol Head Neck Surg. 139 (2013) 828–833.
[71] H. Petersen, F. Tavakoli, S. Kruber, A. Münscher, A. Gliese, N.O. Hansen, S. Uschold, D. Eggert, W.D. Robertson, T. Gosau, S. Sehner, M. Kwiatkowski, H. Schlüter, U. Schumacher, R. Knecht, R.J. Miller, Comparative study of wound healing in rat skin following incision with a novel picosecond infrared laser (PIRL) and different surgical modalities, Lasers Surg. Med. 48 (2016) 385–391.
[72] K. Franjic, D. Miller, Vibrationally excited ultrafast thermodynamic phase transitions at the water/air interface, Phys. Chem. Chem. Phys. 12 (2010) 5225–5239.
[73] D. Calligaris, D. Caragacianu, X. Liu, I. Norton, C.J. Thompson, A.L. Richardson, M. Golshan, M.L. Easterling, S. Santagata, D.A. Dillon, F.A. Jolesz, N.Y. Agar, Application of desorption electrospray ionization mass spectrometry imaging in breast cancer margin analysis, Proc. Natl. Acad. Sci. U.S.A. 111 (2014) 15184–15189.
[74] S. Guenther, L.J. Muirhead, A.V. Speller, O. Golf, N. Strittmatter, R. Ramakrishnan, R.D. Goldin, E. Jones, K. Veselkov, J. Nicholson, A. Darzi, Z. Takats, Spatially resolved metabolic phenotyping of breast cancer by desorption electrospray ionization mass spectrometry, Cancer Res. 75 (2015) 1828–1837.
[75] A. Tata, A. Gribble, M. Ventura, M. Ganguly, E. Bluemke, H. Ginsberg, D. Jaffray, D. Ifa, A. Vitkin, A. Zarrine-Afsar, Wide field tissue polarimetry allows accelerated tumor margin assessment by ambient mass spectrometry imaging, Chem. Sci. 7 (2016) 2162–2169.
[76] B. Fatou, P. Saudemont, E. Leblanc, D. Vinatier, V. Mesdag, M. Wisztorski, C. Focsa, M. Salzet, M. Ziskind, I. Fournier, In vivo real-time mass spectrometry for guided surgery application, Sci. Rep. 6 (2016) 25919.
[77] B. Fatou, P. Saudemont, M. Duhamel, M. Ziskind, C. Focsa, M. Salzet, I. Fournier, Real time and in vivo pharmaceutical and environmental studies with SpiderMass instrument, J. Biotechnol. 281 (2018) 61–66.

CHAPTER 6

Probe electrospray ionization/ mass spectrometry and its applications to the life sciences

Tasuku Murata[1], Kei Zaitsu[2]
[1]MS Business Unit, Life Science Business Department, Analytical & Measuring Instruments Division, Shimadzu Corporation, Nakagyo-ku, Kyoto, Japan; [2]In Vivo Real-time Omics Laboratory, Institute for Advanced Research, Nagoya University, Chikusa-ku, Nagoya, Japan

Contents

6.1 Principle of probe electrospray ionization and the development of instruments

6.1.1 Principle of probe electrospray ionization

It is well known that J. Fenn and coworkers developed electrospray ionization (ESI) in 1989 [1], for which he received the Nobel Prize in 2002. ESI is an ionization method which ionizes molecules in the liquid

Ambient Ionization Mass Spectrometry in Life Sciences
ISBN 978-0-12-817220-9
https://doi.org/10.1016/B978-0-12-817220-9.00006-0

phase. Although similar phenomena to ESI were confirmed in the early 1700s, Fenn et al. demonstrated that ESI was a powerful technique for producing intact ions from large and complex species in solution. Needless to say, ESI is one of the most innovational techniques not only in mass spectrometry but also in biological chemistry as well as soft laser desorption ionization, which was developed by Fenn's Nobel Prize cowinner K. Tanaka [2]. However, the detailed mechanism of ESI has not yet been elucidated completely, but in general, the mechanism of ESI can be explained as follows. A sample solution enters an ionization source thorough a capillary needle, to which a high voltage is continuously applied. The solution is atomized with a nebulizer gas such as nitrogen, producing charged droplets that also evaporate. After desolvation of the droplets in the self-generated electric fields, quasi-molecular ions are produced, and the ions are analyzed using a mass spectrometer. ESI is ordinarily used for conventional liquid chromatography/mass spectrometry (LC/MS) or LC/MS/MS, but the ionization efficiency of ESI needs to be improved to reach higher sensitivity, especially for the analysis of trace-level peptides. To achieve this, nano-ESI was developed [3,4], where the flow rate of the LC and the inner diameter of the column were downsized to acquire higher ionization efficiency; a low flow rate of the sample solution increases the excess charge per volume of the solution, leading to an improved ionization efficiency. Therefore, nano-ESI achieves a higher ionization efficiency than that of conventional flow-ESI. In other words, ionization efficiency will be dramatically improved when we minimize the volume of the sample solution and apply a high voltage to the solution. There is, however, a limitation for downsizing the inside diameter of the capillary because a narrow capillary can be easily clogged, and thus, new solid-substrate-based ESI techniques were developed.

The Shiea group first reported solid-substrate-based ESI [5,6]; they used a fine cooper wire as the ionization unit instead of the narrow capillary used in conventional ESI or nano-ESI, and they succeeded in producing sufficient ions of the target compounds. They further demonstrated that a surface-modified optical fiber or nanostructured tungsten oxide surfaces were applicable for solid-substrate-based ESI [7,8]. These studies were the first-ever reports of solid-substrate-based ESI.

Hiraoka and coworkers, as well as the abovementioned studies regarding solid-substrate-based ESI, came up with an idea to use an ultrathin metal needle probe, which was ordinarily used as an acupuncture needle, as the ionization unit. This idea was based on the concept of nano-ESI, and the method has become known as "probe electrospray ionization (PESI)" [9].

As shown in Fig. 6.1, PESI ordinarily uses an ultrafine needle, with a submicrometer tip diameter in general, and it works as an ionization unit like the capillary needle used in capillary-based ESI. As explained later in more detail, the needle probe also works as a sampling unit, achieving direct sampling of biological samples such as tissues [10].

In principle, an extremely small volume (c. a few picoliter) of the sample solution is attached onto the surface of the needle, and such a small volume of the solution is easily ionized, leading to a high ionization efficiency [9,11]. Furthermore, the direct voltage impression to the attached solution via the metal probe strongly contributes to the dramatic improvement in the ionization efficiency. It should also be noted that the adsorption of target compounds onto the surface of the probe plays an important role in sampling, as mentioned later [12,13].

Fig. 6.2 shows how electrospray is generated from the tip of the needle probe [11]; a high voltage impression to the probe immediately causes electrospray, followed by the production of charged droplets. In general, however, a solution of an organic solvent and water, such as a 50% ethanol aqueous solution, must be supplied to the needle tip to enhance ionization efficiency.

6.1.2 PESI-single quadrupole mass spectrometer: DPiMS™-2020

Following the invention of PESI by Hiraoka et al., a proposal for the development of a PESI-single quadrupole mass spectrometer (PESI-MS)

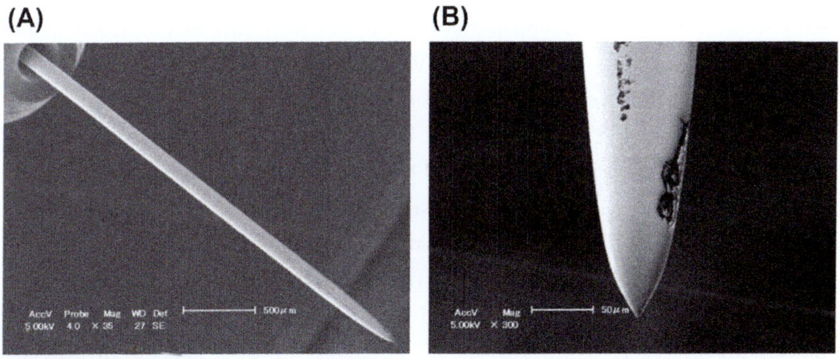

Figure 6.1 Scanned electron microscope images of the probe needle. (A) Low magnification and (B) high magnification.

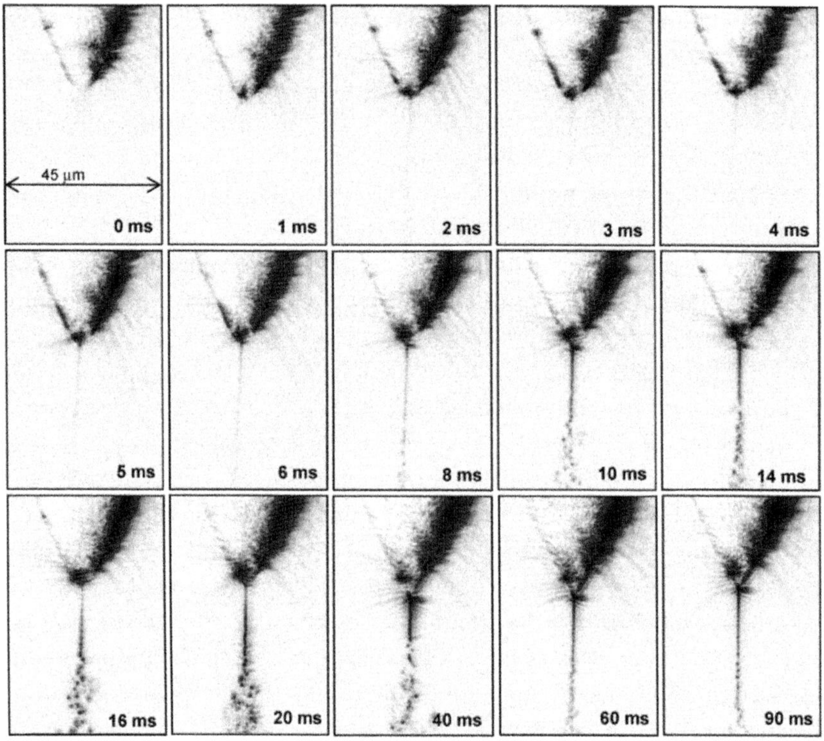

Figure 6.2 Electrospray on the tip of a solid needle probe. *(Reprinted with permission from L.C. Chen, K. Nishidate, Y. Saito, K. Mori, D. Asakawa, S. Takeda, T. Kubota, H. Hori, K. Hiraoka, Characteristics of probe electrospray generated from a solid needle, J. Phys. Chem. B 112(35) (2008) 11164–11170. Copyright 2008 American Chemical Society.)*

was adopted by Japan Science and Technology Agency (JST). The proposal's main aim was to develop a system to support cancer diagnosis using PESI-MS. Yamanashi University, Waseda University, and Shimadzu Corporation jointly participated in the project. The Japanese Red Cross Society Hospital and Yokohama City University contributed to a database to determine cancer using PESI-MS. During the project, the Takeda group reported that PESI-MS was able to support the pathological diagnosis of some cancers [14–16]. Interestingly, this system uses artificial intelligence (AI) software to automatically learn mass spectral profiles obtained from both cancer and noncancer tissue samples; the AI helps in discriminating whether an analyzed tissue sample is cancerous or not. Recent improvements to the system have made it more sophisticated.

Following the results mentioned above, Shimadzu Corporation successfully released a PESI-MS named DPiMS^{TM}-2020 in July 2017, shown in Fig. 6.3. This instrument is based on an existing LC-single quadrupole mass spectrometer, LCMS^{TM}-2020, and the difference between the two instruments is only the ionization source.

As shown in Fig. 6.3, DPiMS^{TM}-2020 has a plate holder, which is for fixing a sample plate, in front of the inlet of the mass spectrometer. Sample plates are available for either liquid or solid samples; there are two types of sample plate for liquid and tissue samples. As will be explained later, PESI needs an organic solvent in an aqueous solution, such as 50% ethanol, supplied to the probe tip in order to enhance ionization efficiency. Thus, the sample solution is ordinarily diluted with an appropriate organic solution prior to analysis. The sample plate for the liquid sample has a small hole, which acts as the reservoir for the liquid sample.

On the other hand, the sample plate for a tissue sample is foldable; a small portion of the tissue sample is set onto the plate, which is then closed. As described above, an appropriate organic solution is added onto the tissue. The capacity of the liquid sample plate is up to 10–15 μL, and c. 2–3 mm^2 of the tissue sample is required for the solid sample plate, although most of the sample is not consumed because PESI requires only an extremely small amount of sample; analytical results can be obtained from small samples, which meets the demands of clinical and forensic toxicology.

The probe holder is placed in the ion source, and a needle probe is set into the holder. When PESI-MS is working, the needle probe moves vertically up and down. At the bottom position, the probe acts as the sampling unit without voltage impression, whereas at the upper position, it

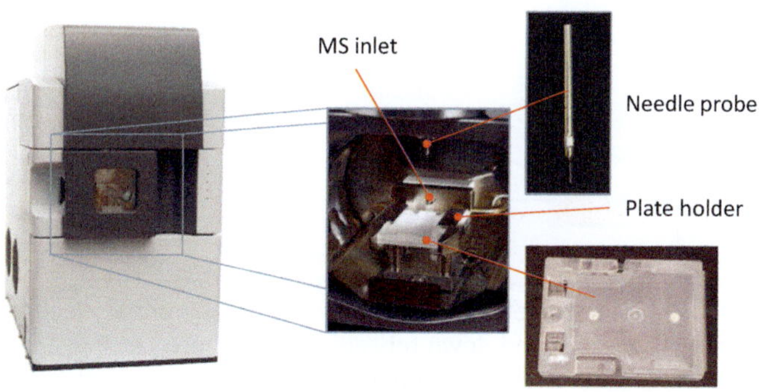

Figure 6.3 Photographs of DPiMS^{TM}-2020 and its attachments.

acts as the ionization unit with a high voltage impression (Fig. 6.4). The probe movement is generally so rapid that analysis is finished in a short time; PESI/MS makes analysis on a very short timeframe possible.

The combinational use of PESI with specific mass spectrometric techniques such as MS/MS or high-resolution mass spectrometry is preferable, although some applications are achievable with PESI-MS because PESI allows for the direct analysis of samples and sample consumption is almost negligible. Here, we will briefly introduce two applications: (1) continuous monitoring of a chemical reaction and (2) the rapid analysis of cyanide ions in blood serum for forensic purposes.

The continuous monitoring of a chemical reaction and the calculation of the reaction rate are necessary to precisely understand the chemical reaction. In this case, PESI/MS was used to continuously monitor the deprotection reaction of a peptide protecting group in a sample plate [17]. Fmoc-Tyr (tBu)-OH (*N*-(9-fluorenylmethoxycarbonyl)-*O*-tert-butyl-L-tyrosine) was dissolved in DMF, the mixture was placed into the small hole of the sample plate, and a piperidine and DMF mixture solution (1:4, v/v) was then added into the plate hole to start the deprotection reaction. It is well known that the deprotection reaction of the Fmoc group is rapid under alkaline conditions at room temperature. As shown in Fig. 6.5, we can monitor the deprotection reaction for 5 minutes using PESI/MS. The intensity of the ion at *m/z* 458, which corresponds to $[M-H]^-$ of Fmoc-Tyr (tBu)-OH, clearly decreased as the reaction proceeded, and the intensity of the ion at *m/z* 474, which is derived from the solvent, did not

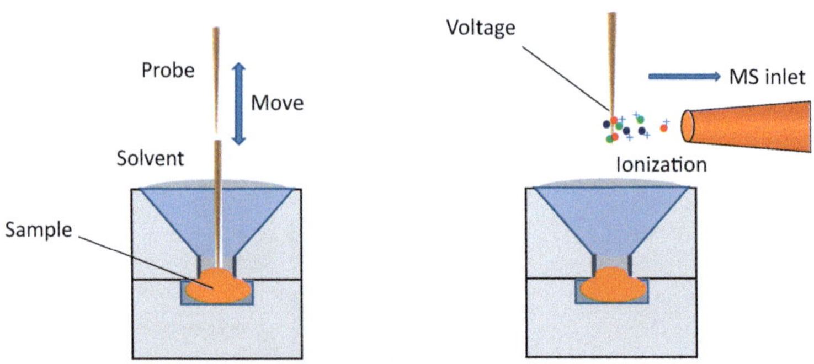

Figure 6.4 Schematic images of sampling and ionization by probe electrospray ionization. The probe moves up and down repeatedly and is inserted into the sample at the bottom position (left). Following sampling, the probe moves up to the highest position, and a high voltage is applied to the probe to ionize any compounds attached to the surface of the probe (right).

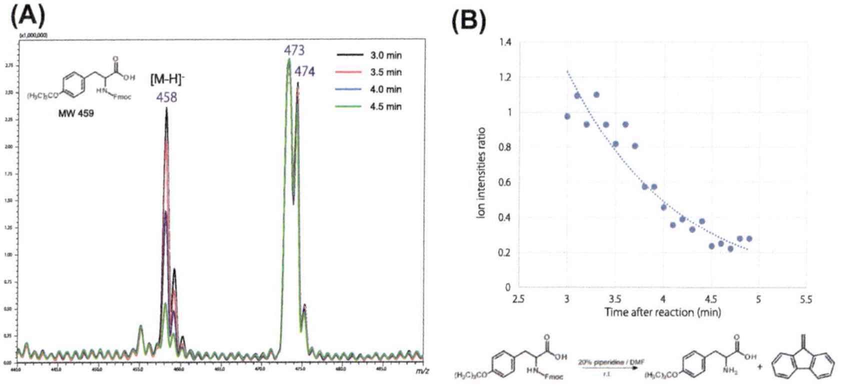

Figure 6.5 Continuous monitoring of the Fmoc deprotection reaction by probe electrospray ionization/mass spectrometry. (A) Changes in peak intensities with reaction time, (B) time-course changes of the ion intensity ratio, and (C) scheme of the Fmoc deprotection reaction.

change during the reaction. As explained earlier, when the reaction occurs in the sample plate at room temperature, the reaction solution may evaporate because the sample plate is exposed to the desolvation line, which is ordinarily kept at a high temperature. However, the intensity of the ion at *m/z* 474 did not change during the reaction, proving that there was no remarkable effect of solvent evaporation on the concentration of Fmoc-Tyr (tBu)-OH; PESI/MS was therefore able to monitor the deprotection reaction continuously. In addition, we estimated the reaction rate of the deprotection of Fmoc-Tyr(tBu)-OH using the intensity ratio of *m/z* 458 to *m/z* 474, where *m/z* 474 was used as a quasi-internal standard. As shown in Fig. 6.5, we successfully observed the deprotection reaction with PESI/MS.

There have been some reports on the forensic applications of PESI/tandem mass spectrometry (PESI/MS/MS) [18,19], which will be discussed later, but we have also applied PESI/MS to forensic toxicological analysis. Here, we will introduce a rapid screening method for the detection of cyanide ions in blood serum without the need for tedious sample pretreatment. Needless to say, the cyanide ion has a high toxic effect and is of great importance in forensic toxicology [20,21]. In general, however, it is impossible to identify the intact cyanide ions in biological specimens using mass spectrometry. We therefore used the in situ derivatization of cyanide ions in serum and investigated the direct detection of the cyanide derivative with PESI/MS [22,23], where we utilized in-source collision-induced dissociation (CID) to improve the qualitative response. Briefly, potassium cyanide (10 μg/mL) was added to blank human serum that had been obtained from healthy volunteers. According to previous reports [22,23], the cyanide ions are directly derivatized with taurine and naphthalene dialdehyde (NDA) in serum at room temperature. The scheme of the derivatization is shown in Scheme 6.1.

We applied 80 V to both the desolvation line and the Q-array bias in order to cause in-source CID. Owing to in-source CID, quasi-product ions at m/z 191 and 81, as well as protonated ions at m/z 299, were observed in the mass spectrum, demonstrating a high potential to identify cyanide ions in serum with PESI/MS. The product ion spectrum obtained with PESI/MS was in good agreement with that obtained with LC-MS/MS. As shown in Fig. 6.6, PESI/MS showed a unique mass chronogram, where each peak appeared at each ionization point. Therefore, quantitative analysis was executed using a calibration curve constructed by calculating the total area under the peaks observed in the chronogram.

6.1.3 Development of PESI-MS/MS (series DPiMS-8060)

PESI-MS (DPiMS™-2020) shows a high potential for the direct analysis of target compounds in biological specimens, despite a strong demand to improve mass spectrometric selectivity by combining PESI with a tandem mass spectrometer. The commercial demand for such an instrument was met by Shimadzu Corporation, who has developed a PESI-MS/MS, the DPiMS™-8060 series, as shown in Fig. 6.7.

The DPiMS™-8060 series is a category name for PESI-MS/MS, where it is possible to combine a PESI unit with either a middle-spec mass spectrometer, LCMS™-8045, or high-spec mass spectrometers, LCMS™-8050/8060, all of which are tandem mass spectrometers supplied by Shimadzu Corporation. Unlike the PESI-MS (DPiMS™-2020), ESI and PESI units are easily interchangeable with each other; both LC/MS/MS and PESI/MS/MS are achievable with one machine. Needless to say, the improvement in the signal-to-noise ratio (S/N) by tandem mass

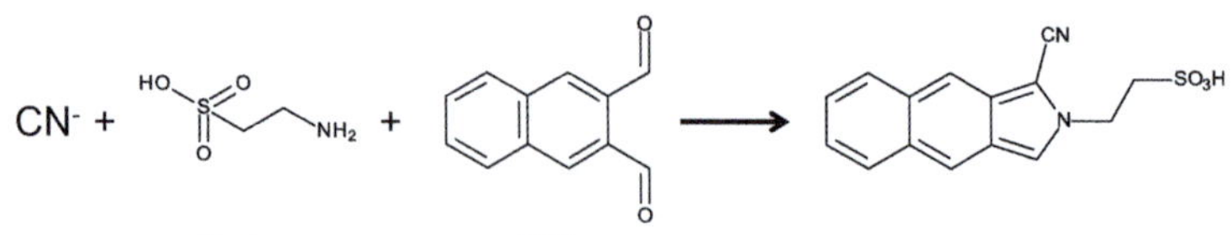

Scheme 6.1 Scheme for the derivatization of cyanide.

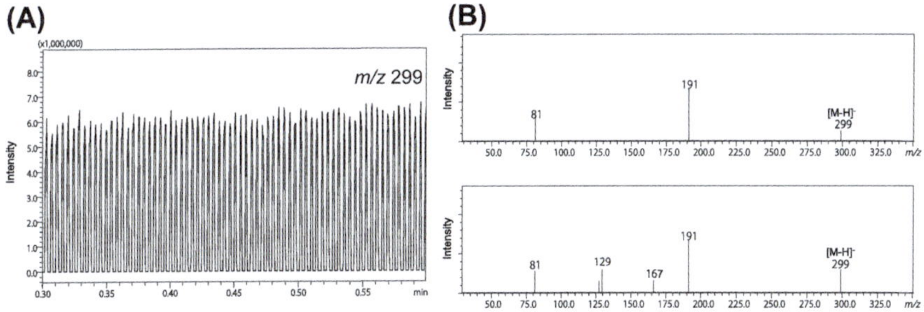

Figure 6.6 Rapid analysis of cyanide ion derivatives by probe electrospray ionization/mass spectrometry (PESI/MS). (A) Extracted ion chronogram for the ions at *m/z* 299 and (B) product ion spectra for cyanide ion derivatives obtained from PESI/MS (upper) and LC/MS/MS (lower).

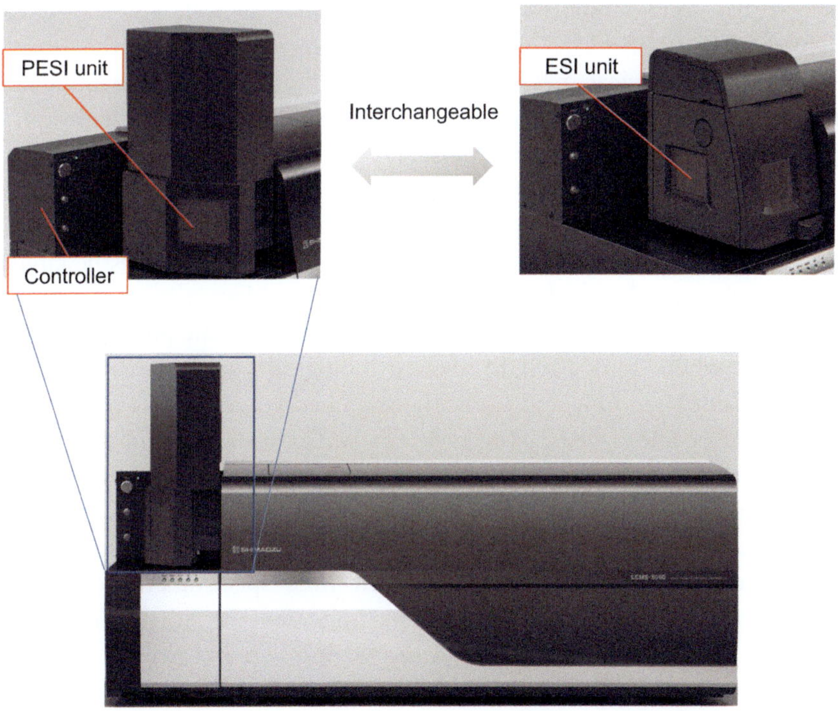

Figure 6.7 Photographs of DPiMS^{TM}-8060 and its ionization units. probe electrospray ionization/mass spectrometry unit (upper left) and ESI unit (upper right) are interchangeable with each other.

spectrometry allowed for more sensitive analysis than that by PESI-MS [24]. The development of PESI/MS/MS succeeded in opening the door to various new applications, which will be introduced in the next section.

6.2 Applications of PESI to life sciences

6.2.1 Introduction

In the second section of this chapter, we will mainly review some interesting applications that use PESI with mass spectrometric techniques in the life sciences. To date, there have been a lot of studies on using PESI in single-cell analysis [25–29], the direct analysis of biological samples [10,24,30], real-time monitoring of metabolites in a living mouse [30,31], imaging mass spectrometry of a mouse brain slice [32], and direct drug-screening analysis in forensic toxicology [18,19]. In addition, applications using other solid-substrate-based ESI techniques as well as PESI have increased [12,13,33–36]. As described above, PESI enables the direct sampling of biological specimens without tedious sample pretreatments, such as extraction and evaporation. This is mainly due to the high ionization efficiency of PESI like nano-ESI, even for samples with a high salt concentration, as explained in the first section of this chapter.

As mentioned above, the principle behind the sampling method in PESI has not been completely understood, although the adsorption of the target compound onto the probe surface undoubtedly plays a pivotal role in sampling. The surface conditions or the modification of the probe surface is important when using PESI to analyze biological samples, even though they have not been fully and explicitly examined yet, especially in the case of metallic needle probes. However, the surface modification of probes changes the adsorption profiles in wooden-tip ESI [7,12,13], and the same is true for PESI.

Since PESI does not require chromatographic separation, specific mass spectrometric techniques such as tandem mass spectrometry or high-resolution mass spectrometry are mandatory for the identification of compounds in biological samples. In this section, we will mainly focus on the applications of PESI/MS/MS and PESI/high-resolution mass spectrometry in the life sciences.

In particular, an extremely thin needle probe is able to achieve low-invasive sampling, allowing real-time monitoring of biogenic compounds in a living animal. In addition, PESI has been expanded not only to analyze the local distribution of endogenous compounds in tissues but also to include single-cell analysis. Needless to say, the real-time monitoring of biogenic compounds in a living animal and mass spectrometric single-cell analysis are attracting scientific attention [37—43]. It is expected that the combinational use of PESI and specific mass spectrometry will become mainstream in the next generation of analytical techniques. In this section, the key features of some representative applications of PESI and the future perspectives of PESI will be demonstrated.

6.2.2 Application of PESI/high-resolution mass spectrometry to single-cell analysis

Gong in the Zhang group reported the application of PESI/high-resolution mass spectrometry to botanical single-cell analysis [25]. They reported that the distinctive features of single-cell analysis by PESI are as follows: (1) direct sampling in a living cell, (2) enrichment of target compounds, which is mainly due to the adsorption of the target compounds onto the surface of the needle probe, and (3) a relatively simple device without the need for LC. They used a tungsten probe with a 1 μm tip diameter and an Orbitrap mass spectrometer (Q-Exactive, Thermo Scientific) as the detector. They applied PESI/high-resolution mass spectrometry to an *Allium cepa* cell, demonstrating that there are significant differences in the metabolites of the inner and outer epidermal cells. Moreover, they successfully proved that their method is capable of subcellular detection using methylene blue to stain the nuclei of the cells, which was not detected in the cytoplasm but only in the nuclei. Finally, they were able to discriminate between the metabolic diversity of two different kinds of *A. cepa* cells (Fig. 6.8). In addition to these new findings, they demonstrated that the ionization efficiency of PESI is higher than that of the conventional nano-ESI method. They compared the intensities of the protonated ions and the sodium adduct ions of angiotensin II and concluded that PESI/high-resolution mass spectrometry showed a 30-fold higher sensitivity than that of conventional nano-ESI. Interestingly, the signal intensity of angiotensin II increased with enrichment time, i.e., the penetration time of the probe into the cell. Consequently, the signal intensity reached a plateau. This result suggested that the adsorption of target compounds onto the probe surface plays an important role in sampling by PESI, and the surface modification of the probe has a large effect on the sampling property.

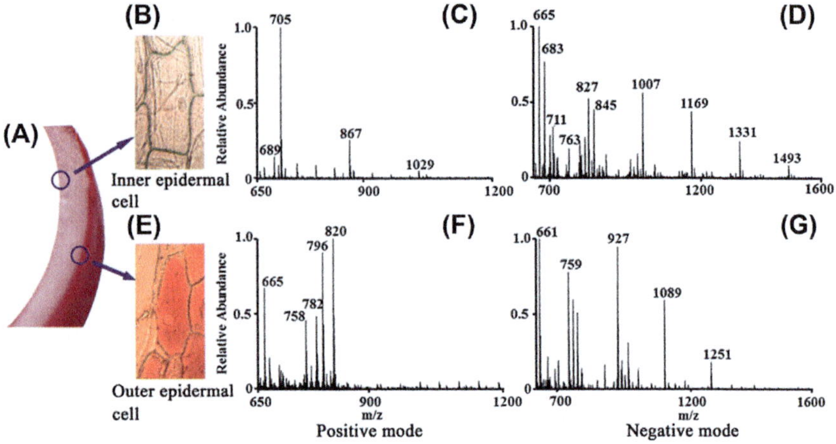

Figure 6.8 Metabolite diversity of two different kinds of *A. cepa* cells. (A) Actual picture of an *A. cepa* bulb layer. The microscopic image and corresponding mass spectra of (B)–(D) an inner epidermal cell and (E)–(G) an outer layer cell. *(Reprinted with permission from X. Gong, Y. Zhao, S. Cai, S. Fu, C. Yang, S. Zhang, X. Zhang, Single cell analysis with probe ESI-mass spectrometry: detection of metabolites at cellular and subcellular levels, Anal. Chem. 86(8) (2014) 3809–3816. Copyright 2014 American Chemical Society.)*

Yu in the Nonami group reported that piezoelectric-inkjet-assisted PESI can be applied to the analysis of a single plant cell [27]. In their work, samples were mounted on a glass slide, and a needle probe was controlled by a piezomicro-manipulator. They were able to take samples from single cells using a long-working distance microscope, as reported by Gong et al. [25], and precisely inserted the needle probe tip into the inside of the cell. Following sampling, the probe was pulled out from the cell and set in front of the orifice of TOFMS (AccuTOF, JEOL). The organic solvent (acetonitrile: water, 50:50, v/v) was supplied to the probe tip by a piezoelectric inkjet system using an inkjet head. In their study, an etched stainless-steel needle (tip diameter: c. 150 nm) was used as the sampling probe; its surface was chemically modified to be hydrophilic and afford greater wettability. The voltage applied to the probe needle was 1.7 kV, and the flow rate of the solvent using the inkjet system was a less than 1 nL/s, which contribute to the enhanced sensitivity. They applied their system to analyze single cells from pink and purple rhododendron petals and successfully detected anthocyanins, saccharides, and flavonoids as the major components. Although they did not confirm each product ion spectrum, they observed that the dominant anthocyanins in the rhododendron petal cells were cyanidin and its glycosides, and delphinidin was observed in the pink and purple cells.

The Nonami group also developed internal electrode capillary pressure PESI (IEC-PPESI) [29]. In their previous work [28], the wire electrode was externally coiled around the capillary probe. In the IEC-PPESI system, the modified quartz microcapillary (two to five μm opening diameter) was used as the sampling unit, and the capillary was filled with 0.01% (v/v) ionic liquid/silicone oil mixture, as shown in Fig. 6.9. An internal wire electrode, which was pre-embedded into a capillary holder, was inserted into the capillary, resulting in an airtight fit. The capillary was operated by a manipulator and was precisely inserted into a glandular cell of a potted plant under microscope observation, and a small amount of the cell sap was

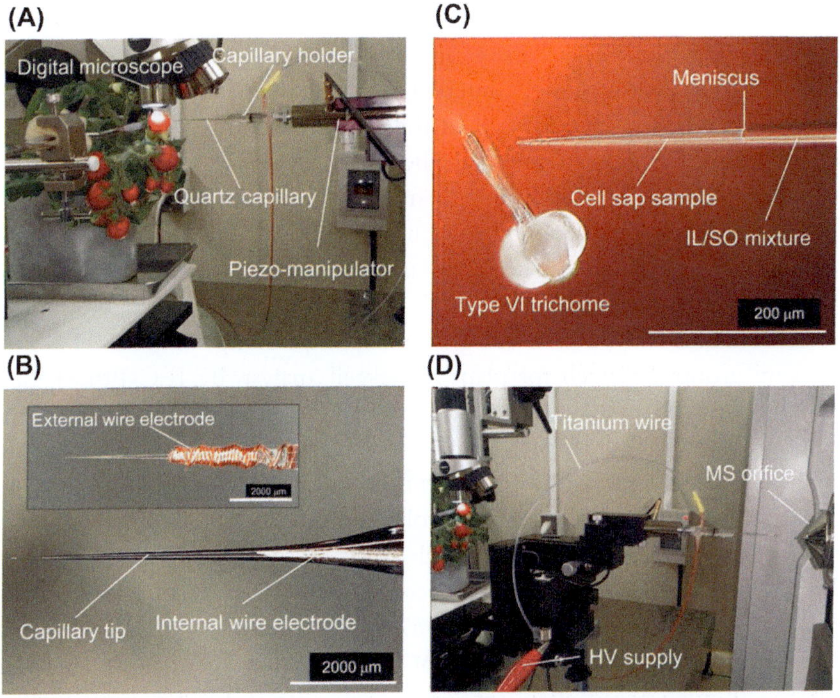

Figure 6.9 Photographs of a pressure probe system. (A) Mechanical setup of the pressure probe system, (B) quartz capillary tip with an internal titanium wire electrode used in the current study, (C) capillary tip containing sampled sap from type VI stalk cell, and (D) subsequent ionization/detection by Orbitrap MS. The inset photograph in (B) shows the external copper wire electrode, which was used in a previous study. *(Reprinted with permission from T. Nakashima, H. Wada, S. Morita, R. Erra-Balsells, K. Hiraoka, H. Nonami, Single-cell metabolite profiling of stalk and glandular cells of intact trichomes with internal electrode capillary pressure probe electrospray ionization mass spectrometry, Anal. Chem. 88(6) (2016) 3049–3057. Copyright 2016 American Chemical Society.)*

sucked into the capillary. After sampling the cell sap from the living plant cell, the capillary tip was oriented toward the orifice of an Orbitrap mass spectrometer (Exactive Plus, Thermo Fisher Scientific Inc.), and then, the capillary was electrified via the internal electrode. The applied voltage was −4 kV, and a scan analysis in negative ionization mode with 35,000 resolution was performed.

As compared to their previous setup, the new system showed a higher sensitivity for standard metabolites and 30- and 50-fold increases in sensitivity for positive and negative ionization modes, respectively. They applied the system to individual stalks and the adjacent glandular cells of an intact type IV trichome. The obtained sample volume of the cell sap ranged from subpicoliter to a few picoliters. Amino acids, organic acids, carbohydrates, phenolic and flavonoid precursors, and flavonoids were detected in the stalk cell, although it was strikingly different from the mass spectrometric profile obtained from the glandular cell of the same trichome. This result highlighted the significant differences in the abundant metabolites between the neighboring stalk and glandular cells from the same trichome.

PESI has also been applied to animal single-cell analysis. Chen et al. reported that a combination of PESI and Q-TOFMS (microTOF-Q mass spectrometer, Bruker Daltonics) with a drop-on-demand inkjet printing device could be applied to single-cell lipid profiling [26]. Unlike the Nonami group approach for plant single-cell analysis by IEC-PPESI, the inkjet printing device was used not to supply the organic solvent but to push a "single-cell-droplet" onto a tungsten probe tip, as shown in Fig. 6.10. In their strategy, a 2 μm tip diameter tungsten probe was used as an ionization unit, and the applied voltage was 2.5 kV. They adjusted the cell concentration ($\sim 2.1 \times 10^6$ cells/mL) to produce a "single-cell-droplet," and 43.8% of the droplets contained a single-cell under optimal conditions. The target cells were U87 (human glioblastoma cell), U251 (human brain glioblastoma cell), HepG2 (human liver hepatocellular cell), MCF-7 (human breast adenocarcinoma cell), 293 (HEK293, human embryonic kidney cell), Caco-2 (human colon adenocarcinoma cell), HUVEC (human umbilical vein endothelial cell), and 3T3 (murine embryo fibroblast cell). Before analysis, the cells were fixed with paraformaldehyde. This analytical system was able to detect 11 kinds of phospholipids on the cell surface, and four different kinds of cells (MCF-7, Caco-2, HUVEC, and 3T3) were clearly separated in principal component analysis (PCA) score plots by their cell classification experiment. To verify the performance of their system, they analyzed an MCF-7 single-cell marked with rhodamine

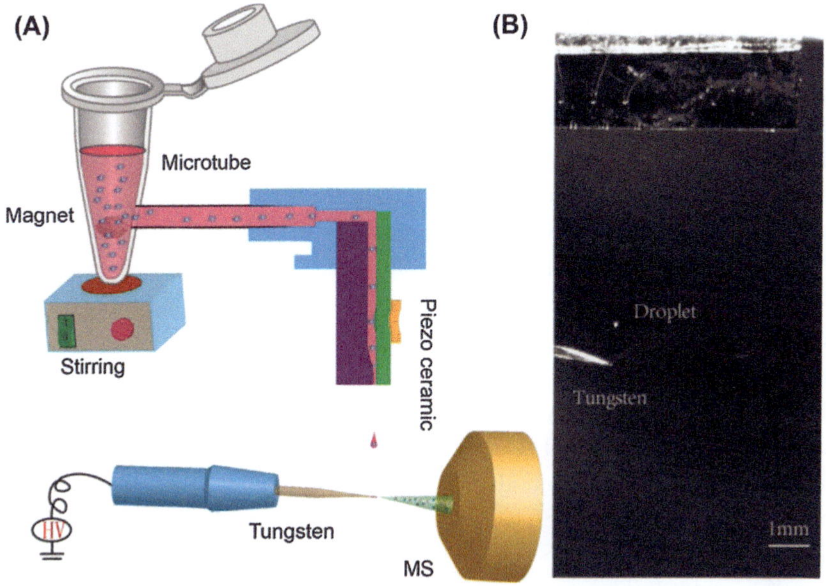

Figure 6.10 Schematic diagrams of the setup for drop-on-demand inkjet cell printing and probe electrospray ionization mass spectrometry (A) Inkjet printing of cells onto the tungsten tip for MS analysis and (B) image of the inkjet printing of cell-containing droplets. *(Reprinted with permission from F. Chen, L. Lin, J. Zhang, Z. He, K. Uchiyama, J.-M. Lin, Single-cell analysis using drop-on-demand inkjet printing and probe electrospray ionization mass spectrometry, Anal. Chem. 88(8) (2016) 4354–4360. Copyright 2016 American Chemical Society.)*

6G and an MCF-7 native single-cell. The ion corresponding to rhodamine 6G (*m/z* 443.3304) was only detected in the marked cell, demonstrating that the system could detect cell markers at the single-cell level.

These studies demonstrated that PESI could be useful in the qualitative analysis of a single-cell for both plants and animals. Although there are still some difficulties in performing quantitative analysis of a single-cell by PESI, it is envisaged that this will be achieved in the near future.

6.2.3 Application of PESI/MS/MS for the analysis of intact metabolites of mouse tissue and in the real-time analysis of metabolites in a living mouse

6.2.3.1 Direct analysis of biological samples

Chen et al. reported the direct qualitative analysis of various biological samples by PESI/TOFMS (AccuTOF, JEOL) [10]. They directly analyzed human urine, human milk obtained from a breast-feeding mother, commercially available bovine milk, mouse brain and liver, salmon egg, and various fruits (orange, banana, etc.). Prior to the analysis of the biological samples, they demonstrated that PESI/TOFMS was able to detect gramicidin S, which is a polypeptide antibiotic, and insulin at 0.1 and 6 μM, respectively, in a 100 mM NaCl aqueous solution. These results indicate that PESI has the potential to detect analytes even against a high salt concentration background. They also detected urea and creatinine in human urine, and disaccharide and lipid clusters were detected in human and bovine milk. Mass spectral profiles using PESI/TOFMS were directly obtained from a mouse brain, a mouse liver, a salmon egg, and fruit, as well as from the liquid biological samples mentioned above, although the biogenic components in each sample were not identified in detail. However, they successfully showed that PESI was able to ionize some of the components in various biological samples.

Zaitsu et al. first combined PESI with a triple quadrupole tandem mass spectrometer (LCMS^{TM}-8040 tandem mass spectrometer, Shimadzu Corporation) and applied PESI/MS/MS to analyze the intact metabolites in a mouse liver [30]. The needle tip diameter was c. 700 nm, and the voltage applied to the probe was ±2.5 kV. The liver samples were obtained by dissection of 129/sv male mice, and each mouse liver was cut into c. 3 mm^2 samples. These samples were frozen immediately after dissection and were stored at −80°C until analysis. A frozen liver sample was set onto a sample plate (Shimadzu Corporation), and to enhance ionization efficiency, 30 μL of 50% aqueous ethanol solution (v/v) was poured onto the sample. Selected reaction monitoring (SRM) mode was used for analysis, and the SRM transition for each metabolite was optimized using the corresponding authentic standard solutions. As explained above, the needle probe moves up and down repeatedly, and thus, the loop time of each SRM transition for mass spectrometry and the cycle time of the probe movement need to be synchronized to avoid missing data acquisition. Consequently, PESI/MS/MS was able to directly detect 26 metabolites, including sugars, amino acids, and organic acids, in the mouse liver without the need for any tedious sample pretreatment, such as homogenization and extraction. To evaluate the practicality of their method, the CCl_4-induced acute hepatic injury model and control mice liver samples were analyzed using PESI/MS/MS in the same way. PCA was applied to the metabolic data, and the two groups could be clearly distinguished by their PCA score plots, as shown in Fig. 6.11. The PCA loading plots indicated that taurine played a pivotal role in group separation, and the hepatic taurine levels were significantly

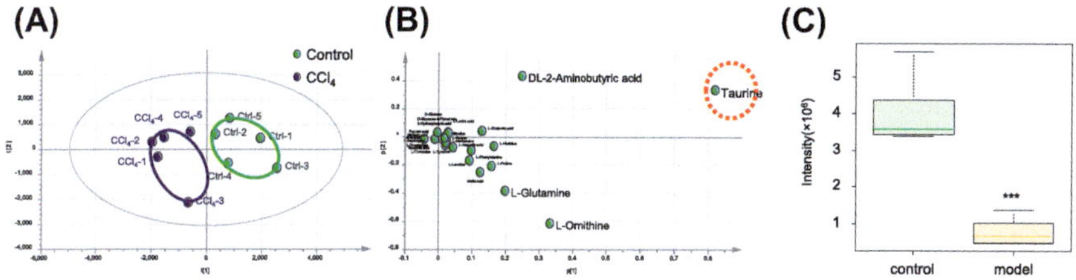

Figure 6.11 Results of the principal component analysis (PCA) applying the metabolite data obtained by PESI/MS/MS (A) Score plots, (B) loading plots, and (C) box-whisker plot of the hepatic levels of taurine (***$P < .01$). The ellipse in the PCA score plots shows Hotelling's T^2 (<95%). *(Reprinted with permission from K. Zaitsu, Y. Hayashi, T. Murata, T. Ohara, K. Nakagiri, M. Kusano, H. Nakajima, T. Nakajima, T. Ishikawa, H. Tsuchihashi, A. Ishii, Intact endogenous metabolite analysis of mice liver by probe electrospray ionization/triple quadrupole tandem mass spectrometry and its preliminary application to in vivo real-time analysis, Anal. Chem. 88(7) (2016) 3556–3561. Copyright 2016 American Chemical Society.)*

different between the control and model groups ($P < .001$, Welch's t-test), with the hepatic taurine levels of the model group being significantly lower.

As is well known, CCl_4 is metabolized by the metabolic enzyme CYP2E1 to the active CCl_3 radical, which is the main cause of CCl_4-induced acute hepatic injury [44]. In general, taurine acts as a scavenger against oxidants such as the CCl_3 radical in the body, and thus, a significant decrease in the hepatic taurine levels in CCl_4-induced acute injury model mice was a reasonable result. In addition, to further verify the effectiveness of PESI/MS/MS, its results for taurine and glutamic acid were compared with those from an established GC/MS/MS-based metabolome analysis that employed a stable isotope-based internal standard (glutamic acid-$^{13}C_5$, $^{15}N_1$) correction method. The hepatic taurine and glutamic acid levels were well correlated between the two methods, proving that PESI/MS/MS was able to capture metabolic profiles directly.

Hayashi in the Zaitsu group expanded intact metabolite analysis by PESI/MS/MS by applying it to mouse brain samples [24]. Since the brain contains a lot of lipids and fats, sample pretreatments such as the Bligh and Dyer method or protein precipitation are generally mandatory for brain metabolome analysis [45–48]. PESI/MS/MS was able to directly detect 26 metabolites in mouse brain, which is mainly due to the high ionization efficiency of PESI. In addition, they investigated the potential of PESI/MS/MS to perform local distribution analysis of the brain. In brief, the frontal cortex and hippocampus of the mouse brain were obtained by dissection under isoflurane anesthesia and were rapidly frozen by liquid nitrogen. The frozen sample was cut into a c. 3 mm^2 sample, which was set onto the sample plate. After 30 μL of cold 50% aqueous ethanol solution (v/v) was poured onto the sample, it was directly analyzed using PESI/MS/MS. The metabolic profiles of the frontal cortex and hippocampus were compared by PCA sore plots, resulting in a clear separation of both groups (Fig. 6.12). This observation demonstrated that slight metabolic differences between the frontal cortex and hippocampus could be detected using PESI/MS/MS. The matrix effect differences between the frontal cortex and hippocampus were also evaluated using isotope-labeled L-glutamic acid-$^{13}C_5$, $^{15}N_1$, which was spiked to either the frontal cortex or hippocampus at the same concentration levels.

The results indicated that there was no significant difference in matrix effects between the frontal cortex and hippocampus. Thus, PESI/MS/MS has a high potential for matrix-free imaging mass spectrometry. In fact, imaging mass spectrometry using PESI was demonstrated by the Hiraoka group, as described later.

6.2.3.2 *In vivo real-time monitoring of metabolites in a living mouse*

As described above, the in vivo real-time monitoring of biogenic compounds in a living animal has been one of the most difficult challenges not only in analytical chemistry but also in biological chemistry. There is a

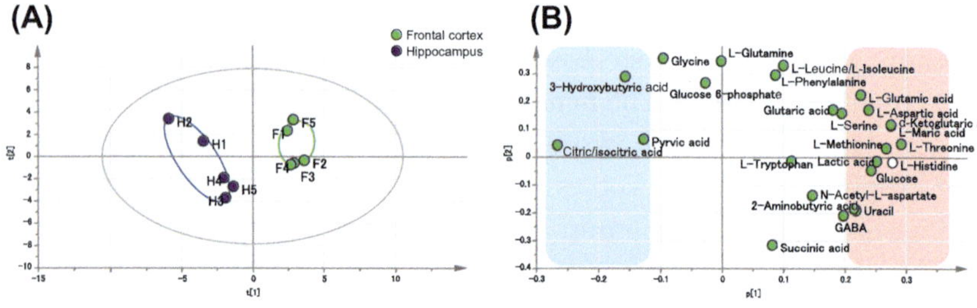

Figure 6.12 (A) Results of the principal component analysis (PCA) for the frontal cortex and hippocampus samples of the control group. (B) Blue or red zones shown in the loading plot correspond to decreased or increased metabolite levels in the frontal cortex in comparison with those in the hippocampus. *(Reprinted from Analytical Chimica Acta, 987, Y. Hayashi, K. Zaitsu, T. Murata, T. Ohara, S. Moreau, M. Kusano, H. Tanihata, H. Tsuchihashi, A. Ishii, T. Ishikawa, Intact metabolite profiling of mouse brain by probe electrospray ionization/triple quadrupole tandem mass spectrometry (PESI/MS/MS) and its potential use for local distribution analysis of the brain, 160–165, Copyright 2017, with permission from Elsevier.)*

difficulty in low-invasive sampling from living animals, which has been overcome by approaches such as the microdialysis technique [49–52]. As explained earlier, the very thin needle probe used in PESI has a high potential to achieve low-invasive sampling from living animals, which strongly matches real-time monitoring.

In fact, Zaitsu et al. used a c. 700 nm diameter tip needle probe as both sampling and ionization units and achieved real-time monitoring of endogenous metabolites in a living mouse liver or brain [30,31]. Fig. 6.13 shows their experimental work flow for in vivo real-time monitoring of a living mouse liver.

In this experiment, a 129/sv male mouse was anesthetized with isoflurane, and laparotomy was performed. The mouse's liver surface was exposed, and a special solvent cup was attached onto the surface with an adhesive suitable for biotissues. As explained earlier, PESI needs an organic solvent, such as ethanol, to be supplied to the probe tip in order to enhance ionization efficiency; however, contact of the organic solvent with the tissue surface can lead to tissue injury. To avoid this, Zaitsu et al. invented the special solvent cup shown in Fig. 6.13. This solvent cup has a double-layer structure, which supplies an organic solvent to the probe tip without contacting the tissue surface. In their study, 150 μL of 50% aqueous ethanol solution with heparin (0.5 mg/mL) was supplied to the solvent cup; the heparin was used to prevent blood coagulation on the probe surface during analysis. Owing to the newly invented solvent cup, they achieved real-time monitoring of two metabolites (α-ketoglutaric and succinic acids), both of which are intermediates of the tricarboxylic acid (TCA) cycle in a living mouse liver. In this experiment, they injected pyruvic acid via the hepatic

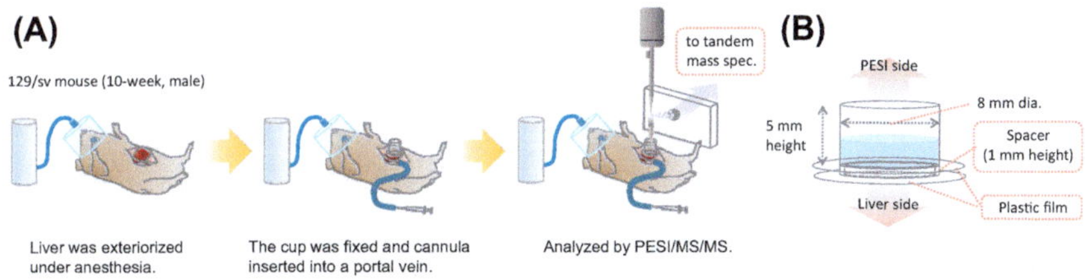

Figure 6.13 (A) Schematic images of the experimental workflow for in vivo real-time monitoring of mouse liver and (B) details of the special solvent cup. *(Reprinted with permission from K. Zaitsu, Y. Hayashi, T. Murata, T. Ohara, K. Nakagiri, M. Kusano, H. Nakajima, T. Nakajima, T. Ishikawa, H. Tsuchihashi, A. Ishii, Intact endogenous metabolite analysis of mice liver by probe electrospray ionization/triple quadrupole tandem mass spectrometry and its preliminary application to in vivo real-time analysis, Anal. Chem. 88(7) (2016) 3556–3561. Copyright 2016 American Chemical Society.)*

portal vein to intentionally activate the TCA cycle activity. This is because pyruvic acid can easily penetrate through the outer membrane of the mitochondrion, which directly leads to enhance TCA cycle activity. As shown in Fig. 6.14, the intensities of α-ketoglutaric and succinic acids were dramatically increased immediately after pyruvic acid injection, demonstrating that activation of the TCA cycle caused by pyruvic acid could be successfully monitored using PESI/MS/MS in real time.

In order to control the puncture position correctly and to analyze other tissues, especially the brain, Zaitsu et al. developed an in vivo real-time monitoring system, which is composed of a semiautomatic stage, a PESI unit, and a tandem mass spectrometer (LCMS^{TM}-8040 tandem mass spectrometer, Shimadzu), as shown in Fig. 6.15 [31]. They applied the newly developed system to the real-time monitoring of eight brain metabolites in a living mouse. The experimental workflow was as follows: an ICR male mouse was anesthetized under isoflurane and fixed to a stereotaxic frame. The hair was removed, and the scalp was carefully opened. A small hole (c. 4 mm dia.) was drilled at the center of the skull, paying

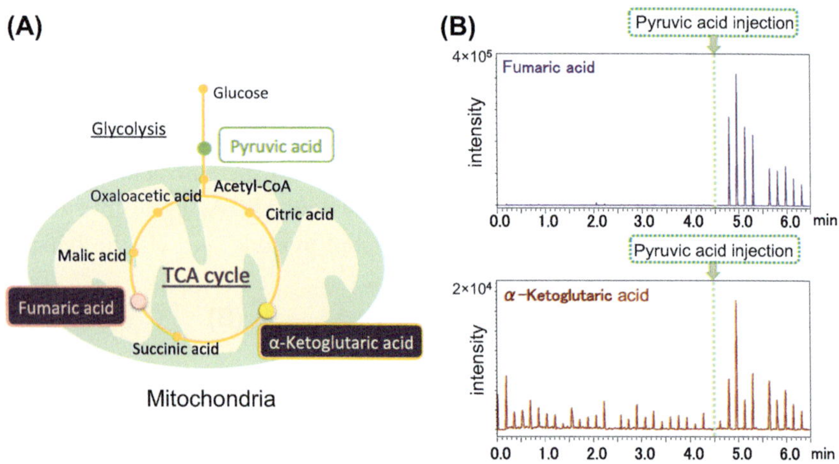

Figure 6.14 (A) Schematics of the tricarboxylic acid (TCA) cycle and its mitochondorial localization (B) Chronograms of the monitoring of fumaric and α-ketoglutaric acids. A dramatic increase in the hepatic levels of these metabolites after the administration of pyruvic acid (monitoring time at 4.5 min) via the portal vein was observed. *(Reprinted with permission from K. Zaitsu, Y. Hayashi, T. Murata, T. Ohara, K. Nakagiri, M. Kusano, H. Nakajima, T. Nakajima, T. Ishikawa, H. Tsuchihashi, A. Ishii, Intact endogenous metabolite analysis of mice liver by probe electrospray ionization/triple quadrupole tandem mass spectrometry and its preliminary application to in vivo real-time analysis, Anal. Chem. 88(7) (2016) 3556–3561. Copyright 2016 American Chemical Society.)*

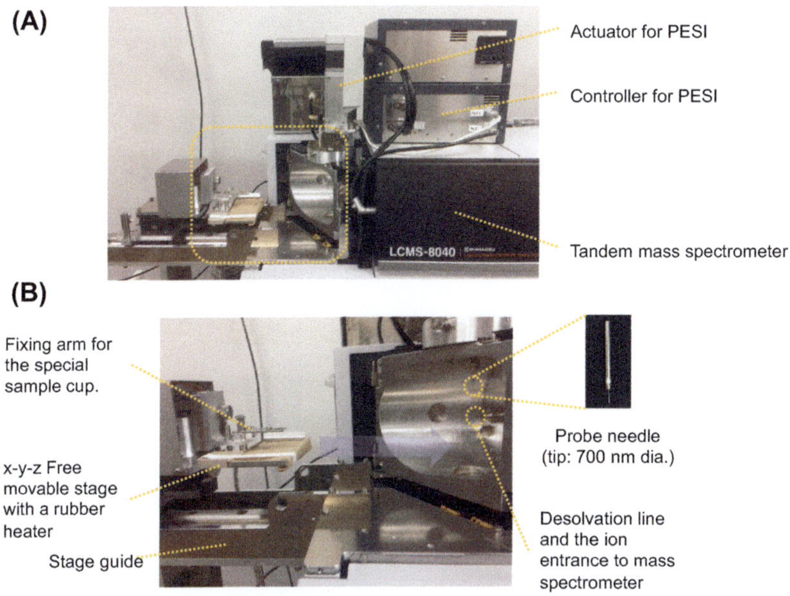

Figure 6.15 In vivo real-time monitoring system (A) Photograph of the overall system. The system consists of a free-movable stage, PESI actuator, and conventional tandem mass spectrometer. (B) Close-up photograph of the orange dotted square section in (A) and a picture of the PESI probe. The free-movable stage can be manually moved to the front of the mass spectrometer (*purple arrow*). *(Reprinted with permission from K. Zaitsu, Y. Hayashi, T. Murata, K. Yokota, T. Ohara, M. Kusano, H. Tsuchihashi, T. Ishikawa, A. Ishii, K. Ogata, H. Tanihata, In vivo real-time monitoring system using probe electrospray ionization/tandem mass spectrometry for metabolites in mouse brain, Anal. Chem. 90(7) (2018) 4695–4701. Copyright 2018 American Chemical Society.)*

attention not to injure the brain. Following the exposure of the brain surface, a special solvent cup was attached to the brain surface with an adhesive, and the mouse brain was directly analyzed with PESI/MS/MS. During the analysis, the 50% aqueous solution with heparin in the sample cup was replaced with a fresh solution every 2 minutes.

Owing to the semiautomatic stage, they achieved real-time monitoring of brain metabolites in a living mouse for 3 hours (Fig. 6.16). In their report, the sampling interval was set at 20 seconds to monitor long-time metabolic fluctuations, though a higher temporal resolution is also possible. It is noted that the mouse was still alive, and there was no noticeable injury on the brain surface after the analysis, proving the low-invasive nature of PESI. In this study, the cannabinoid type 1 receptor agonist, which is known to suppress mitochondrial activity, was administered 5 minutes after the analysis started.

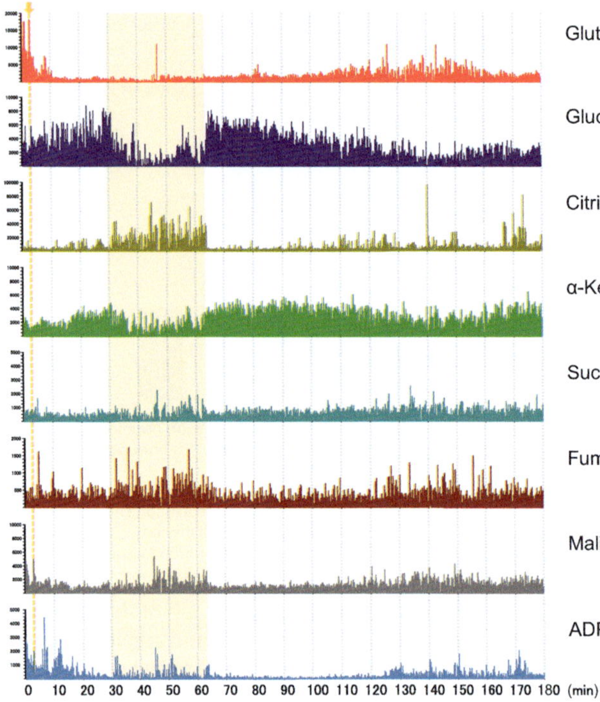

Figure 6.16 In vivo real-time monitoring of the brain of a CB_1R agonist-administered mouse Yellow arrow: CB_1R agonist injection point. Light yellow zone: period in which dynamic changes of energy metabolism were observed. Run time: 180 minutes. *(Reprinted with permission from K. Zaitsu, Y. Hayashi, T. Murata, K. Yokota, T. Ohara, M. Kusano, H. Tsuchihashi, T. Ishikawa, A. Ishii, K. Ogata, H. Tanihata, In vivo real-time monitoring system using probe electrospray ionization/tandem mass spectrometry for metabolites in mouse brain, Anal. Chem. 90(7) (2018) 4695–4701. Copyright 2018 American Chemical Society.)*

As shown in Fig. 6.16, some metabolite levels synchronized with each other. On the other hand, these phenomena were not observed in a control mouse, demonstrating that their system was able to capture metabolic dynamics in a living mouse brain. In addition, they proved that contamination from the solvent cup was negligible.

Interestingly, the changes in glutamic acid levels obtained by PESI/MS/MS showed a different tendency from the experimental results using microdialysis, which can monitor extracellular changes in the mouse brain. Although there is a need to verify the technique using an animal cell, PESI/MS/MS may capture both intracellular and extracellular metabolic dynamics.

6.2.4 Application of PESI/MS to imaging mass spectrometry

Imaging mass spectrometry based on ambient mass spectrometry such as DESI and LAESI has been widely used not only in pharmaceutical fields but also in medical applications [53–57]. As detailed in Chapter 4, DESI enables matrix-free imaging mass spectrometry. PESI, as well as other ambient ionization techniques used in mass spectrometric imaging, also has a potential in imaging mass spectrometry.

To date, Chen in the Hiraoka group applied PESI/TOFMS (AccuTOF, JEOL) to imaging mass spectrometry of a C57/BL6 male mouse brain [32]. Prior to the analysis, the mouse was fixed by paraformaldehyde under anesthesia, and the brain was removed. After washing, the brain was sliced using a razor blade into a 7 × 5 × 2 mm sample. The needle probe used in their experiment had a c. 120 μm diameter, with a submicrometer tip radius. The probe was L-shaped in order to monitor the sampling process using a long-working distance microscope, and the needle was attached to the actuator. The applied voltage was set to around 2–2.5 kV, and the sampling and ionization times were 50 and 200 ms, respectively. An auxiliary sprayer, which was made by modifying an ESI sprayer, was used to supply uncharged solvents for enhancing the ionization efficiency. This is because the brain slice sample quickly dried out during analysis, resulting in failure of sampling by the needle probe. Also, the detectable analytes were fully ionized with the sprayer, and thus, contamination between sampling spots was eliminated. A sliced brain sample was set onto the sample stage, which could be moved along the X-, Y-, and Z-axes in 25 nm steps, and the sampling depth was adjusted to 100 μm, in spite of the surface roughness of the sliced sample. The spatial resolution was set at 60 μm, and data acquisition was repeated for 8 seconds at each sampling point. Mass spectrometric images of the mouse brain slice (30 × 25 pixels) were successfully obtained by PESI/TOFMS, as shown in Fig. 6.17.

PESI imaging mass spectrometry was performed in positive ionization mode, and some peaks were derived from sodium- or potassium-adduct ions of phosphatidylcholines (PCs). For instance, $[PC34:1 + Na]^+$ (*m/z* 782.6, Fig. 6.17A) and $[PC32:0 + Na]^+$ (*m/z* 756.6, Fig. 6.17D) were distributed everywhere in the imaging area. On the other hand, sodium adduct ions of galactosylceramide $[GalCer\ 24h:0 + Na]^+$ (*m/z* 850.7, Fig. 6.17G), $[GalCer\ 22h:0 + Na]^+$ (*m/z* 822.7, Fig. 6.17F), and $[GalCer\ 30h:0 + Na]^+$ (*m/z* 934.7, Fig. 6.17H) were detected only in one distinctive area (the alveus of the hippocampus). Such a distribution was confirmed by previous studies

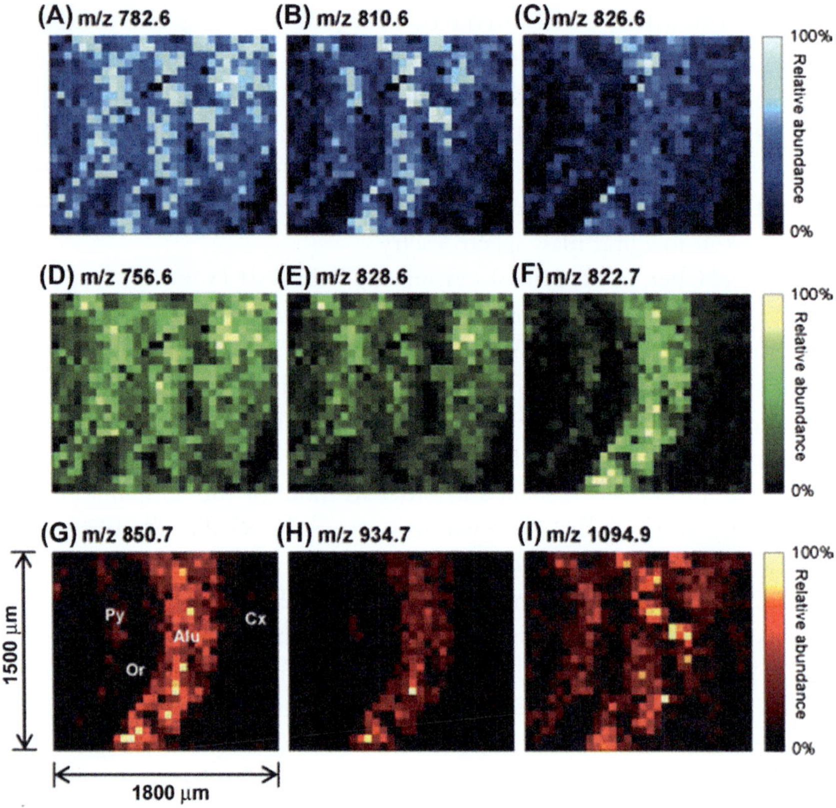

Figure 6.17 Probe electrospray ionization-mass spectrometric imaging of the mouse brain The pixel size is 60 μm, and the imaging area is 1500 × 1800 μm^2. Cx: cerebral cortex, Alu: alveus of the hippocampus, Or: oriens layer, and Py: pyramidal cell layer. *(Reprinted with permission from L.C. Chen, K. Yoshimura, Z. Yu, R. Iwata, H. Ito, H. Suzuki, K. Mori, O. Ariyada, S. Takeda, T. Kubota, K. Hiraoka, Ambient imaging mass spectrometry by electrospray ionization using solid needle as sampling probe, J. Mass Spectrom. 44(10) (2009) 1469–1477. Copyright 2009 John Wiley and Sons.)*

[58,59], where MALDI-MS imaging was performed, in which 2,5-dihydroxybenzoic acid (DHB) for PC and nanoparticles for GalCer were used as the matrix. PESI imaging mass spectrometry showed the same mapping results obtained from MALDI-MS imaging. There is, however, a need to change the matrix (DHB and nanoparticles) for MALDI-MS imaging, and such a tedious treatment is avoided in PESI imaging mass spectrometry. The spatial resolution in PESI was 60 μm, but a higher resolution may be possible by reducing the tip diameter of the needle probe and by an improvement in the ionization efficiency.

6.2.5 Application of PESI/MS/MS to direct drug analysis in forensic toxicology

In forensic toxicology, the quantitative analysis of toxicants in the blood is indispensable for the estimation of acute toxic levels. However, since a lot of toxicants such as stimulants, narcotics, hypnotics, and new psychoactive substances (NPSs) are target compounds, a drug-screening test is also necessary prior to confirmation (qualitative) analysis. A screening test requires high operativity and rapidity to obtain a screening result. Portable kits such as the commercially available Triage kit have their limitations and give both false-positive or false-negative results. Therefore, a mass spectrometric drug-screening test using GC/MS, GC/MS/MS, LC/MS/MS, and LC/Q-TOFMS is commonly accepted by forensic toxicologists [60,61]. In general, however, sample preparation steps such as extraction and evaporation are necessary before the instrumental analysis of drugs in biological specimens. However, it is quite difficult to select the most appropriate extraction procedure, such as liquid—liquid extraction or solid phase extraction, in nontarget drug-screening analysis because the kinds of drugs and metabolites in real forensic specimens are "unknown." Therefore, protein precipitation or QuEChERS methods have been widely accepted as the first choice of pretreatment method in drug-screening tests even if the recovery rates and purification are somewhat low [62—64]. Ultimately, however, direct drug-screening without any extraction is superlative for a drug-screening test, and PESI/MS/MS has a high potential in the direct analysis of drugs in biological specimens.

Zaitsu et al. constructed a preliminary drug-screening method for 160 drugs and metabolites in serum and whole blood, where the screening results were obtained within about 6 minutes [18]. In this preliminary study, despite some of the drugs being only detectable above 10 ng/mL, most of the drugs were detected at 10 ng/mL, demonstrating the practicality of PESI/MS/MS in drug-screening analysis. Quantitative analysis by PESI/MS/MS is also feasible, and this technique can be applied to the screening for NPSs by all ion MS/MS like Sequential Window Acquisition of all THeoretical Mass Spectra (SWATH) analysis [65]. In this instance, all precursor ions existing in the target m/z range are preselected, and a product ion scan for all precursor ions is executed. Although this is a preliminary study, this approach has a high potential as an effective drug-screening method.

Usui et al. demonstrated the rapid drug analysis of postmortem human tissues by the combinational use of PESI/MS/MS by the QuEChERS extraction method [19]. In their study, postmortem human tissues (liver, kidney, lung, brain, and heart) were obtained from a fatal intoxication by an opioid-like psychoactive substance, MT-45. Qualitative analysis by PESI/MS/MS (LCMS™-8040 tandem mass spectrometer, Shimadzu) was performed on all tissues. Briefly, a 1–2 mm^2 tissue sample was placed onto the sample plate (Shimadzu), and 35 μL of ethanol/10 mM ammonium formate aqueous solution (1:1, v/v) was added to the sample. SRM-trigger analysis, where a product ion scan was automatically executed when the SRM ion intensity was over criterion, was performed. MT-45 was detected in all tissues by PESI/MS/MS, where the spectra of MT-45 matched that of the authentic standard for MT-45. In addition, mono- and di-hydroxylated metabolites of MT-45 were detected in the liver, where the chemical structure of the metabolites was estimated from their product ion spectra. Although a direct qualitative analysis of the target drugs in the postmortem tissues was achieved, quantitative analysis was performed by the QuEChERS method. A 0.5 g tissue sample was treated by the QuEChERS method, and a standard addition method was used for quantification. Consequently, the quantitative results obtained by PESI/MS/MS were almost the same as those obtained by LC/MS/MS. Although a direct quantitative analysis of tissues was not achieved, future improvements to the analytical conditions and/or modification of the probe surface may lead to the direct drug quantification of tissues by PESI/MS/MS.

6.2.6 Other applications and ambient ionization techniques similar to PESI

The Hiraoka group developed dipping PESI for on-site food analysis [66]. As mentioned above, PESI requires a solvent supply to the probe tip to enhance its ionization efficiency, and a sophisticated system is needed for supplying the solvent to the probe tip. Following their previous studies, the sampling and ionization processes were considered separately in this study; sampling was manually performed using a needle probe. After drying the probe, a solvent reservoir was used to supply organic solvents to the probe tip, and vegetables (tomato, spinach, and onion), salmon, fresh, cow's milk, yogurt, and soy bean milk were analyzed using PESI-TOFMS (JEOL). In their study, they applied an ionization voltage between +2.5 and +3.0 kV to the probe and successfully obtained mass spectral profiles of all samples.

In general, PESI uses a thin solid needle probe as the ionization unit, although similar approaches have been reported as mentioned above. Hereafter, we will introduce the wooden-tip ESI method [12,13,33,35,67].

Hu et al. reported ESI using a disposable "wooden-tip," which is made from a commercially available wooden toothpick [33,35]. They explained that the hydrophilic and porous nature of wood allows for the adsorption of target compounds on the tip surface. The sharp wooden-tip end (~0.2 mm) was simply dipped into the sample solution for 5 seconds, and 5 μL of solvent (MeOH/H_2O, 1:1, containing 0.1% formic acid) was pipetted onto the tip surface. Subsequently, the tip was mounted onto a capillary holder for analysis. Q-TOFMS (Q-TOF2, Waters, or QStar, ABI) or a tandem mass spectrometer (Quattro Ultima, Waters) was used as a detector. The applied voltage to the tip was adjusted in the range 2.5–3.5 kV. The wooden-tip ESI was compared to nano-ESI and paper spray ionization on the basis of sensitivity to reserpine, indicating that the wooden-tip ESI showed a similar sensitivity to paper spray ionization, but a lower sensitivity than that of nano-ESI. The negative ionization mode was also possible using the wooden-tip ESI, and the method was applicable to various compounds/samples such as proteins, powder samples, and forensic case specimens.

Deng et al. developed a surface-coated wooden-tip ESI and used it to analyze antibiotics in tap and river waters; the details of their study are not explained here because the reported applications were not for life sciences [12]. However, their study demonstrated that the surface modification of the wooden-tip strongly affects the sensitivity, suggesting that surface adsorption onto the probe is one of the important factors in PESI. Hu et al. also reported that a surface-modified wooden-tip ESI exhibited enhanced sensitivity [13]. They modified the surface of a wooden-tip with C_{18}- (hydrophobic), H_2N- (basic), and HSO_3- (acidic) functional groups, and direct loading or dipping methods were examined as different sampling ways. The dipping method gave a strong interaction between the modified wooden-tip and the target analyte, whereas the direct loading method gave a weak interaction. There were no significant differences between tips purchased via the internet or those prepared from birch wood. The HSO_3-type surface-modified wooden-tip was used to directly analyze cocaine in raw human oral fluids (0.1 μg/mL). The other types of modified wooden tips were not able to detect cocaine; however, the predominant signal of cocaine was observed by the HSO_3-type wooden-tip ESI, achieving sufficient extraction of cocaine (Fig. 6.18). In addition, a sufficient calibration

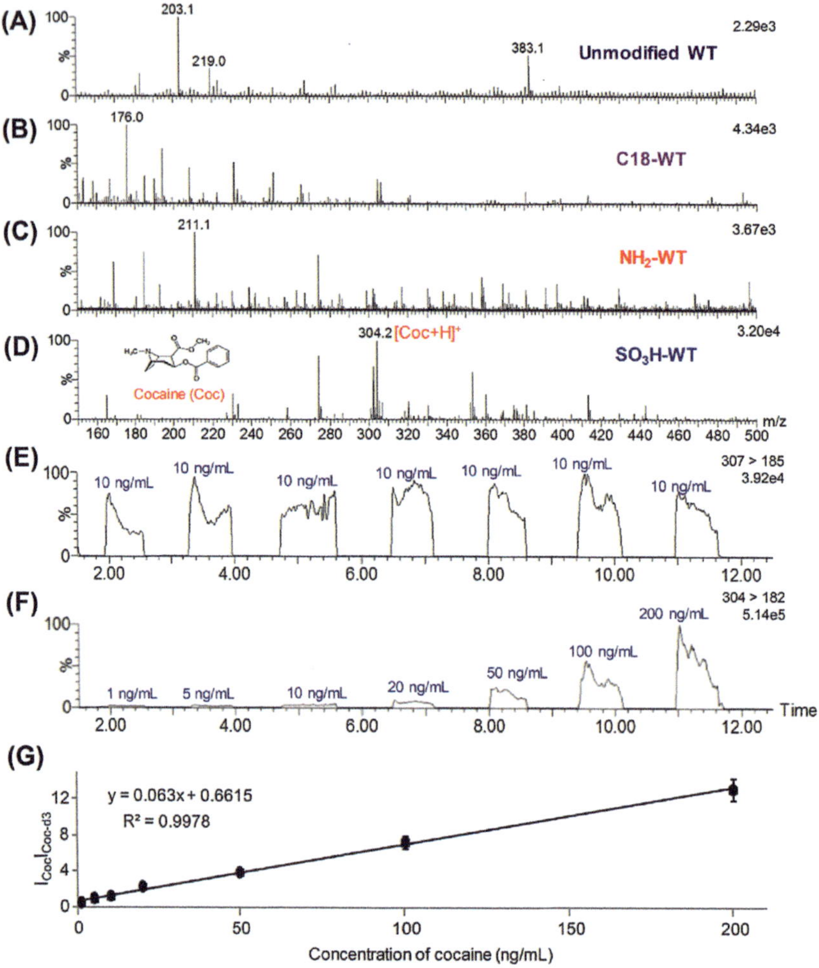

Figure 6.18 Results from wooden-tip (WT) ESI of cocaine in raw oral fluids (A) Unmodified-WT, (B) C18-WT, (C) NH2-WT, (D) SO3-WT, (E) Selected reaction monitoring (SRM) chronograms of cocaine-d3 (10 ppb), (F) SRM chronograms of cocaine at different concentrations, and (G) a calibration curve obtained from the data shown in (E) and (F). *(Reprinted with permission from B. Hu, P.-K. So, Y. Yang, J. Deng, Y.-C. Choi, T. Luan, Z.-P. Yao, Surface-modified wooden-tip electrospray ionization mass spectrometry for enhanced detection of analytes in complex samples, Anal. Chem. 90(3) (2018) 1759–1766. Copyright 2018 American Chemical Society.)*

curve (1–200 ng/mL) with an internal standard (IS) method was constructed, where cocaine-d_3 was used as IS. Also, the C_{18}-type modified wooden-tip was used for the analysis of protein compounds in high concentration of salts and detergents.

Finally, an interesting study that uses a sampling method similar to PESI will be discussed here. The Fernández group reported a new generation of instrument and methods for 3D surface analysis, called the robotic surface analysis (RoSA) system [68].

As shown in Fig. 6.19, the RoSA system was constructed from a robotic arm with 360 degrees rotation, force-sensing capabilities, a control framework, and a 3D line laser scanner, which was attached to the robotic arm's wrist. A spring-loaded semiblunt needle probe was used as the sampling unit, as shown in Fig. 6.20.

Following sampling of the surface, the robot arm moves the probe into an open port interface of a mass spectrometer. In one example, they applied the system to analyze the surface of a rubber football which had been touched by a person who had used *N,N*-diethyl-meta-toluamide (DEET), resulting in the detection of DEET only in the specific area that had been touched (Fig. 6.21). This study strongly encourages the combination of PESI and its similar other ambient ionization techniques with a 3D analytical system such as RoSA.

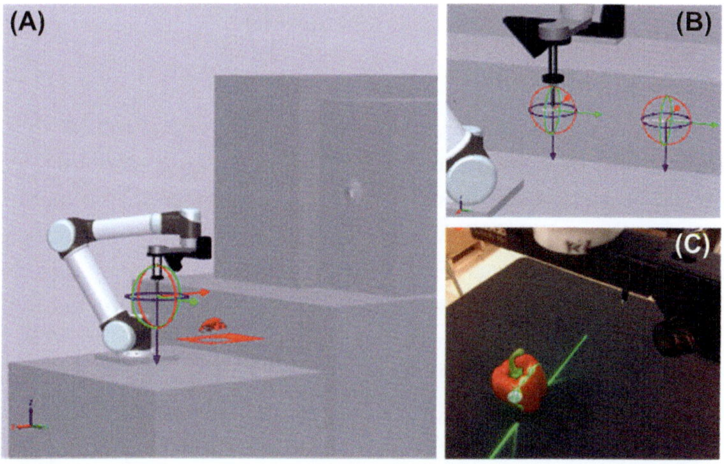

Figure 6.19 Schematic images of robotic surface analysis mass spectrometry (A) The robot and laboratory environment integrated into the Actin GUI, (B) laser scan start and end waypoints being programmed in Actin-MS to define the length of the scan, and (C) a bell pepper being scanned by the laser. *(Reprinted with permission from A. Li, M.R.L. Paine, S. Zambrzycki, R.B. Stryffeler, J. Wu, M. Bouza, J. Huckaby, C.-Y. Chang, M. Kumar, P. Mukhija, F.M. Fernández, Robotic surface analysis mass spectrometry (RoSA-MS) of three-dimensional objects, Anal. Chem. 90(6) (2018) 3981–3986. Copyright 2018 American Chemical Society.)*

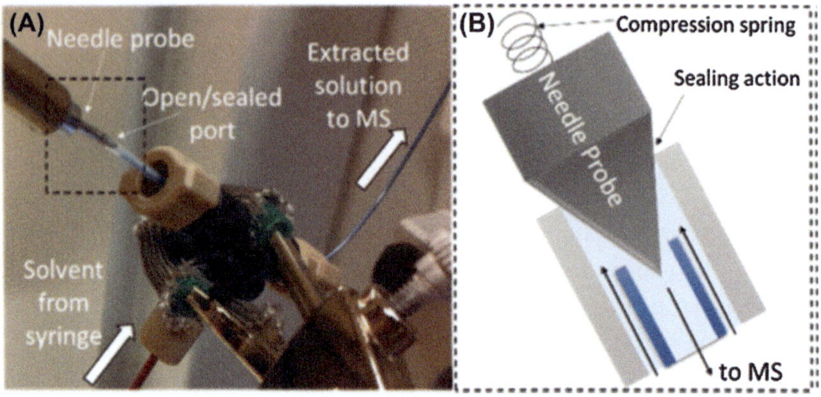

Figure 6.20 Sampling unit of robotic surface analysis mass spectrometry (RoSA-MS) (A) During RoSA-MS analysis, the robot docks the needle probe into an open port-sampling interface. (B) Schematic of the needle probe against the annulus tubing; the capillary dissolves the material on the needle tip. The sealed fluidics system delivers the solubilized sample to the mass spectrometer. *(Reprinted with permission from A. Li, M.R.L. Paine, S. Zambrzycki, R.B. Stryffeler, J. Wu, M. Bouza, J. Huckaby, C.-Y. Chang, M. Kumar, P. Mukhija, F.M. Fernández, Robotic surface analysis mass spectrometry (RoSA-MS) of three-dimensional objects, Anal. Chem. 90(6) (2018) 3981–3986. Copyright 2018 American Chemical Society.)*

6.2.7 Future perspectives

PESI allows the direct analysis, narrow-area observation, and in vivo real-time monitoring of a living animal with low-invasive sampling, and its applications in the life sciences will be expanded more than ever. In fact, applications using PESI/MS/MS in the plant sciences are currently underway by our collaborators. Furthermore, single-cell analysis and in vivo real-time monitoring by PESI will contribute to new scientific fields such as single-cell metabolomics and real-time metabolomics.

The combinational use of PESI and Q-TOFMS will be a powerful tool to push PESI to be used in next generation analysis, and PESI-Q-TOFMS will be developed in the near future. Moreover, there may be a possibility to apply an ion mobility instrument to PESI-MS/MS or PESI-Q-TOFMS to provide additional separation ability to these mass spectrometers. The development of the new instruments and technologies mentioned above will produce new applications to the life sciences; the door to the next scientific stage will soon be opened.

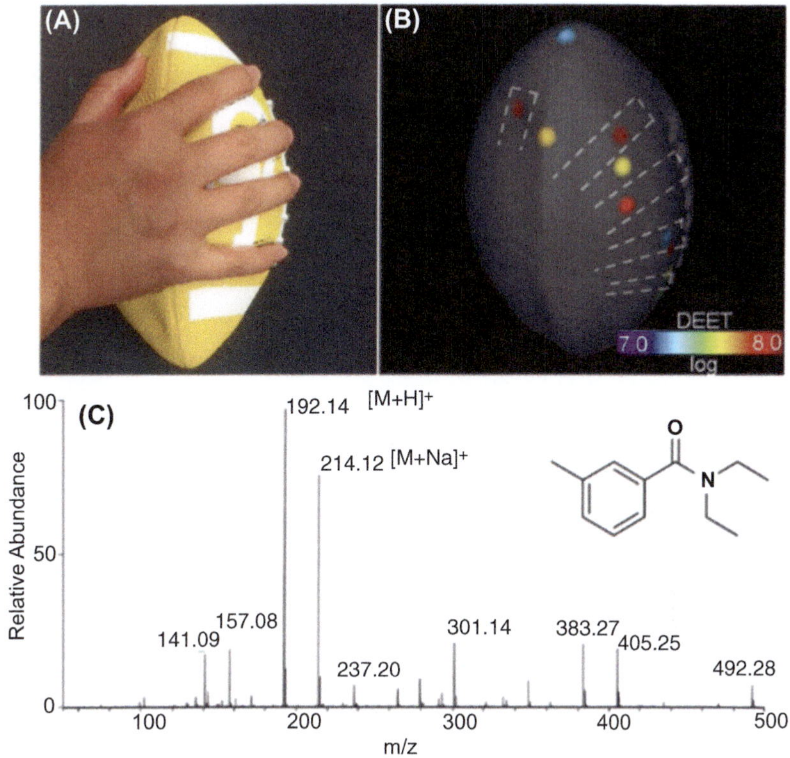

Figure 6.21 (A) A hand that was sprayed with *N,N*-diethyl-meta-toluamide (DEET) holding a football. (B) Several locations on this football were analyzed by RoSA-MS. The summed peak area (*m/z* 192 + 214) for DEET ions was converted to a pseudo color scale for the specific coordinates sampled. The dashed lines indicate the approximate boundary of the fingers when touching the football surface. (C) Full-scan mass spectrum obtained from the coordinates where the thumb contacted the surface. *(Reprinted with permission from A. Li, M.R.L. Paine, S. Zambrzycki, R.B. Stryffeler, J. Wu, M. Bouza, J. Huckaby, C.-Y. Chang, M. Kumar, P. Mukhija, F.M. Fernández, Robotic surface analysis mass spectrometry (RoSA-MS) of three-dimensional objects, Anal. Chem. 90(6) (2018) 3981–3986. Copyright 2018 American Chemical Society.)*

ACKNOWLEDGMENTS

We express our sincere thanks to Dr. Yumi Hayashi for making permission logs.

References

[1] J. Fenn, M. Mann, C. Meng, S. Wong, C. Whitehouse, Electrospray ionization for mass spectrometry of large biomolecules, Science 246 (4926) (1989) 64–71.

[2] K. Tanaka, H. Waki, Y. Ido, S. Akita, Y. Yoshida, T. Yoshida, T. Matsuo, Protein and polymer analyses up to m/z 100 000 by laser ionization time-of-flight mass spectrometry, Rapid Commun. Mass Spectrom. 2 (8) (1988) 151–153.

[3] M.S. Wilm, M. Mann, Electrospray and Taylor-Cone theory, Dole's beam of macromolecules at last? Int. J. Mass Spectrom. Ion Process. 136 (2) (1994) 167–180.
[4] B. Brügger, G. Erben, R. Sandhoff, F.T. Wieland, W.D. Lehmann, Quantitative analysis of biological membrane lipids at the low picomole level by nano-electrospray ionization tandem mass spectrometry, Proc. Natl. Acad. Sci. Unit. States Am. 94 (6) (1997) 2339–2344.
[5] C.-M. Hong, C.-T. Lee, Y.-M. Lee, C.-P. Kuo, C.-H. Yuan, J. Shiea, Generating electrospray from solutions predeposited on a copper wire, Rapid Commun. Mass Spectrom. 13 (1) (1999) 21–25.
[6] C.-P. Kuo, J. Shiea, Application of direct electrospray probe to analyze biological compounds and to couple to solid-phase microextraction to detect trace surfactants in aqueous solution, Anal. Chem. 71 (19) (1999) 4413–4417.
[7] J. Jeng, J. Shiea, Electrospray ionization from a droplet deposited on a surface-modified glass rod, Rapid Commun. Mass Spectrom. 17 (15) (2003) 1709–1713.
[8] J. Jeng, C.-H. Lin, J. Shiea, Electrospray from nanostructured tungsten oxide surfaces with ultralow sample volume, Anal. Chem. 77 (24) (2005) 8170–8173.
[9] K. Hiraoka, K. Nishidate, K. Mori, D. Asakawa, S. Suzuki, Development of probe electrospray using a solid needle, Rapid Commun. Mass Spectrom. 21 (18) (2007) 3139–3144.
[10] L.C. Chen, K. Nishidate, Y. Saito, K. Mori, D. Asakawa, S. Takeda, T. Kubota, N. Terada, Y. Hashimoto, H. Hori, K. Hiraoka, Application of probe electrospray to direct ambient analysis of biological samples, Rapid Commun. Mass Spectrom. 22 (15) (2008) 2366–2374.
[11] L.C. Chen, K. Nishidate, Y. Saito, K. Mori, D. Asakawa, S. Takeda, T. Kubota, H. Hori, K. Hiraoka, Characteristics of probe electrospray generated from a solid needle, J. Phys. Chem. B 112 (35) (2008) 11164–11170.
[12] J. Deng, T. Yu, Y. Yao, Q. Peng, L. Luo, B. Chen, X. Wang, Y. Yang, T. Luan, Surface-coated wooden-tip electrospray ionization mass spectrometry for determination of trace fluoroquinolone and macrolide antibiotics in water, Anal. Chim. Acta 954 (2017) 52–59.
[13] B. Hu, P.-K. So, Y. Yang, J. Deng, Y.-C. Choi, T. Luan, Z.-P. Yao, Surface-modified wooden-tip electrospray ionization mass spectrometry for enhanced detection of analytes in complex samples, Anal. Chem. 90 (3) (2018) 1759–1766.
[14] K. Yoshimura, L.C. Chen, M.K. Mandal, T. Nakazawa, Z. Yu, T. Uchiyama, H. Hori, K. Tanabe, T. Kubota, H. Fujii, R. Katoh, K. Hiraoka, S. Takeda, Analysis of renal cell carcinoma as a first step for developing mass spectrometry-based diagnostics, J. Am. Soc. Mass Spectrom. 23 (10) (2012) 1741–1749.
[15] K. Yoshimura, M.K. Mandal, M. Hara, H. Fujii, L.C. Chen, K. Tanabe, K. Hiraoka, S. Takeda, Real-time diagnosis of chemically induced hepatocellular carcinoma using a novel mass spectrometry-based technique, Anal. Biochem. 441 (1) (2013) 32–37.
[16] S. Takeda, K. Yoshimura, K. Hiraoka, Innovations in analytical oncology - status quo of mass spectrometry-based diagnostics for malignant tumor, J. Anal. Oncol. 1 (2012) 6.
[17] L.A. Carpino, G.Y. Han, 9-Fluorenylmethoxycarbonyl function, a new base-sensitive amino-protecting group, J. Am. Chem. Soc. 92 (19) (1970) 5748–5749.
[18] K. Zaitsu, Cutting-edge mass spectrometric screening methods for new psychoactive substances, in: 15th International Congress of Therapeutic Drug Monitoring and Clinical Toxicology, Kyoto, Japan, 2017.
[19] K. Usui, T. Murata, Y. Fujita, Y. Kamijo, T. Hanazawa, T. Yoshizawa, M. Funayama, Direct detection of the psychoactive substance MT-45 in human tissue samples by probe electrospray ionization-tandem mass spectrometry, Drug Test. Anal. 10 (6) (2018) 1033–1038.

[20] Y. Seto, Peer reviewed: false cyanide detection, Anal. Chem. 74 (5) (2002) 134 A−141 A.

[21] Y. Seto, N. Tsunoda, H. Ohta, T. Shinohara, Determination of blood cyanide by headspace gas chromatography with nitrogen-phosphorus detection and using a megabore capillary column, Anal. Chim. Acta 276 (2) (1993) 247−259.

[22] A. Sano, M. Takezawa, S. Takitani, High performance liquid chromatography determination of cyanide in urine by pre-column fluorescence derivatization, Biomed. Chromatogr. 3 (5) (1989) 209−212.

[23] S. Chinaka, N. Takayama, Y. Michigami, K. Ueda, Simultaneous determination of cyanide and thiocyanate in blood by ion chromatography with fluorescence and ultraviolet detection, J. Chromatogr. B Biomed. Sci. Appl. 713 (2) (1998) 353−359.

[24] Y. Hayashi, K. Zaitsu, T. Murata, T. Ohara, S. Moreau, M. Kusano, H. Tanihata, H. Tsuchihashi, A. Ishii, T. Ishikawa, Intact metabolite profiling of mouse brain by probe electrospray ionization/triple quadrupole tandem mass spectrometry (PESI/MS/MS) and its potential use for local distribution analysis of the brain, Anal. Chim. Acta 983 (2017) 160−165.

[25] X. Gong, Y. Zhao, S. Cai, S. Fu, C. Yang, S. Zhang, X. Zhang, Single cell analysis with probe ESI-mass spectrometry: detection of metabolites at cellular and subcellular levels, Anal. Chem. 86 (8) (2014) 3809−3816.

[26] F. Chen, L. Lin, J. Zhang, Z. He, K. Uchiyama, J.-M. Lin, Single-cell analysis using drop-on-demand inkjet printing and probe electrospray ionization mass spectrometry, Anal. Chem. 88 (8) (2016) 4354−4360.

[27] Z. Yu, L.C. Chen, S. Ninomiya, M.K. Mandal, K. Hiraoka, H. Nonami, Piezoelectric inkjet assisted rapid electrospray ionization mass spectrometric analysis of metabolites in plant single cells via a direct sampling probe, Analyst 139 (22) (2014) 5734−5739.

[28] Y. Gholipour, R. Erra-Balsells, K. Hiraoka, H. Nonami, Living cell manipulation, manageable sampling, and shotgun picoliter electrospray mass spectrometry for profiling metabolites, Anal. Biochem. 433 (1) (2013) 70−78.

[29] T. Nakashima, H. Wada, S. Morita, R. Erra-Balsells, K. Hiraoka, H. Nonami, Single-cell metabolite profiling of stalk and glandular cells of intact trichomes with internal electrode capillary pressure probe electrospray ionization mass spectrometry, Anal. Chem. 88 (6) (2016) 3049−3057.

[30] K. Zaitsu, Y. Hayashi, T. Murata, T. Ohara, K. Nakagiri, M. Kusano, H. Nakajima, T. Nakajima, T. Ishikawa, H. Tsuchihashi, A. Ishii, Intact endogenous metabolite analysis of mice liver by probe electrospray ionization/triple quadrupole tandem mass spectrometry and its preliminary application to in vivo real-time analysis, Anal. Chem. 88 (7) (2016) 3556−3561.

[31] K. Zaitsu, Y. Hayashi, T. Murata, K. Yokota, T. Ohara, M. Kusano, H. Tsuchihashi, T. Ishikawa, A. Ishii, K. Ogata, H. Tanihata, In vivo real-time monitoring system using probe electrospray ionization/tandem mass spectrometry for metabolites in mouse brain, Anal. Chem. 90 (7) (2018) 4695−4701.

[32] L.C. Chen, K. Yoshimura, Z. Yu, R. Iwata, H. Ito, H. Suzuki, K. Mori, O. Ariyada, S. Takeda, T. Kubota, K. Hiraoka, Ambient imaging mass spectrometry by electrospray ionization using solid needle as sampling probe, J. Mass Spectrom. 44 (10) (2009) 1469−1477.

[33] B. Hu, P.-K. So, Z.-P. Yao, Analytical properties of solid-substrate electrospray ionization mass spectrometry, J. Am. Soc. Mass Spectrom. 24 (1) (2013) 57−65.

[34] B. Hu, P.-K. So, Z.-P. Yao, Electrospray ionization with aluminum foil: a versatile mass spectrometric technique, Anal. Chim. Acta 817 (2014) 1−8.

[35] B. Hu, P.-K. So, H. Chen, Z.-P. Yao, Electrospray ionization using wooden tips, Anal. Chem. 83 (21) (2011) 8201−8207.

[36] P. Liu, A. Forni, H. Chen, Development of solvent-free ambient mass spectrometry for green chemistry applications, Anal. Chem. 86 (8) (2014) 4024–4032.
[37] C.-C. Hsu, M.S. ElNaggar, Y. Peng, J. Fang, L.M. Sanchez, S.J. Mascuch, K.A. Møller, E.K. Alazzeh, J. Pikula, R.A. Quinn, Y. Zeng, B.E. Wolfe, R.J. Dutton, L. Gerwick, L. Zhang, X. Liu, M. Månsson, P.C. Dorrestein, Real-time metabolomics on living microorganisms using ambient electrospray ionization flow-probe, Anal. Chem. 85 (15) (2013) 7014–7018.
[38] H. Wen, Y.J. An, W.J. Xu, K.W. Kang, S. Park, Real-time monitoring of cancer cell metabolism and effects of an anticancer agent using 2D in-cell NMR spectroscopy, Angew. Chem. Int. Ed. 54 (18) (2015) 5374–5377.
[39] H. Link, T. Fuhrer, L. Gerosa, N. Zamboni, U. Sauer, Real-time metabolome profiling of the metabolic switch between starvation and growth, Nat. Methods 12 (2015) 1091.
[40] D. Wang, S. Bodovitz, Single cell analysis: the new frontier in 'omics', Trends Biotechnol. 28 (6) (2010) 281–290.
[41] T. Fujii, S. Matsuda, M.L. Tejedor, T. Esaki, I. Sakane, H. Mizuno, N. Tsuyama, T. Masujima, Direct metabolomics for plant cells by live single-cell mass spectrometry, Nat. Protoc. 10 (2015) 1445.
[42] K. Yamamoto, K. Takahashi, H. Mizuno, A. Anegawa, K. Ishizaki, H. Fukaki, M. Ohnishi, M. Yamazaki, T. Masujima, T. Mimura, Cell-specific localization of alkaloids in Catharanthus roseus stem tissue measured with Imaging MS and Single-cell MS, Proc. Natl. Acad. Sci. Unit. States Am. 113 (14) (2016) 3891–3896.
[43] M. Fessenden, Metabolomics: small molecules, single cells, Nature 540 (2016) 153.
[44] L. Jiang, J. Huang, Y. Wang, H. Tang, Metabonomic analysis reveals the CCl4-induced systems alterations for multiple rat organs, J. Proteome Res. 11 (7) (2012) 3848–3859.
[45] E.G. Bligh, W.J. Dyer, A rapid method of total lipid extraction and purification, Can. J. Biochem. Physiol. 37 (8) (1959) 911–917.
[46] K. Zaitsu, Y. Hayashi, K. Suzuki, H. Nakayama, N. Hattori, R. Takahara, M. Kusano, H. Tsuchihashi, A. Ishii, Metabolome disruption of the rat cerebrum induced by the acute toxic effects of the synthetic cannabinoid MAM-2201, Life Sci. 137 (2015) 49–55.
[47] K. Zaitsu, Y. Hayashi, M. Kusano, H. Tsuchihashi, A. Ishii, Application of metabolomics to toxicology of drugs of abuse: a mini review of metabolomics approach to acute and chronic toxicity studies, Drug Metab. Pharmacokinet. 31 (1) (2016) 21–26.
[48] W. Römisch-Margl, C. Prehn, R. Bogumil, C. Röhring, K. Suhre, J. Adamski, Procedure for tissue sample preparation and metabolite extraction for high-throughput targeted metabolomics, Metabolomics 8 (1) (2012) 133–142.
[49] M.R. Emmett, P.E. Andrén, R.M. Caprioli, Specific molecular mass detection of endogenously released neuropeptides using in vivo microdialysis/mass spectrometry, J. Neurosci. Methods 62 (1) (1995) 141–147.
[50] C. Liu, Q. Wu, A.C. Harms, R.D. Smith, On-line microdialysis sample cleanup for electrospray ionization mass spectrometry of nucleic acid samples, Anal. Chem. 68 (18) (1996) 3295–3299.
[51] M. Wang, T. Slaney, O. Mabrouk, R.T. Kennedy, Collection of nanoliter microdialysate fractions in plugs for off-line in vivo chemical monitoring with up to 2s temporal resolution, J. Neurosci. Methods 190 (1) (2010) 39–48.
[52] P. Song, N.D. Hershey, O.S. Mabrouk, T.R. Slaney, R.T. Kennedy, Mass spectrometry "sensor" for in vivo acetylcholine monitoring, Anal. Chem. 84 (11) (2012) 4659–4664.
[53] D.S. Cornett, M.L. Reyzer, P. Chaurand, R.M. Caprioli, MALDI imaging mass spectrometry: molecular snapshots of biochemical systems, Nat. Methods 4 (2007) 828.

[54] M. Stoeckli, P. Chaurand, D.E. Hallahan, R.M. Caprioli, Imaging mass spectrometry: a new technology for the analysis of protein expression in mammalian tissues, Nat. Med. 7 (2001) 493.

[55] Z. Takáts, J.M. Wiseman, B. Gologan, R.G. Cooks, Mass spectrometry sampling under ambient conditions with desorption electrospray ionization, Science 306 (5695) (2004) 471−473.

[56] J.M. Wiseman, D.R. Ifa, Q. Song, R.G. Cooks, Tissue imaging at atmospheric pressure using desorption electrospray ionization (DESI) mass spectrometry, Angew. Chem. Int. Ed. 45 (43) (2006) 7188−7192.

[57] P. Nemes, A. Vertes, Laser ablation electrospray ionization for atmospheric pressure, in vivo, and imaging mass spectrometry, Anal. Chem. 79 (21) (2007) 8098−8106.

[58] S.N. Jackson, M. Ugarov, T. Egan, J.D. Post, D. Langlais, J. Albert Schultz, A.S. Woods, MALDI-ion mobility-TOFMS imaging of lipids in rat brain tissue, J. Mass Spectrom. 42 (8) (2007) 1093−1098.

[59] S. Taira, Y. Sugiura, S. Moritake, S. Shimma, Y. Ichiyanagi, M. Setou, Nanoparticle-assisted laser desorption/ionization based mass imaging with cellular resolution, Anal. Chem. 80 (12) (2008) 4761−4766.

[60] S. Ojanperä, A. Pelander, M. Pelzing, I. Krebs, E. Vuori, I. Ojanperä, Isotopic pattern and accurate mass determination in urine drug screening by liquid chromatography/time-of-flight mass spectrometry, Rapid Commun. Mass Spectrom. 20 (7) (2006) 1161−1167.

[61] K. Zaitsu, H. Nakayama, M. Yamanaka, K. Hisatsune, K. Taki, T. Asano, T. Kamata, M. Katagai, Y. Hayashi, M. Kusano, H. Tsuchihashi, A. Ishii, High-resolution mass spectrometric determination of the synthetic cannabinoids MAM-2201, AM-2201, AM-2232, and their metabolites in postmortem plasma and urine by LC/Q-TOFMS, Int. J. Legal Med. 129 (6) (2015) 1233−1245.

[62] K. Zaitsu, M. Katagi, H.T. Kamata, T. Kamata, N. Shima, A. Miki, H. Tsuchihashi, Y. Mori, Determination of the metabolites of the new designer drugs bk-MBDB and bk-MDEA in human urine, Forensic Sci. Int. 188 (1−3) (2009) 131−139.

[63] K. Usui, Y. Hayashizaki, M. Hashiyada, M. Funayama, Rapid drug extraction from human whole blood using a modified QuEChERS extraction method, Leg. Med. 14 (6) (2012) 286−296.

[64] S. Matsuta, K. Nakanishi, A. Miki, K. Zaitsu, N. Shima, T. Kamata, H. Nishioka, M. Katagi, M. Tatsuno, K. Tsuboi, H. Tsuchihashi, K. Suzuki, Development of a simple one-pot extraction method for various drugs and metabolites of forensic interest in blood by modifying the QuEChERS method, Forensic Sci. Int. 232 (1) (2013) 40−45.

[65] K. Arnhard, A. Gottschall, F. Pitterl, H. Oberacher, Applying 'sequential windowed acquisition of all theoretical fragment ion mass spectra' (SWATH) for systematic toxicological analysis with liquid chromatography-high-resolution tandem mass spectrometry, Anal. Bioanal. Chem. 407 (2) (2015) 405−414.

[66] D.T. Usmanov, M.K. Mandal, K. Hiraoka, S. Ninomiya, H. Wada, M. Matsumura, S. Sanada-Morimura, H. Nonami, S. Yamabe, Dipping probe electrospray ionization/mass spectrometry for direct on-site and low-invasive food analysis, Food Chem. 260 (2018) 53−60.

[67] Y. Yang, J. Deng, Z.-P. Yao, Field-induced wooden-tip electrospray ionization mass spectrometry for high-throughput analysis of herbal medicines, Anal. Chim. Acta 887 (2015) 127−137.

[68] A. Li, M.R.L. Paine, S. Zambrzycki, R.B. Stryffeler, J. Wu, M. Bouza, J. Huckaby, C.-Y. Chang, M. Kumar, P. Mukhija, F.M. Fernández, Robotic surface analysis mass spectrometry (RoSA-MS) of three-dimensional objects, Anal. Chem. 90 (6) (2018) 3981−3986.

CHAPTER 7

Design and construction of paper-spray ionization/mass spectrometry

Cheng-Huang Lin, Yea-Wenn Liou
Department of Chemistry, National Taiwan Normal University, Taipei, Taiwan

Contents

Ambient Ionization Mass Spectrometry in Life Sciences
ISBN 978-0-12-817220-9
https://doi.org/10.1016/B978-0-12-817220-9.00007-2

7.1 Introduction

The use of a triangular-shaped section of chromatography paper, so-called paper spray—mass spectrometry (PS-MS) [1], since it debuted in 2010, has created new insights and opportunities in the field of mass spectrometric analysis. It is a variant of electrospray ionization—mass spectrometry (ESI-MS) and, since the ionization process can be achieved at ambient conditions by using a simple piece of paper, this permits the experimental operation process to be simplified thus reducing the time required for a mass spectral analysis. Furthermore, it is relatively easy to use and it has now been realized that the phenomenon of the original ESI can not only be achieved by a metal needle but even the use a wet chromatography paper is possible. Thus, when a high voltage is applied to a "paper" to create an aerosol and to produce ions, a mass spectrometric analysis can be completed. For this reason, a number of readily available materials have been proposed for use and a variety of similar approaches have been developed, including leaf spray [2], tissue spray [3], wooden tip spray [4], conductive polymer spray, and fiber-spray [5—7]. Related to this, high-throughput PS-MS [8], nib-assisted PS-MS [9], hydrophobic PS-MS, and laser-assisted PS-MS [10,11] have also been reported. Even a decade before, it was hard to imagine that a part of an MS analysis could be performed by simply using a piece of paper, including sample concentration, transformation, desorption, and ionization. The technique is particularly efficient when used in conjunction with complex mixtures extracted from blood, urine, saliva, or tissues and has now become increasingly popular and important. In this chapter, the mechanism, methodology, and applications of PS-MS analysis, which are currently in use, are described.

7.2 Paper spray ionization/mass spectrometry

7.2.1 Fundamentals of paper spray

Fig. 7.1 shows a schematic diagram of paper spray ionization. A triangular section of chromatography paper is held in place by a clip that is connected to a high voltage power supply. The sample solution drops on the paper and is then wetted by a solvent, usually a MeOH:H_2O solution. Since the sharp portion of the wet paper is positioned in front of a mass spectrometer and toward the MS inlet, once a high direct current voltage is applied, the charged sample solution is immediately vaporized and then ionized, and as a result, the ions can be analyzed with the mass spectrometer.

Cooks et al. proposed a mechanism for paper spray because they were interested in how that actually functions [12]. They found that there were two distinct spray modes during the process of paper spray. Multiple Taylor cone-jets are initially created in the solvent-rich paper surface, resulting in the production of a range of droplet sizes. Meanwhile, at low solvent flow rates and higher currents a corona discharge is produced, leading to the production of cone-jets. The magnitude of these cone-jets was dependent on the applied voltage and solvent composition, such as viscosity or the ratio of water to organic solvent. The use of 100% water is difficult in paper spray because the sample molecules volatilize with the organic solvent. The vapor pressures of liquid methanol and ethanol (at 20°C) are 13.02 and 5.95 kPa, respectively. The rate of evaporation of methanol is greater than ethanol. Thus, the use of an additional organic solvent, usually methanol, is preferred to ethanol. Finally, there is no observable deformation when the chromatography paper is wetted with a solution, and, after the spray solution was dried, the paper triangle could be reused. Reproducible spectra can be obtained, even when the solution is reloaded.

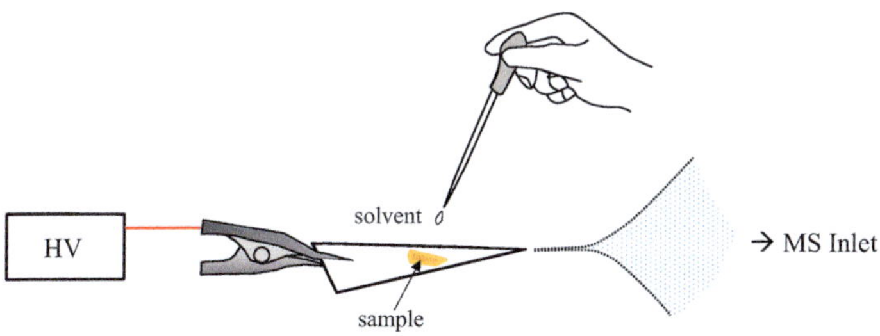

Figure 7.1 Schematic diagram of the paper-spray ionization.

7.2.2 Characteristic of the type of paper

The characteristic of the paper itself is also an important factor. Cooks et al. investigated the optimal type of papers [13]. In their study, four types of filter paper, glass fiber paper, and chromatography paper were examined, in which a MeOH:H_2O solution of cocaine (10 μL, 200 ng/mL) was selected as the model sample solution. The findings indicated that the optimal performance occurred when a chromatography paper was used, whereas the poorest performance was observed for glass fiber paper.

In Fig. 7.2, the inset photographs show scanning electron microscope (SEM) images of gampi paper (A) and (B) chromatography paper, respectively, and show that the surface comprised of disorganized fibers. In fact, wood is the major raw material for the production of paper-based products, including chromatography paper, in the pulp and paper industry. The Japanese have preserved traditional papermaking methods and many types of traditional Japanese paper, the so-called "Washi" are still available. "Washi" is produced by a process that is similar to that of ordinary paper, but relies heavily on manual methods. In making "Washi" various raw materials (such as hemp, ramie, and bamboo) are used, depending on the

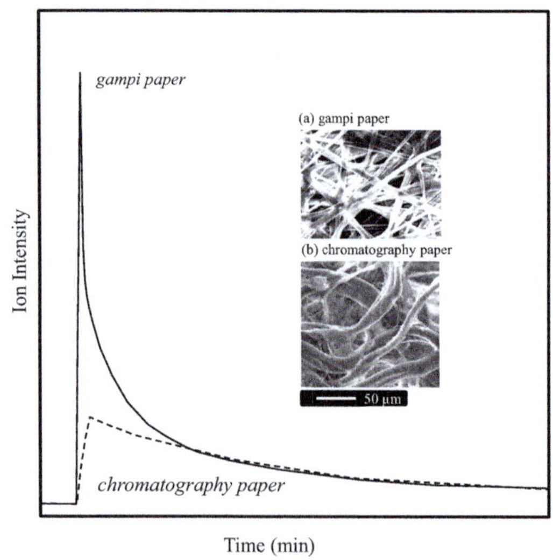

Figure 7.2 Relationship between ion intensity and the period of successive ion occurring (time; s) for chromatography paper (*dashed line*) and gampi paper (*solid line*) were used for paper spray–mass spectrometry, respectively. The inset photographs show scanning electron microscope images of a gampi paper (A) and (B) chromatography paper, respectively.

intended use of the paper. For this reason, "Washi" is generally tougher than ordinary paper made from wood pulp. In ancient times, one types of "Washi," gampi paper, was known to be the finest art paper available. It has a smooth, shiny surface and has a natural cream color (light yellow), with a unique desirable luster. Fig. 7.2 shows the relationship between ion intensity and the period of successive ion occurring (time; s) when chromatography paper (dashed line) and gampi paper (solid line) are used for PS-MS, respectively, although these numbers will be different when the solvent or voltage are changed. In the case of chromatography paper, the ion intensity decreased very slowly. About ~100 seconds was required for the ion intensity to drop to one half of the maximum, after the sample solution was applied to the paper. However, when gampi paper was used, the value for the half maximum time was only 1.5 seconds. It is clear that, after applying the sample molecules on a thin and tough paper, they are translated and evaporated immediately, and, as a result, more sample molecules can be detected within a very short period of time. Various types of paper, including cicada paper, tengujou paper, and glassine paper, were investigated and the findings show the gampi paper proved to be superior to the others [14]. Gampi paper is comprised of more uniform, directional, straighter fibers that are smaller in diameter and more compact. However, gampi paper is made by hand in the traditional manner, which may result in differences in the final product, depending on the location where it is manufactured. Gampi paper can be purchased from PAPER NAO (Japan; http://www.papernao.com). However, it may not be easily obtained in other countries. Instead of gampi paper, the use of synthetic "paper", i.e., fibers, was also investigated to confirm whether they result in an effect similar to gampi paper.

7.2.3 Tip angle

The angle of the paper tip plays a key role in the generation of the spray [15]. The sharpness of the portion of the tip of the triangular paper has a substantial effect on the ionization efficiency. It has been reported that, if the angle of the paper tip is changed, the total spray current and the intensity of the electric field at the tip would all vary correspondingly, leading to differences in signal intensity. It was found that when the degree is sharper than 30 degrees, the S/N ratios are improved dramatically, indicating that the corona discharge also plays an important role. The optimized degree is 10- 20 degrees [16]. It is also noteworthy that hollow fiber spray MS was recently developed [6]. It is very surprising that a cut flat end on a porous hollow fiber caused the production of a Taylor cone. It should be noted that the electrospray is usually formed from a sharp tip, such as a needle, pipette tip, or the tip of a triangular-shaped chromatography paper, where the opposing forces of surface tension and electric field produce a cusp-shaped Taylor cone. Once it was discovered that a wet porous hollow fiber could be used for an ESI material, many applications become possible, such as the use of a frit, a wooden tip, a porous material, or even a sponge.

7.2.4 Optimized distance

The ESI process requires an appropriate distance for evaporation and ionization and there is a specific position that provides maximum ion intensity. Tolerance to positioning was also reported, since this is an important factor for judging the ease of use of an ionization source [13], where a paper triangle mounted on a 2D moving stage was used. In that case, the translation distances, in the y-direction and the x-direction, were 8 and 3 cm, respectively, with a 2 mm increment for each step. Solvent was continuously fed to the paper from the bottom of the triangle to maintain the spray. In the meantime, when a high voltage was applied, an intense peak corresponding to the protonated ion appeared and was recorded as a function of the position of the triangle paper tip. The findings show that the peak intensities were normalized and then plotted as a contour map. It was found that a stable high intensity can be obtained for PS when the paper tip is located anywhere in an area of about 5×10 mm (x−y plane). It therefore appears that accurate positioning may not be required for implementing paper spray. After this report, a similar study appeared [17]. Using a 4-chloroamphetamine solution and chromatography paper, the tolerance of paper spray ionization with respect to the position of the paper tip was investigated. As a result, a stable high intensity can be obtained for PS when the paper tip is positioned in an area of about 6×4 mm (x by y), which was in agreement with data reported in the literature. However, two specific spots were found, located on both sides of the mass inlet, respectively, that provided the most appropriate results. The limit of detection (LOD) can be further improved when the above two hot spots are selected.

7.3 Modifications

As mentioned in the section of "Tip angle," many wetted porous materials can be used for PS-MS. A series of materials were investigated or developed for PS-MS and are summarized below.

7.3.1 Leaf spray

Instead of a paper, the use of a piece of a leaf to investigate the chemical constituents of intact plant materials was a clever idea. Fig. 7.3A shows a schematic diagram of leaf spray ionization.

Cooks' group proposed this method, in which the chemical constituents of intact plant material, including living plants, could be directly analyzed. This is a simple spray method that provides real-time information regarding sugars, amino acids, fatty acids, lipids, and alkaloids [2]. They confirmed that a leaf spray spectrum can be acquired directly from a green onion leaf and sucrose and glucose ions were identified. As a result, leaf spray MS has now become a rapid and simple method for the direct analysis of components in fresh leaves, without the need for any sample pretreatment, such as sweet glycosides in an untreated Stevia leaf [18]. Furthermore, this method is particularly useful for the analysis of important plants that serve as medicinal sources, such as Tulsi or Holy Basil (Ocimum sanctum Linn) [19]. Leaf spray is also extremely easy to implement with no need for a nebulizing gas or sample preparation and is suited to the semiquantitative determination of complex materials.

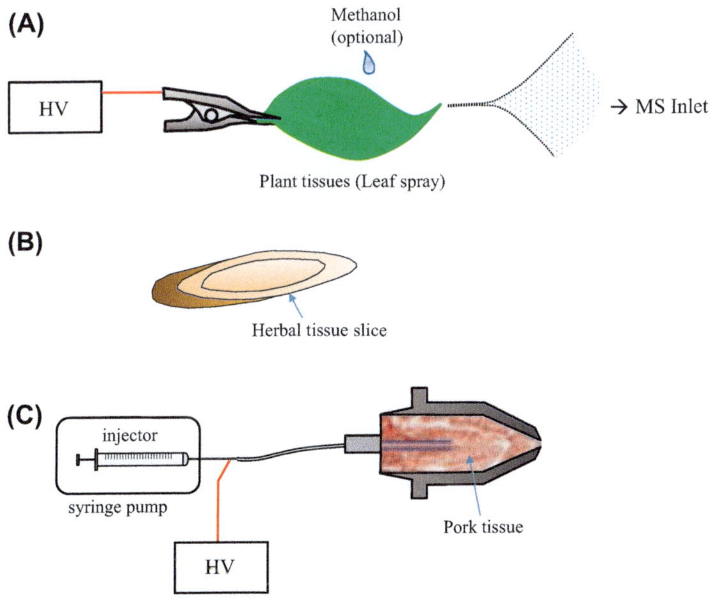

Figure 7.3 (A) Schematic diagram of leaf spray ionization. (B) tissue spray. (C) pork tissue spray.

7.3.2 Tissue spray

Instead of a leaf, an alternate option is the use of an herbal tissue slice, so-called tissue spray, as shown in Fig. 7.3B [20]. For chemical profiling, this is a simple, fast, and very useful method for the in situ analysis of solid tissue samples, such as American ginseng. A raw herb analysis has been reported based on this method. However, a dry slice does not absorb solvent as efficiently as a chromatography paper, so that the distance from the tip of the dry slice to the MS inlet is closer than that for a regular paper spray. In spite of this, analyte molecules can be directly sprayed and ionized from the solvent-wetted slices when a high electrical voltage was applied. It should be very carefully done since a short distance between the ion inlet and the tissue tip or too high an electrical voltage could easily induce an electrical discharge, leading to the production of an unstable ion signal. Using this approach, the sample extraction process can be simplified and the sample transfer step can be reduced. It is quite interesting that differences were found between wild-type and cultivated-type American ginseng. On the other hand, direct spray ionization from a biopsy needle (or fine-needle aspiration biopsy) for the analysis of biological tissue was also reported [21]. A needle biopsy is a diagnostic procedure that is used to investigate superficial lumps or masses. Fig. 7.3C shows a schematic diagram of pork tissue spray MS. In this study, tissue is extracted from the organ by inserting the needle into the organ. The extracted tissue is then partially released from the needle by pushing the plunger along the barrel, and it is retained on the needle tip. A high voltage and spray solvent are applied to the needle and the exposed tissue, respectively, to generate charged droplets carrying chemicals from the tissue into the MS inlet. Thus, any type of tissue, either a harder ginseng slice or soft biological tissue can all be used in the tissue spray ionization MS method.

7.3.3 Fiber spray

In contrast to these "real" papers (chromatography paper or office A4 paper), a series of synthetic "papers" were prepared by means of an electrospinning technique [22–28] and used in PS-MS. These synthetic "papers" are thin synthetic fibers, including a microarray membrane and nanofibers, which are prepared from polycarbonate, polylactic acid, and poly-L-lactic acid (PLLA), respectively, by means of a coaxial electrospinning technique. The findings show that, compared to regular chromatography paper, the LOD can also be improved when a PLLA nanofiber is used; in this study, 3,4-methylenedioxy-*N*-methylamphetamine (MDMA), caffeine, and ketamine were selected as the model samples, respectively [14]. This is because this type of paper-like nanofiber is also very thin, tough, and hydrophobic, which permits ionization to occur within a very short period. Among the synthesized papers, the PLLA nanofiber was superior to polylactic acid, polycarbonate, polyvinylidene fluoride or nylon.

7.3.4 Wooden tip spray

As described in Chapter 6, the use of a disposable wooden tip (toothpick) for the loading and ionization of samples, i.e., a wooden tip spray, was developed [4].

Fig. 7.4A shows a schematic diagram of the wooden tip spray. The sampling method can be done either by normal pipetting onto the tip or by simply dipping the tip into the sample solution. The photograph (A) in Fig. 7.4 shows an image of the tip of a wooden tip. The porous nature and hydrophilic property of the wood tip allows the sample solution to effectively adhere to the tip. When a high voltage is applied, durable ion signals can be produced and desirable mass spectra can be obtained. It is particularly useful for slurried or powdered samples. Another advantage is that hard

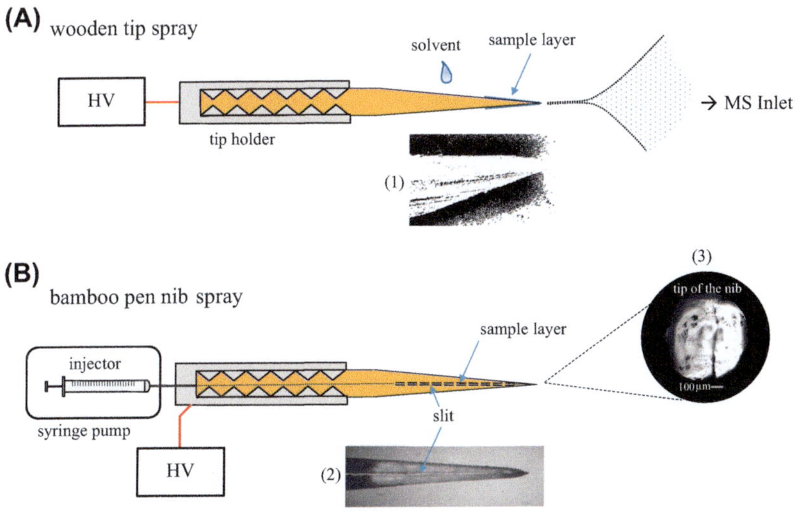

Figure 7.4 (A) Schematic diagram of the wooden tip spray; photograph (1) shows an image of the tip of a wooden tip. (B) schematic diagram of the bamboo pen nib spray; photographs (2) and (3) show the scanning electron microscope image of the tip and the top view of the tip, respectively.

portions, which enable sampling from specific locations, such as meat or vegetables, can be included in the slim and hard properties of the wooden tip. In order to enhance the analytical capability of a solid substrate, several methods for surface modification have recently been reported [29], including the treatment of wooden tips with hydrophobic ($-C_{18}$), basic ($-NH_2$), and acidic ($-SO_3H$) functional groups. Various sampling methods, i.e., extractive sampling and direct loading, were examined in an attempt to understand the analytical properties of solid-substrate ESI-MS. For the direct loading method, the results showed that analytes with weak interactions with the solid-substrate surface could be readily incorporated in the spray for detection. In contrast to this, for the extractive sampling method, analytes that are strongly retained on solid-substrate surface could be selectively enriched and detected. After sample loading, a washing step can be used to remove unbound components, thus effectively reducing the level of interfering substances.

7.3.5 Bamboo pen nib

Attempts have also been made to physically improve the wooden tip. In fact, if a slit in the middle of a toothpick is created, as a nib, the use of bamboo is better than wood. Fig. 7.4B shows a schematic diagram of a bamboo pen nib spray [30]. Herein, a bamboo pen nib was shaped and used for sample loading and ionization; the sample solution was loaded onto the slit by normal pipetting or simply dipping the nib into a sample solution. The sample solution enters the slit of the nib by capillary action. The photographs (b) and (c) in Fig. 7.4B show SEM images of the tip of a wooden tip (b) and the top view of the tip, respectively. It can be seen that a slit exists and the slit functions as a micro reservoir that prevents the sample solution from spreading to the surface. When a high voltage is applied to the bamboo nib, the sample solution is rapidly ejected and ionized toward the mass inlet. Since ionization occurs within a very short period of time, an abundance of ions is formed, leading to a dramatic improvement in the LOD.

7.3.6 Touch spray

A teasing needle is used for teasing tissues for microscopic examination. Fig. 7.5A shows a schematic diagram of touch spray, where a teasing needle (i.e., a small probe) is used to acquire a sample.

In this case, a field-induced droplet emission occurs when a high voltage is applied. This method is very useful for testing the disease state of tissue,

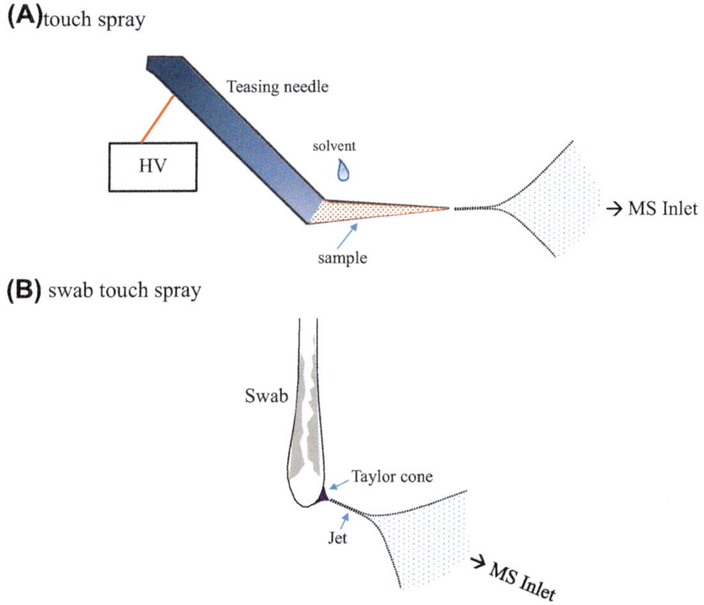

Figure 7.5 (A) Schematic diagram of touch spray, where a teasing needle is used to acquire the sample. (B) Schematic diagram of the medical swab touch spray, where a teasing needle is replaced by medical swab for scraping the sample.

microorganism identification, and therapeutic drug quantitation. This method can be considered to be a variant of probe electrospray ionization (PESI) [31], in which a solid needle or wire is used as sampling probe and ESI emitter instead of capillaries as described in Chapter 6. The distance between the top and bottom position of the needle is usually about 10 mm and the needle or wire is moved up to the highest position, in which a high voltage of about 2.0–3.0 kV is applied to it. Touch spray allows nearly simultaneous chemical derivatization and ionization, generating increased signal intensities for derivatives and enhance qualitative identification. It is useful for in situ analyses of complex samples, such as the detection of lipids in biological tissue [32].

7.3.7 Medical swab touch spray

Fig. 7.5B shows a schematic diagram of medical swab touch spray, where a teasing needle is replaced by a medical swab for scraping the sample [33]. Since a medical swab is a wad of absorbent material wound around one end of a small stick, solvent can be delivered via a fused silica capillary or a syringe pump to keep the swab continuously wet. The Taylor cone can be observed when the portion of swab tip is expanded. This method has been successfully used for the direct drug analysis from an oral fluid. Compounds can be analyzed directly from medical swabs using touch spray MS, and 14 common drugs of abuse were identified in a spiked oral fluid (ng/mL levels). Furthermore, it has been applied for the rapid analysis of an organic gunshot residue from a human hand and various surfaces using commercial and fieldable MS systems [34]. Organic gunshot residues, including methyl centralite (1,3-dimethyl-1,3-diphenylurea) and ethyl centralite (1,3-diethyl-1,3-diphenylurea) can be identified by this method. These characteristic compounds can be considered as forensic analyst tests to determine if an individual has discharged a firearm.

7.3.8 Sponge spray

There are many blood collection devices currently in use since this is an easy way to achieve microsampling. In the clamshells type, it is available in 2- or 4-sampler configurations for flexibility and 10, 20, or 30 μL collection volumes can be selected. Customer uses include preclinical studies, academic research, assisted specimen collection, etc. The cartridge type is a user-friendly format and it is well suited for unassisted remote collection. A tip of a volumetric absorptive microsampling device ("sponge") was reportedly applied to paper spray. This is because the "sponge" is useful for sample preparation, especially for the analysis of clinical samples, including blood, plasma, and urine. As a result, although significant matrix effects were observed, it was possible to quantitate penicillin G, a common antimicrobial, in plasma and in urine, with essentially no sample preparation [35].

7.3.9 Brush spray

A part of a brush removed from a watercolor pen was used for PS-MS. This idea originated from an ink cartridge of a calligraphy brush.

Fig. 7.6A shows a schematic diagram of the brush spray derived from a calligraphy-brush-style synthetic hair pen [36]. Fig. 7.6B shows a schematic drawing of the brush kit; the seal cap, fixed cover, solvent cell, and the body were all prepared by means of a 3D printer. The kit is economical, disposable, and makes sampling easy. When the seal cap is installed, it can be used for the purpose of sampling to measure the nature and level of any type of chemical contaminant. The solvent cell for loading can hold 0.4 mL of methanol. In fact, it was possible to continuously rinse the nylon brush when methanol (at a rate of 6 μL/min), as an auxiliary liquid, was passed

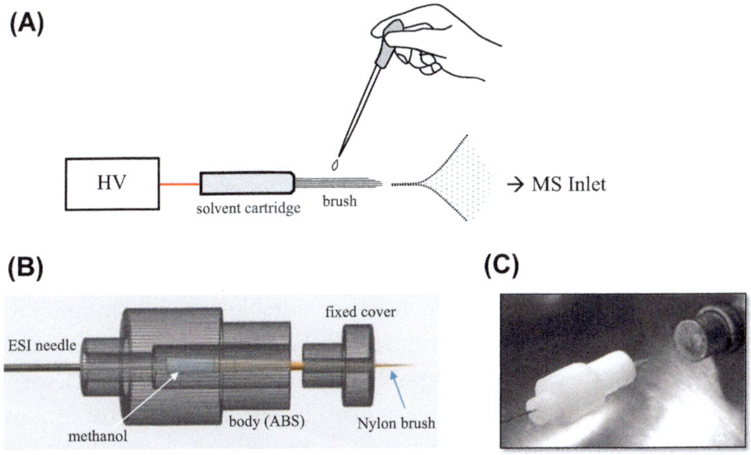

Figure 7.6 (A) Schematic diagram of the brush spray derived from a calligraphy-brush-style synthetic hair pen. (B) Schematic drawing of the brush kit; the seal cap, fixed cover, solvent cell, and the body were all prepared by a 3D printer. (C) Photograph of the actual situation at the mass inlet when a sampling/ionization kit was used.

through the ESI needle. An appropriate distance for the brush tip to acquire maximum ion intensity is an important factor in this method. Fig. 7.6C shows a photograph of an actual mass inlet when the sampling/ionization kit was used. The kit is very light and can be held by an ordinary stainless ESI needle in which a high voltage is applied. The ESI needle was attached to a metric XYZ translation stage, and for this reason, axial alignment was readily achieved. To investigate the sampling and ionization effect, various types of hair samples, i.e., nylon, from horse and weasel, respectively, were examined, and the findings indicated that a nylon brush provided the most satisfactory results. This is because nylon brush hairs are very smooth, tough, and hydrophobic, which permit ionization to occur within a very short period of time. It was found that, when a 3 μL sample solution was loaded onto the brush by normal pipetting, a major peak appeared during ~0.1 seconds, when a high voltage was applied. In contrast to this, when chromatography paper was used, the ion intensity was weaker and decreased very slowly (up to ~5 minutes). When vitamin B2 was used as the test sample, the linear range of vitamin B2 was found from 0.1 to 100 ppm and a correlation coefficient of 0.98 was found. The optimum ESI voltage for the procedure was also investigated. The ideal distance between the tip of the nylon brush and the mass inlet was determined to be 8 mm. This arrangement, however, depends on different conditions, and the actual value would likely change, depending on the analytical parameters in use. Another type of brush spray MS was developed based on modifying thermal desorption electrospray ionization MS (TD-ESI) [37], as shown in Fig. 7.7A [38].

In the case of TD-ESI, a metallic sampling probe (60 mm long, 2.5 mm in diameter) was used. The sampling probe was dipped into the sample solution, then removed rapidly and then inserted promptly into the TD-ESI source, so that alignment skills are not required. However, the probe is composed of metal tubing, which is not suitable for collecting a sample from soft body tissue. Furthermore, it is not applicable for use in conjunction with thermal decomposition products because the desorption temperature of the TD-ESI source is usually set at 280°C. In contrast to TD-ESI, a watercolor pen (brush) was used for brush spray. The analytes are generated by vaporization with the solvent. In the beginning, the watercolor pen (brush) was rinsed with ethanol or methanol, and the

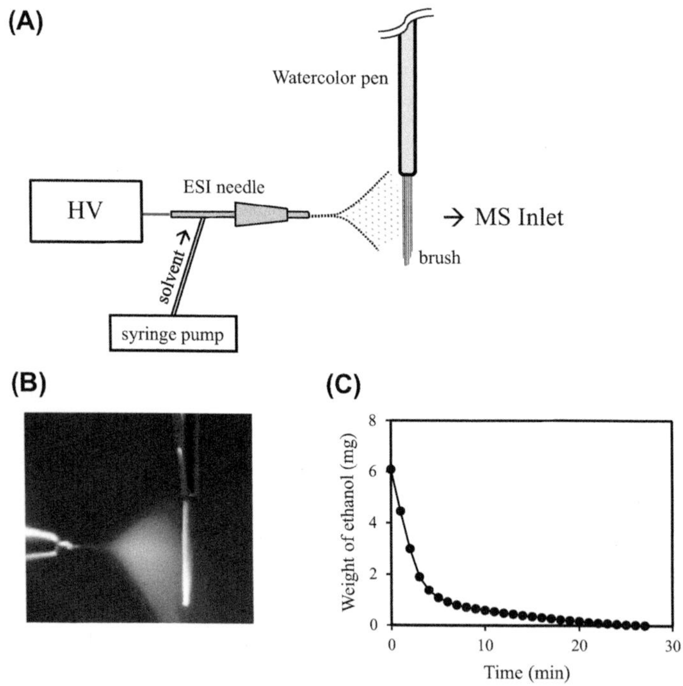

Figure 7.7 (A) Schematic diagram of the brush spray. (B) Actual position of the watercolor pen located between the electrospray ionization needle and the mass inlet (right side). (C) Relationship between time (min, X-axil) and the weight of ethanol absorbed by a watercolor pen brush (mg, Y-axil), respectively.

analytes were collected by sweeping them onto the solvent-saturated brush. The collected analytes, along with the solvent, rapidly evaporate and escape from the brush surface and are then ionized by an ESI plume. The ESI needle tip approaches from the other direction to produce electrosprayed/charged droplets from a regular ESI stainless needle. The watercolor pen can provide both functions, i.e., sampling and volatilization of the samples for ionization. Once the molecules (mostly neutral) evaporate from the surface of the brush, they encounter the electrospray plume and ionization occurs under ambient conditions. Photograph (B) in Fig. 7.7 shows the actual position of the watercolor pen located between the ESI needle and the mass inlet (right side). The position of the pen brush is very important in acquiring sufficient ions for detection. Fig. 7.7C shows the relationship between time (min, X-axil) and the weight of ethanol absorbed by the watercolor pen brush (mg, Y-axis), respectively (conditions: temperature, 26°C; relative humidity, 41%). It was found that ethanol evaporates very rapidly from the brush surface, so that we were able to routinely complete MS experiments within 3 minutes after sampling, thus permitting better ion signals to be acquired.

7.3.10 Hollow fiber spray

Thus far, all of the PS-MS methods are difficult to combine with an online concentration technique. An online microextraction kit for PS-MS was recently developed, in which a hollow fiber was used [7].

Fig. 7.8 shows a schematic diagram of this system. The inset photograph shows the actual position of the kit located between the ESI needle and the mass inlet. The kit consists of a section of a hollow polypropylene (PP) fiber (3.5 cm in length), a glass serum bottle (2 mL), a seal cap with a gasket, and an electromagnetic stirrer. The use of a stirrer is very useful for improving the extraction efficiency of the setup. The end of the PP hollow fiber was sealed with a needle-nose plier to prevent the extraction solvent from leaking. In order to demonstrate the efficiency of online microextraction, a 2 mL sample of a glyphosate solution was used as the model sample and was placed in the 2 mL serum bottle for extraction. In this case, the optimized pH value was 12.0. The concentrated analytes are vaporized and escape, along with the mixed solvent, through the porous surface. Electrosprayed/charged droplets can also be produced from a standard ESI stainless needle. Once the materials that evaporate from the surface of the hollow fiber (mostly neutral) meet the electrospray plume, ionization occurs under

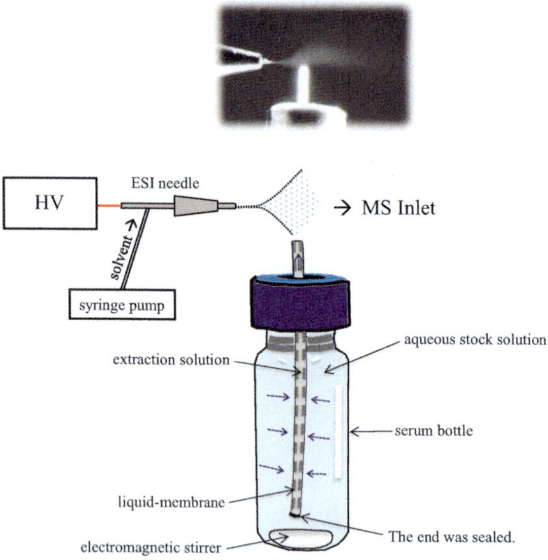

Figure 7.8 Schematic diagram of the online microextraction kit, in which a hollow fiber was used.

ambient conditions. The methodology is simple and economical and is suitable for use in screening for low-level pesticides or drugs.

7.3.11 High-throughput paper spray

In PS-MS, cutting the chromatography paper into a small triangle piece for each experiment is somewhat of a task. Frequently and routinely clipping the triangle paper is not very convenient.

Fig. 7.9 shows a schematic diagram of a high-throughput device based on the PS-MS [39]. The paper substrate contains multiple triangles for loading samples and, as a result, this system represents a rapid method for the analysis of chemicals in complex mixtures, including therapeutic drug monitoring and food safety screening. In the case of bovine blood analysis, a blood sample (spiked with sunitinib and nicotine) was used as a model sample. Under optimized conditions, a relatively rapid analysis speed of 7 seconds/sample was achieved with relatively good limit of quantitation (LOQ) (1 ng/mL).

7.3.12 Catalytic paper spray

In order to improve the ionization efficiency for compounds that are difficult to ionize in the ESI process, a catalytic paper spray method using Pt-nanoparticle-decorated nanotubes (Pt NP-NTs) was examined [40]. Pt

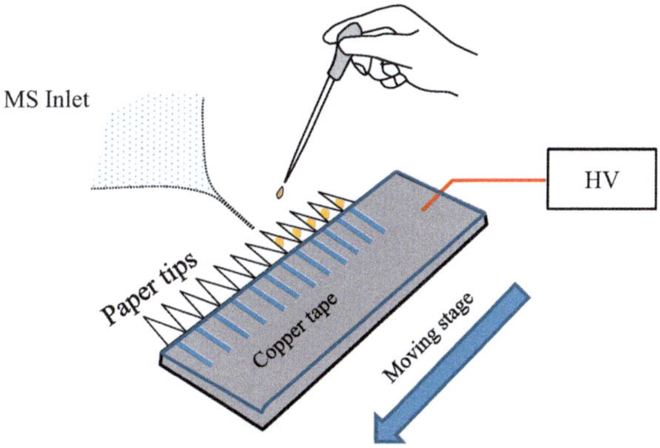

Figure 7.9 Schematic diagram of a high-throughput device based on paper spray technology.

NP-NT is a type of highly active catalyst that is used for the reduction of 2,4,6-trinitrotoluene (TNT) to 2,4,6-triaminotoluene (TAT) and the subsequent easy detection of the product by in situ ambient ionization MS.

Fig. 7.10A shows a schematic of the process for accumulating TNT vapor on PtNT-coated paper. The roughness on the surface provides more surface area and hence more catalytically active areas. Fig. 7.10B shows a

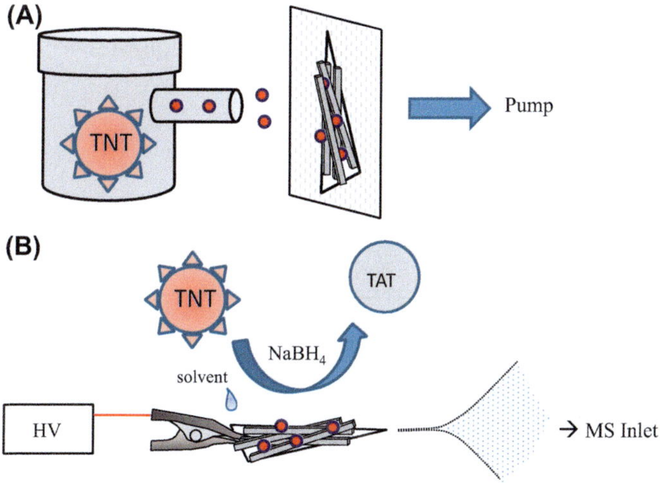

Figure 7.10 (A) schematic of the process for accumulation of 2,4,6-trinitrotoluene (TNT) vapor on Pt-decorated nanotube-coated paper. (B) Schematic of a typical paper spray ionization process, showing that, after eluting the TNT-adsorbed Pt-nanoparticle-decorated nanotube-coated paper with $NaBH_4$ solution, it is converted into triaminotoluene.

schematic of a typical paper spray ionization process, showing that, after eluting the TNT-adsorbed Pt NP-NT-coated paper with a $NaBH_4$ solution, it is transformed into TAT, in which a typical mass spectrum can be observed with a peak corresponding to TAT at m/z 138. A sensitivity in the range of 4.5 ng was demonstrated.

7.3.13 Hydrophobic paper spray

While the majority of paper used in paper spray absorbs solvents rapidly, the use of a hydrophobic paper was also investigated [10].

Fig. 7.11 shows a schematic diagram of this device. In this case, the hydrophobic paper can be obtained by modifying the surface by exposure to trichloro (3,3,3-trifluoropropyl) silane vapor, leading to a hydrophobic oligomeric silylated layer. The reaction by-product (gaseous HCl) can be removed by vacuum in situ. The difference between an untreated and a treated (hydrophobic) paper can easily be observed based on the interaction of a dye solution. The use of the hydrophobic paper spray prevents the organic solvent from spreading beyond the point where the biological sample (e.g., blood) was deposited. As a result, an online liquid/liquid extraction of small organic compounds from a drop of biological fluid placed on the dry hydrophobic paper can be achieved. This method is rapid and reaches completion under 1 minute, and in situ ESI can be used for the more efficient detection of extracted molecules. Furthermore, with no sample pretreatment, the direct analysis of small volumes of biofluids is

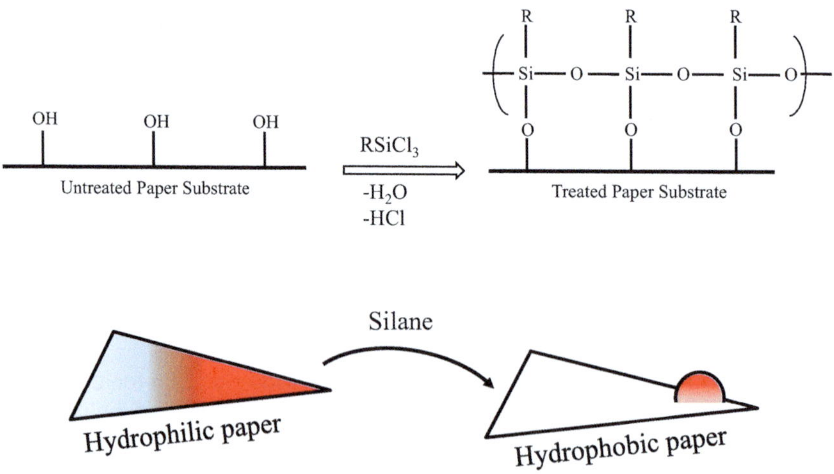

Figure 7.11 Schematic diagram of the hydrophobic paper spray.

possible. High sensitivity and quantitative accuracy was achieved for the direct analysis of illicit drugs in 4 μL of raw blood, serum, and whole urine. This study has extended to monitor the activity of alanine transaminase enzyme, a key biomarker for the detection of liver injury in patients (with HIV and tuberculosis) who typically take several types of drugs.

7.3.14 Conductive polymer spray

A conductive polymer spray ionization method was recently reported [5]. It is a very useful method for the direct mass spectrometric analysis of hydrophilic drugs, saccharides, peptides, and proteins in biofluids.

Fig. 7.12 shows a schematic diagram of this device. In this study, carbon nanotubes (CNTs) are introduced into poly(methyl methacrylate) (PMMA) to fabricate a conductive composite substrate. Since the conductive polymer is hydrophobic, it absorbs less hydrophilic compounds. When aqueous biofluid samples are loaded, they also diffuse less on this nonporous surface. Only 1.0–2.0 μL of solvent is needed to extract the components in a dried biofluid spot and to form charged microdroplets (4.5 kV high voltage applied). As a result, the ion intensity of the target compound from the conductive polymer support is significantly higher than that for paper spray, typically by a factor of 20–100 when sunitinib, acarbose, melamine, and angiotensin II were tested. These results suggest that the CNTs/PMMA conductive polymer spray has great potential in the analysis of hydrophilic drugs, saccharides, peptides, and proteins in biofluids.

7.3.15 Coupling liquid-phase microextraction

Fig. 7.13 shows a schematic diagram of the steps that are involved in liquid-phase microextraction for the paper spray method [41].

In this case, a 10 mL Hamilton microsyringe (Reno, Nevada) was used in the liquid-phase microextraction. The extraction of analytes was performed in 4 mL vials with magnetic screw caps containing 2 mL aqueous samples. The needle of the syringe containing 2 mL of a water immiscible

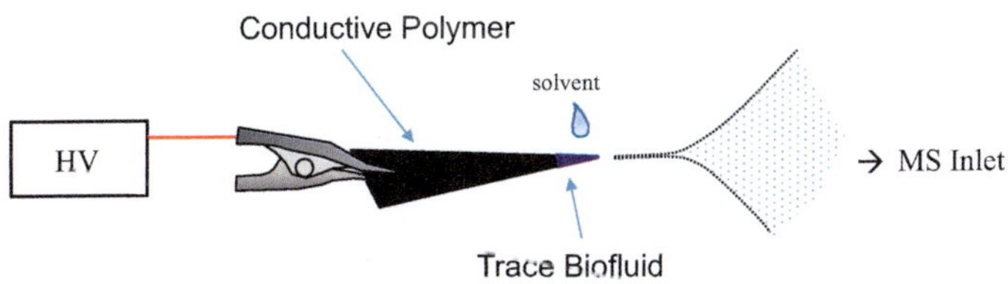

Figure 7.12 Schematic diagram of the conductive polymer spray.

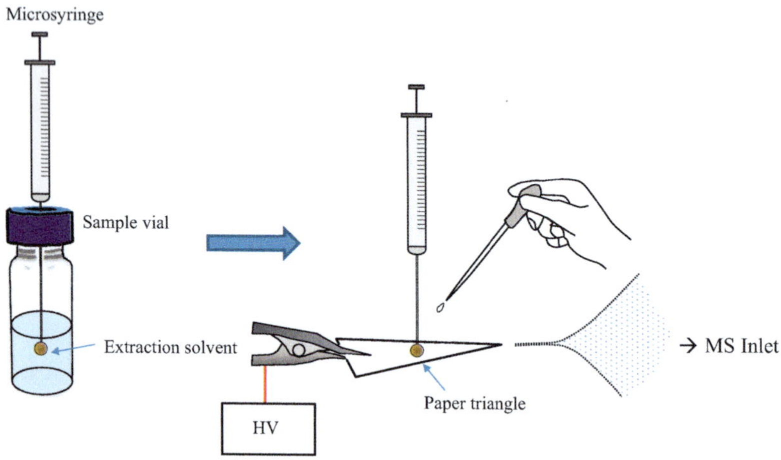

Figure 7.13 Schematic diagram showing the steps of liquid-phase microextraction used for paper spray.

organic solvent was penetrated through the septum of the vial until the tip protruded below the meniscus of the sample solution. The plunger was depressed to form a drop which was suspended from the needle tip. After extraction for a predetermined period with continuous magnetic stirring, the drop was drawn back into the syringe and the needle was then removed from the sample vial for subsequent analysis.

7.3.16 A cartridge with solid-phase extraction

Fig. 7.14 shows a schematic diagram of the cartridge design for use in solid-phase microextraction used in PS-MS (A, sample extraction; B, ionization) [42].

It was designed and fabricated as a prototype paper spray cartridge with an integrated solid-phase extraction (SPE) column. Basically, the cartridge consists of two parts. The bottom part contains two separate recessed regions to hold an absorbent waste pad and the paper spray substrate. In the original studies, the volumes of plasma samples ranged from 10 μL up to hundreds of microliters and were added to the hole in the top part of the cartridge containing the SPE column. The sample would then wick through the SPE column and subsequently onto the absorbent pad located in the bottom part of the cartridge. The target analytes were retained on the SPE column after the sample passed through it. The sample was then allowed to dry, during which a cover was placed over the top part of the cartridge to prevent sample evaporation from the top of the SPE column.

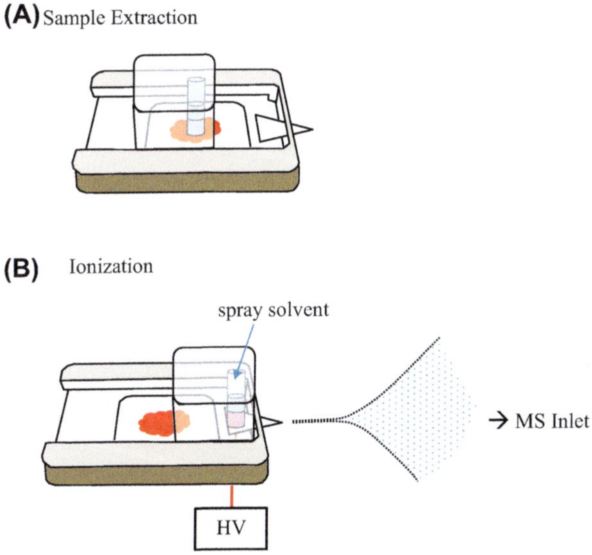

Figure 7.14 Diagram depicting the workflow for paper spray analysis with integrated on-cartridge solid phase extraction.

The cartridge was positioned in front of the inlet to the mass spectrometer, and the extraction/spray solvent was added to the top of the SPE column. The solvent wicked through the SPE column, recovering the analytes in the process and onto the spray substrate by passive capillary action. Ionization was induced directly from the paper substrate by applying a high voltage to the paper through a wire inserted in the bottom part. By spiking the plasma sample with an isotopically labeled internal standard, a linear calibration curve for the drugs was obtained from the LOD to 1 μg/mL, indicating that this method would be useful for quantitative analysis.

7.3.17 Laser assisted paper spray

Polycyclic aromatic hydrocarbons (PAHs) are ubiquitous environmental pollutants that are primarily generated by the incomplete combustion of organic materials (e.g., coal, oil, petrol, and wood). Many of them have toxic, mutagenic, and/or carcinogenic properties and have been linked to cancers in well-established animal model studies. PAHs possess very characteristic UV absorbance and fluorescence spectra, and in general, gas chromatography/mass spectrometry (GC/MS) or liquid chromatography/mass spectrometry (LC/MS) is also suitable for their analysis. Laser assisted paper spray ionization MS was recently reported in this context [11]. This

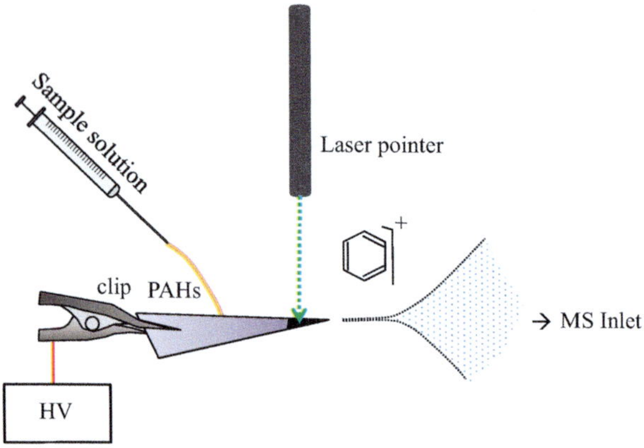

Figure 7.15 Schematic diagram of laser assisted paper spray.

method is a new technique for the analysis of PAHs based on the paper spray MS. A low power laser pointer (532 $\pm$ 10 nm, $\leq$10 mW) was used in this study.

Fig. 7.15 shows a schematic diagram of this method. When the laser is focused on a triangular-shaped piece of paper along with a high voltage, it facilitates ionization. Field-induced distortion of the potential well can be large in paper spray since the fibers that make up the paper are separated from one another by tens of nanometers, and consequently, the PAH molecules are subjected to very large electric fields in the order of 10^7 V/cm. It was speculated that the mechanism for ionization involves field assisted photoionization. Ionization occurs from their distorted electronic states of reduced ionization energy caused by the laser. The solution containing the PAHs can be continuously passed through a fused silica capillary, using a syringe pump, at a preferred infusion rate. PAHs do not typically undergo ionization when ionization methods involving solvent sprays are used; however, the use of a laser promises to result in enhanced ionization and a high signal intensity. A clear enhancement in signal intensity was seen in the total ion chronogram for most analytes when a laser was used.

7.3.18 MS combined with Raman spectroscopy

The use of a paper substrate, the surface of which had been inkjet printed with silver nanoparticles to enhance Raman spectroscopy, was reported [43]. This is because traditional forensic tests involve complex separations, such as LC and GC or various types of extraction techniques, are typically coupled with MS, all of which are time-consuming. Even though some ambient ionization techniques have been adopted by the forensic community to speed up analysis time, there is still significant resistance to discarding chromatography, since most forensic analyses require both an identification and a confirmation technique. To overcome these, a study of paper spray MS combined with a surface enhanced Raman spectroscopy was completed. The paper substrate was inkjet printed with silver nanoparticles and the same substrate can also act as the paper substrate for paper spray MS. The coupling of surface-enhanced Raman scattering (SERS) and paper spray ionization indeed creates a rapid, forensically feasible combination.

7.4 Applications

7.4.1 Dried blood spots

The first application of paper spray was reported by Cooks and Ouyang et al., dealing with an analysis of therapeutic drugs in whole blood [1]. It was quite astonishing to see that such a complicated procedure in MS analysis could be solved in such a simple manner. The authors successfully analyzed a blood spot, in which 0.4 μL of whole blood containing 4 μg/mL of imatinib was identified and quantified by the selected reaction monitoring (SRM) transition at *m/z* 494 and *m/z* 394. Furthermore, a quantitative analysis was performed by spiking with imatinib (62.5 ng/mL ~ 4 μg/mL) and its isotopomer [D_8] imatinib (1 μg/mL). These investigators also reported on another type of illicit drug detection [13]. The PS-MS method was applied to therapeutic drug monitoring with whole blood samples and to illicit drug detection in raw urine samples. As a result, limits of detection of 50 ng/mL (or 20 pg absolute) were achieved for atenolol in bovine blood. A 100 pg amount of heroin distributed on a surface and agrochemicals on fruit peels were also detectable. Furthermore, online derivatization with a preloaded reagent was demonstrated for the analysis of cholesterol in human serum. Regarding in-field sample analysis, PS-MS has considerable advantages in that blood from a finger prick can be used to measure drugs in whole blood at low cost and in a short time. Some studies may require specialized blood collection devices that can be used in remote areas. Based on research by Cooks et al., four different devices for handling dried blood were successfully tested for use in rapidly quantifying imatinib and *N*-desmethyl imatinib [44]. Hence, clinical samples can be collected and analyzed in a remote location, and these data can be considered as proof for in-field blood collection and subsequent rapid laboratory analysis. On the other hand, although PS-MS is useful for the rapid analysis of small molecules from blood spots with minimal sample preparation, plasma is often the preferred matrix for bioanalysis. A paper spray cartridge was developed to meet this need [45]. This cartridge contains a plasma fractionation membrane that can perform automatic on-cartridge plasma fractionation from whole blood samples. Three different membranes were evaluated for lysis and drug binding, resulting in the development of a disposable paper spray MS cartridge. It can be used to analyze a dried plasma sample obtained from a drop of whole blood. Reports have appeared on the analytical characteristics of paper spray as related to analyses of dried blood spots [46–48]. One such example involved the rapid quantitation of acylcarnitine derivatives in serum and whole blood [47]. Dried serum and whole blood samples containing a mixture of 10 acylcarnitine derivatives at various concentrations were directly analyzed without any sample pretreatment, separation, or derivatization. The findings showed that the composition of the spray solvent used was a critical factor. For blood samples that contain more matrix components, acetonitrile/water (90:10, v/v) was a much more suitable spray solvent. For serum samples, a spray solvent composed of methanol/water/formic acid (80:20:0.1, v/v/v) was used because this mixture gave the best signal intensity. Another case involves an investigation of protein–drug interactions by comparing the ratio of propranolol and atenolol [46]. The weaknesses and strengths of several procedures for implementing paper spray ionization for the quantitation of small molecule pharmaceuticals in dried blood spots were recently examined [47]. Studies of eight oncology drugs, including pazopanib, tamoxifen, imatinib, cyclophosphamide, paclitaxel, irinotecan, docetaxel and topotecan, were also reported [48], in which the limits of detection ranged from 0.5 to 17 ng/mL.

7.4.2 Tissue sample analysis

The most significant benefit of tissue spray MS is that a very small tissue sample can directly provide highly specific information concerning the molecule contained by it. A quantitative determination of the bulk molecular concentrations of β-agonists in pork tissue samples was recently reported [49]. This was completed by direct internal extractive ESI-MS, without any sample pretreatment. Salbutamol, clenbuterol, ractopamine, terbutaline, tulobuterol, and brombuterol were investigated in pork tissue samples. A single sample analysis required only 1 minute and the LOD varied from 0.002 μg/kg for salbutamol to 0.006 μg/kg for tulobuterol. The relative standard deviation (RSD) of quantitation was in the range 6.5%–11.3%. The analytical results were validated by GC/MS and LC/MS, showing the accuracy rates of 92–105%.

7.4.3 Studies of metabolites

The use of a molecularly imprinted polymer-coated paper substrate for paper spray MS was reported [50]. With its specificity, this type of new substrate was applied to the detection and quantification of dopamine, sarcosine, and butyric acid in synthetic human urine without derivatization or a complex sample pretreatment. This is important in terms of evaluating several pathological conditions including heart disease, stress, neurological disorders, cancerous tumors, and AIDS based on the urinary levels of these metabolites.

7.4.4 Biomarkers

Conventional chromatography-MS methods are useful for quantitating biomarkers in untreated biological samples, even though it is time-consuming in terms of sample preparation and chromatography separation processes. However, in clinical diagnosis a method for the rapid quantitation of biomarkers in freshly obtained biological samples would be highly desirable. An ultrasensitive biomarker quantitation method based on membrane ESI-MS was recently reported [51]. This method provides both vertical and horizontal chemical separations that could be achieved in real time. As a result, disregarding the matrix effects from the presence of salts, small molecules, and macromolecules, the ultrasensitive detection of cytochrome C (>500-fold sensitivity improvement) in untreated urine samples was demonstrated.

7.4.5 Signature ink analysis

The analysis of signature ink on important documents or checks is important. However, ballpoint pen inks are viscous, thus prohibiting the use of HPLC and GC for their analysis. Furthermore, they are manufactured from a wide variety of materials that exhibit different chemical properties, which introduce additional complexities to the experimental work. The use of paper spray MS for the direct analysis of black ink was recently reported [52]. In that study, some commercially available ballpoint pens were classified based on the ink used, and six commonly worldwide utilized brands were then selected as the model samples, including Pilot, Paper Mate, Faber Castell, Pentel, Compactor, and Bic. After exposure to light (artificial aging), changes in the ink chemical composition were monitored by paper spray MS. As a result, it was possible to determine whether the fresh ink lines had overlapped. The analysis of old ink from archived documents and the detection of alterations (simulated forgeries) on archived documents were also accomplished.

7.4.6 Security

The detection of TNT in environmental samples is an important issue due to its toxicity, carcinogenic, and mutagenic properties that pose a high environmental risk. In the past, photofragmentation-laser-induced fluorescence [53] or homogeneous fluoroimmunoassay methodology was used to accomplish this [54]. In fact, TNT is not easily ionized by ESI due to the lack of easily ionized functional groups. However, methods have recently been developed to overcome this, including desorption atmospheric pressure chemical ionization [55], desorption electrospray ionization [56], and direct analysis in real time (DART) [57]. All of these methods generate M^- ions of TNT. As described above, a vapor-phase detection of TNT by modifying the paper with Pt NP-NTs was recently reported [40]. By using the Pt NP-NT as a heterogeneous catalyst in paper spray MS, the nitro ($-NO_2$) groups can be converted to easily ionized amine ($-NH_2$) groups in situ, leading to TNT being readily detectable by MS. After eluting the TNT-adsorbed Pt NP-NT-coated paper with a $NaBH_4$ solution, TNT was transformed into TAT and the peak for TAT at m/z 138 could be easily observed in a typical mass spectrum. Sensitivity in the range of 4.5 ng can be achieved.

7.4.7 Food analysis

In food analysis, one of the most challenges is sample preparation, especially when MS is used as an analytical method. This is because foods are very complex matrices in which some natural components can negatively influence the analysis of the targeted compounds, especially in the detection of contaminants. In addition to food safety, food quality is one of the main concerns of the modern food industry. Food quality encompasses multiple factors, since food authentication and the adulteration of food include food ingredients, such as lipids, proteins, oligosaccharides, vitamins, and carbohydrates, as well as additives. Consequently, the evaluation of food quality is usually a very complex task in which multiple aspects need to be considered to achieve an appropriate food quality. Food composition, aroma, flavor, or nutritional properties are among the most important features that need to be evaluated in food quality assessments. Considering analytes prior to the use of MS, traditional methods, including liquid–liquid extraction or SPE cannot be avoided. More recently, the use of solid-phase microextraction, pressurized liquid extraction, and QuEChERS (quick, easy, cheap, effective, rugged, safe) has become popular [58]. Paper spray MS would be predicted to provide an easy way for accomplishing this. A high-throughput mass spectrometric method for the simultaneous detection of Sudan I, II, III, IV, and Para-Red azo-dyes in foodstuffs was reported [59]. Based on the use of paper spray MS and a deuterium-labeled internal standard, Sudan dyes can be determined at threshold values of above 1 ppm. Paper spray combined with tandem MS is a more powerful approach to determining various contaminants, including clenbuterol, melamine, plasticizers, and Sudan Red in various foodstuffs (e.g., meat, milk, sports drinks, and chili powder). Illegal beverage dyes were also detected by paper spray MS [60]. Quantification of six illegal colorants in beverages, i.e., soft beverages, energy beverages, alcoholic beverages, teas, and fruit juices, was achieved by paper spray ionization combined with tandem mass spectrometry (MS/MS). The samples were diluted in methanol, and isotopically labeled internal standards were added, which allowed detection of particular dyes with analysis times of less than 1 minute per sample. The LODs were below 1.5 ng/mL and LOQs were below 5 ng/mL.

7.4.8 Caffeine

Many people cannot do without caffeine in their daily life. It would be quite interesting to have a method for the rapid analysis of caffeine in beverages and drugs. A simple and rapid method based on paper spray MS for the determination of caffeine in commercial beverages and drugs was developed, in which multiple reaction monitoring and isotope dilution method were applied. Based on monitoring the SRM transitions m/z 195 → m/z 138 for caffeine and m/z 198 → m/z 140 for the labeled internal standard, caffeine-(trimethyl-$^{13}C_3$), the LOQ and LOD values were calculated to be 1.2 and 1.6 μg/mL, respectively. Both beverages and drugs were also analyzed with the classic analytical method based on LC-UV measurements, and the results between the two approaches were consistent, thus confirming the reliability of the paper spray MS. In addition, a novel and noninvasive sampling method was recently developed [38]. As the sampling tool, a watercolor pen (brush) was used. Prior to the analysis, the brush was rinsed with ethanol and then used to collect analytes from the eyelids of volunteers. Following this, the brush was quickly moved to the front of the mass inlet, where the ethanol as well as the analytes evaporated very fast and escaped from the brush surface. When the analytes make contact with the ESI plume, which arises from the ESI needle tip, they are ionized and then detected by a mass spectrometer. The findings show that this new technique can be reliably used for the analysis of caffeine and its metabolites in samples obtained from the eyelids of volunteers after they consumed coffee. The methodology is simple and economical, noninvasive, and suitable for use in monitoring the levels of caffeine and its metabolites over an extended period of time.

7.4.9 Drugs of abuse

From the point of view of the screening and confirmation of drugs of abuse on the illicit market, more detailed detection and separation information would be highly desirable. At this time, GC/MS and LC/MS continue to be the most popular and powerful techniques for the analysis of illicit drugs and analogs thereof. Although each of these above methods has certain unique advantages and disadvantages with respect to sensitivity, precision, and simplicity of use, they can be time-consuming when a separation is required. Hence, a rapid and highly accurate screening method would be highly desirable in this area. The capabilities of PS-MS for this were recently demonstrated. It was possible to identify and elucidate the chemical structure of drugs of abuse by Orbitrap and product ion spectra. Heroin, cocaine, methamphetamine, oxycodone, and buprenorphine (therapeutic drug) were characterized by PS-MS. The ranges of quantitation are sufficient to cover the screening range needed for abused drugs and metabolites with good linearity ($R^2 > 0.99$) and acceptable reproducibility (RSD ~ 12%).

7.4.10 Pesticide analysis

Glyphosate, a pesticide that was once considered to be highly safe since its debut in 1974 under the trade name Roundup, has been very effective in killing a wide variety of weeds. It is now one of the most widely used herbicides in the agricultural, horticulture, viticulture, and silviculture sectors, as well as garden maintenance (including home use). However, it has recently been proposed that glyphosate might be associated with certain diseases, including autism, caused by oxidative damage and neurotoxicity, Parkinson's disease, Alzheimer's disease, and various types of cancer. The use of paper spray in this area would be highly desirable because it is a simple and straightforward method for detecting low concentrations of glyphosate. An online microextraction kit based on fiber spray/MS has recently been designed and tested for this purpose [6]. The kit consists of a section of a porous-PP hollow fiber, a serum bottle, a seal cap with a gasket, and an electromagnetic stirrer. The kit is placed between the mass inlet and the ESI needle. Using this configuration, glyphosate can be extracted from a very dilute solution and then evaporates and escapes from the fiber surface. When glyphosate molecules make contact with the ESI plume, which arises from the ESI needle tip, they are ionized and then detected by a mass spectrometer. Using the setup, it was possible to improve the LOD for glyphosate after microextraction by ~1000-fold, resulting in a LOD of 10 ppb. These studies confirm that the paper spray MS is actually a simple, rapid, and robust methodology.

7.4.11 Accelerated chemical synthesis

Derivatization is a common method for making an analyte more easily detected either by fluorescence spectrometry or by MS. On-column or postcolumn derivatization is also a standard method in separation science, such as GC/MS or capillary electrophoresis, especially when a laser is used to induce fluorescence for the analysis. Derivatization is typically performed off-line and is time-consuming. However, it was recently reported that derivatization and other chemical reactions can be accelerated when the reactions occur in electrosprayed droplets [61]. On the other hand, it has also been reported that the haloform reaction was successfully performed based on the paper spray MS [62]. This is a chemical reaction where a haloform is produced by the exhaustive halogenation of a methyl ketone under an alkaline environment. In this study, chemical analysis and accelerated synthesis can be accomplished in a simple paper spray experiment with haloform oxidation.

7.5 Future perspectives

By applying a high voltage, analyte ions can be generated from a wet paper or porous substrate and only a small volume of solvent is needed. Paper spray ionization is now recognized as a rapid MS analysis method for the direct analysis of raw biological and chemical samples. Matrix effects are minimal, and samples obtained by chromatographic separation extraction or other techniques used to purify samples can all be easily applied. It is noteworthy that the method is operated at ambient conditions that this is very useful in terms of sample preparation prior to an MS analysis. The entire procedure originated from a piece of chromatography paper no one thought or believed would lead to leaf spray, tissue spray, fiber spray, wooden tip spray (swab), touch spray, sponge spray, brush spray, hollow fiber spray, all of which were developed as the result of this initial discovery. Methods using modified materials, including catalytic paper, hydrophobic paper, and conductive polymer, have also been established. The use of high-throughput paper spray can speed up the time needed for an analysis. The limits of detection can be dramatically improved by the use of off-line liquid-phase, on-line microextraction by hollow fiber spray, or a cartridge designed for SPE. In some studies, a laser was introduced either for improving ionization efficiency or for use combined with SERS. The scope of applications of this method includes blood (especially dried blood spots), biofluids, and tissue sample analysis. Furthermore, applications in the fields of life science and studies of metabolites, biomarkers, signature ink analysis, security issue, food analysis, caffeine, drugs of abuse, pesticide analysis, and even accelerated chemical synthesis have all been reported. Further applications in the future will clearly be expected.

References

[1] H. Wang, J. Liu, R.G. Cooks, Z. Ouyang, Paper spray for direct analysis of complex mixtures using mass spectrometry, Angew. Chem. Int. Ed. Engl. 122 (2010) 889–892.

[2] J. Liu, H. Wang, R.G. Cooks, Z. Ouyang, Leaf spray: direct chemical analysis of plant material and living plants by mass spectrometry, Anal. Chem. 83 (2011) 7608–7613.

[3] Y. Wei, L. Chen, W. Zhou, K. Chingin, Y. Ouyang, T. Zhu, H. Wen, J. Ding, J. Xu, H. Chen, Tissue spray ionization mass spectrometry for rapid recognition of human lung squamous cell carcinoma, Sci. Rep. 11 (2015) 10077.

[4] B. Hu, P.K. So, H. Chen, Z.P. Yao, Electrospray ionization using woode tips, Anal. Chem. 83 (2011) 8201–8207.

[5] X. Song, H. Chen, R.N. Zare, Conductive polymer spray ionization mass spectrometry for biofluid analysis, Anal. Chem. 90 (2018) 12878–12885.

[6] Y.W. Liou, J.S. Wang, C.C. Chen, C.H. Lin, Development of an on-line microextraction method for use in fiber-spray/mass spectrometry, Int. J. Mass Spectrom. 421 (2017) 178–183.

[7] Y.W. Liou, S.H. Wang, C.H. Hsu, Y.T. Hsu, C.H. Chen, C.C. Chen, C.H. Lin, Rapid on-line microextraction method for the analysis of glyphosate in soy and pepper based on fiber-spray/mass spectrometry, Int. J. Mass Spectrom. 430 (2018) 104–109.

[8] J.E. Keating, J.T. Minges, S.H. Randell, G.L. Glish, Paper spray mass spectrometry for high-throughput quantification of nicotine and cotinine, Anal Methods 10 (1) (2018) 46–50.

[9] H. Lee, C.S. Jhang, J.T. Liu, C.H. Lin, Rapid screening and determination of designer drugs in saliva by a nib-assisted paper spray-mass spectrometry and separation technique, J. Sep. Sci. 35 (2012) 2822–2825.
[10] D.E. Damon, K.M. Davis, C.R. Moreira, P. Capone, R. Cruttenden, A.K. Badu-Tawiah, Direct biofluid analysis using hydrophobic paper spray mass spectrometry, Anal. Chem. 88 (2016) 1878–1884.
[11] P. Basuri, D. Sarkar, G. Paramasivam, T. Pradeep, Detection of hydrocarbons by laser assisted paper spray ionization mass spectrometry (LAPSI MS), Anal. Chem. 90 (2018) 4663–4668.
[12] R.D. Espy, A.R. Muliadi, Z. Ouyang, R.G. Cooks, Spray mechanism in paper spray ionization, Int. J. Mass Spectrom. 325–327 (2012) 167–171.
[13] J. Liu, H. Wang, N.E. Manicke, J.M. Lin, R.G. Cooks, Z. Ouyang, Development, characterization, and application of paper spray ionization, Anal. Chem. 82 (2010) 2463–2471.
[14] P.H. Lai, P.C. Chen, Y.W. Liao, J.T. Liu, C.C. Chen*, C.H. Lin, Comparison of gampi paper and nanofibers to chromatography paper used in paper spray-mass spectrometry, Int. J. Mass Spectrom. 375 (2015) 14–17.
[15] Q. Yang, H. Wang, J.D. Maas, W.J. Chappell, N.E. Manicke, R.G. Cooks, Z. Ouyang, Paper spray ionization devices for direct, biomedical analysis using mass spectrometry, Int. J. Mass Spectrom. 312 (2012) 201–207.
[16] H. Lee, C. S Jhang, J.T. Liu, C.H. Lin, Rapid screening and determination of designer drugs in saliva by a nib-assisted paper spray-mass spectrometry and separation technique, J. Sep. Sci. 0 (2012) 1–4.
[17] C.S. Jhang, H. Lee, Y.S. He, J.T. Liu, C.H. Lin, Rapid screening and determination of 4-chloroamphetamine in saliva by paper spray-mass spectrometry and capillary electrophoresis-mass spectrometry, Electrophoresis 33 (2012) 1–6.
[18] S.A. Oradu, R.G. Cooks, Multistep mass spectrometry methodology for direct characterization of polar lipids in green microalgae using paper spray ionization, Anal. Chem. 84 (2012) 10576–10585.
[19] D. Sarkar, A. Srimany, T. Pradeep, Rapid identification of molecular changes in tulsi (Ocimum sanctum Linn) upon ageing using leaf spray ionization mass spectrometry, Analyst 137 (2012) 4559–4563.
[20] S.L.F. Chan, M.Y.M. Wong, H.W. Tang, C.M. Che, K.M. Ng, Tissue-spray ionization mass spectrometry for raw herb analysis, Rapid Commun. Mass Spectrom. 25 (2011) 2837–2843.
[21] J. Liu, R.G. Cooks, Z. Ouyang, Biological tissue diagnostics using needle biopsy and spray ionization mass spectrometry, Anal. Chem. 83 (2011) 9221–9225.
[22] J.T. McCann, D. Li, Y. Xia, Electrospinning of nanofibers with core-sheath, hollow, or porous structures, J. Mater. Chem. 15 (2005) 735–738.
[23] S. Sakuldao, T. Yoovidhya, S. Wongsasulak, Coaxial electrospinning and sustained release properties of gelatin-cellulose acetate core-shell ultrafine fibres, Sci. Asia 37 (2011) 335–343.
[24] Y.S. Huang, C.C. Kuo, Y.C. Shu, S.C. Jang, W.C. Tsen, F.S. Chuang, C.C. Chen, Highly aligned and single-layered hollow fibrous membranes prepared from polyurethane and silica blends through a two-fluid coaxial electrospun process, Macromol. Chem. Phys. 215 (2014) 879–887.
[25] J.C. Yang, S.Y. Lee, W.C. Tseng, Y.C. Shu, J.C. Lu, H.S. Shie, C.C. Chen, Formation of highly aligned, single-layered, hollow fibrous assemblies and the fabrication of large pieces of PLLA membranes, Macromol. Mater. Eng. 297 (2012) 115–122.
[26] A.V. Bazilevsky, A.L. Yarin, C.M. Megaridis, Co-electrospinning of core-shell fibers using a single-nozzle technique, Langmuir 23 (2007) 2311–2314.

[27] H. Wu, L. Hu, M.W. Rowell, D. Kong, J.J. Cha, J.R. McDonough, J. Zhu, Y. Yang, M.D. McGehee, Y. Cui, Electrospun metal nanofiber webs as high-performance transparent electrode, Nano Lett. 10 (2010) 4242–4248.
[28] H. Qu, S. Wei, Z. Guo, Coaxial electrospun nanostructures and their applications, J. Mater. Chem. 1 (2013) 11513–11528.
[29] B. Hu, P.K. So, Y. Yang*, J. Deng, Y.C. Choi, T. Luan, Z.P. Yao, Surface-modified wooden-tip electrospray ionization mass spectrometry for enhanced detection of analytes in complex samples, Anal. Chem. 90 (2018) 1759–1766.
[30] H.K. Chen, C.H. Lin, J.T. Liu, C.H. Lin, Electrospray ionization using a bamboo pen nib, Int. J. Mass Spectrom. 356 (2013) 37–40.
[31] D.T. Usmanov, S. Saha, L.C. Chen, S. Ninomiya, M.K. Mandal, K. Hiraoka, Probe electrospray ionization mass spectrometry with discontinuous atmospheric pressure interface, Eur. J. Mass Spectrom. 21 (2015) 327–334.
[32] K.S. Kerian, A.K. Jarmusch, R. Graham Cooks, Touch spray mass spectrometry for in situ analysis of complex samples, Analyst 139 (2014) 2714–2720.
[33] V. Pirro, A.K. Jarmusch, M. Vincenti, R.G. Cooks, Direct drug analysis from oral fluid using medical swab touch spray mass spectrometry, Anal. Chim. Acta 861 (2015) 47–54.
[34] P.W. Fedick, R.M. Bain, Swab touch spray mass spectrometry for rapid analysis of organic gunshot residue from human hand and various surfaces using commercial and fieldable mass spectrometry systems, Forensic Chemistry 5 (2017) 53–57.
[35] M. Hecht, H. Evard, K. Takkis, R. Veigure, R. Aro, R. Lohmus, K. Herodes, I. Leito, K. Kipper, Sponge spray reaching new dimensions of direct sampling and analysis by MS, Anal. Chem. 89 (2017) 11592–11597.
[36] J.Y. Liu, P.C. Chen, Y.W. Liou, K.Y. Chang, C.H. Lin, Development and application of a brush-spray derived from a calligraphy-brush-style synthetic hair pen for use in ESI/MS, J. Mass Spectrom. 6 (2017) S0058.
[37] C.W. Lee, H. Su, Y.D. Cai, M.T. Wu, D.C. Wu, J. Shiea, Rapid identification of psychoactive drugs in drained gastric lavage fluid and whole blood specimens of drug overdose patients using ambient mass spectrometry, J. Mass Spectrom. 6 (2017) S0056.
[38] Y.W. Liou, K.Y. Chang, C.H. Lin, Sampling and profiling caffeine and its metabolites from an eyelid using a watercolor pen based on electrospray ionization/mass spectrometry, Int. J. Mass Spectrom. 422 (2017) 51–55.
[39] L. Shen, J. Zhang, Q. Yang, N.E. Manicke, Z. Ouyang, High throughput paper spray mass spectrometry analysis, Clin. Chim. Acta 420 (2013) 28–33.
[40] D. Sarkar, A. Som, T. Pradeep, Catalytic paper spray ionization mass spectrometry with metal nanotubes and the detection of 2,4,6-trinitrotoluene, Anal. Chem. 89 (2017) 11378–11382.
[41] L. Fang, J. Deng, Y. Yu, Y. Yang, X. Wang, H. Liu, T. Luan, Coupling liquid-phase microextraction with paper spray for rapid analysis of malachite green, crystal violet and their metabolites in complex samples using mass spectrometry, Anal. Methods 8 (2016) 6651–6656.
[42] C. Zhang, N.E. Manicke, Development of a paper spray mass spectrometry cartridge with integrated solid phase extraction for bioanalysis, Anal. Chem. 87 (2015) 6212–6219.
[43] P.W. Fedick, B.J. Bills, N.E. Manicke, R.G. Cooks, Forensic sampling and analysis from a single substrate: Surface enhanced Raman spectroscopy followed by paper spray mass spectrometry, Anal. Chem. 89 (2017) 10973–10979.
[44] K.E. Yannell, K.R. Kesely, H.D. Chien, C.B. Kissinger, R.G. Cooks, Comparison of paper spray mass spectrometry analysis of dried blood spots from devices used for in-field collection of clinical samples, Anal. Bioanal. Chem. 409 (2017) 121–131.

[45] B.J. Bills, N.E. Manicke, Development of a prototype blood fractionation cartridge for plasma analysis by paper spray mass spectrometry, Clin. Mass Spectrom. 2 (2016) 18–24.
[46] Q. Yang, N.E. Manicke, H. Wang, C. Petucci, R.G. Cooks, Z. Ouyang, Direct and quantitative analysis of underivatized acylcarnitines in serum and whole blood using paper spray mass spectrometry, Anal. Bioanal. Chem. 404 (5) (2012) 1389–1397.
[47] P.A. Demirev, Dried blood spots: analysis and applications, Anal. Chem. 85 (2) (2013) 779–789.
[48] R.D. Espy, N.E. Manicke, Z. Ouyang, R.G. Cooks, Rapid analysis of whole blood by paper spray mass spectrometry for point-of-care therapeutic drug monitoring, Analyst 137 (10) (2012) 2344–2349.
[49] J. Xu, S. Xu, Y. Xiao, K. Chingin, H. Lu, R. Yan, H. Chen, Quantitative determination of bulk molecular concentrations of β-agonists in pork tissue samples by direct internal extractive electrospray ionization-mass spectrometry, Anal. Chem. 89 (2017) 11252–11258.
[50] T.P.P. Mendes, I. Pereira, M.R. Ferreira, A.R. Chaves, B.G. Vaz, Molecularly imprinted polymer-coated paper as a substrate for highly sensitive analysis using paper spray mass spectrometry: quantification of metabolites in urine, Anal. Methods 9 (2017) 6117–6123.
[51] M. Zhang, F. Lin, J. Xu, W. Xu, Membrane electrospray ionization for direct ultrasensitive biomarker quantitation in biofluids using mass spectrometry, Anal. Chem. 87 (2015) 3123–3128.
[52] V.S. Amador, H.V. Pereira, M.M. Sena, R. Augusti, E. Piccin, Paper spray mass spectrometry for the forensic analysis of black ballpoint pen inks, J. Am. Soc. Mass Spectrom. 28 (2017) 1965–1976.
[53] J. Wormhoudt, J.H. Shorter, J.B. McManus, P.L. Kebabian, M.S. Zahniser, W.M. Davis, E.R. Cespedes, C.E. Kolb, Tunable infrared laser detection of pyrolysis products of explosives in soils, Appl. Optics 35 (1996) 3998–4003.
[54] E.R. Goldman, T.J. Cohill, C.H. Patterson, G.P. Anderson, A.W. Kusterbeck, J.M.M. Environ, Detection of 2,4,6-trinitrotoluene in environmental samples using a homogeneous fluoroimmunoassay, Sci. Technol. 37 (20) (2003) 4733–4736.
[55] Z. Takáts, J.M. Wiseman, R.G. Cooks, Ambient mass spectrometry using desorption electrospray ionization (DESI): instrumentation, mechanisms and applications in forensics, chemistry and biology, J. Mass Spectrom. 40 (2005) 1261–1275.
[56] Z. Takáts, I. Cotte-Rodriguez, N. Talaty, H. Chen, R.G. Cooks, Direct, trace level detection of explosives on ambient surfaces by desorption electrospray ionization mass spectrometry, Chem. Commun. (2005) 1950–1952. Cambridge, U. K.
[57] V. Sládek, P. Galeta, D. Sosna, Measuring human remains in the field: grid technique, total station, or MicroScribe, Forensic Sci. Int. 221 (2012) 84–91.
[58] F. Cacciola, P. Donato, M. Beccaria, P. Dugo, L. Mondello, Advances in LC-MS for food analysis, LC-GC Eur. 25 (2012) 15–24.
[59] F. Mazzotti, L.D. Donna, L. Maiuolo, A. Napoli, R. Salerno, A. Sajjad, G. Sindona, Assay of the set of all Sudan azodye (I, II, III, IV, and para-red) contaminating agents by liquid Chromatography–Tandem mass spectrometry and isotope dilution methodology, J. Agric. Food Chem. 56 (1) (2008) 63–67.
[60] T. Guo, Z. Zhang, K.E. Yannell, Y. Dong, R.G. Cooks, Paper spray ionization mass spectrometry for rapid quantification of illegal beverage dyes, Anal. Methods 9 (2017) 6273–6279.
[61] R.D. Espy, M. Wleklinski, X. Yan, R.G. Cooks, Beyond the flask: reactions on the fly in ambient mass spectrometry, Trends Anal. Chem. 57 (2014) 135–146.
[62] R.M. Bain, C.J. Pulliam, S.A. Raab, R.G. Cooks, Chemical synthesis accelerated by paper spray: the haloform reaction, J. Chem. Educ. 93 (2016) 340–344.

CHAPTER 8

Rapid evaporative ionization mass spectrometry

Thanai Paxton
Nihon Waters K.K., Shinagawa-ku, Tokyo, Japan

Contents

8.1 Introduction

Rapid Evaporative Ionization Mass Spectrometry (REIMS) is a technique that allows the in vivo and in situ analysis and real-time identification of intact biological tissue [1,2]. REIMS was exclusively developed for intraoperative tissue identification and margin assessment by combining MS with electrosurgical tools. Application has since expanded beyond the surgical arena to cover the characterization and identification of microorganisms (see Section 8.4.2) through to the analysis of food and environmental samples (see Section 8.4.3).

MS has been used to analyze tissues since the 1970s [3] and the traditional workflow of homogenization and metabolite extraction remains the benchmark today [4]. However, one of the holy grails of MS has always been to minimize or eliminate sample preparation altogether. Advances were made with the advent of desorption ionization methods in the 1980s [5–10] that culminated in the introduction of matrix-assisted laser desorption ionization (MALDI) [11,12]. However, it was not until the late 1990s that mass spectrometry imaging (MSI) gained widespread use, when

Ambient Ionization Mass Spectrometry in Life Sciences
ISBN 978-0-12-817220-9
https://doi.org/10.1016/B978-0-12-817220-9.00008-4

MALDI was demonstrated to be able to visualize biomolecules, distinguish small histological differences and its variations with disease [13]. MSI methods commonly used today to analyze biomolecules include secondary ion mass spectrometry (SIMS) [14], MALDI [15], and desorption electrospray ionization (DESI) [16].

Within the intraoperative environment, the use of SIMS is limited due to the time-consuming nature of sample preparation and analysis [17]. In comparison to SIMS, MALDI has a relatively straightforward workflow and has already been shown to be useful in the analysis of tumor margins, though it is complicated by the requirement of a matrix to be deposited prior to analysis [18]. By contrast, DESI allows samples to be analyzed directly without the necessity for a matrix, making it potentially the most suitable for intraoperative deployment allowing analysis times of individual specimens to be reduced to the 10 min range from the 20—30 min required by intraoperative histology [19—21]. Nevertheless, these MSI techniques remain restricted to in vitro samples operating within the limits of the histology workflow. The development of REIMS overcomes this barrier allowing in vivo and in situ tissue analysis [1,2].

REIMS is based on the discovery that a variety of tools that already existed in surgery are a rich source of biological information, specifically in electrosurgery and laser surgery. In both instances, the cauterization or dissection of tissue produces smoke that traditionally has been discarded. This surgical smoke is found to contain a treasure of gaseous ions including the biological signature of the specific tissue area being vaporized. It is this rapid evaporation of the samples that the term REIMS is derived. In doing so it is possible to increase the rate of molecular evaporation to overcome any thermal degradation and yield gaseous molecular ions.

8.2 REIMS instrumentation

The original experimental setup for REIMS analysis employed an electrosurgical electrode as the ion source to apply a high-frequency electric current directly to the tissue. The resultant smoke containing both positive and negative ions from the rapid heating and evaporation of the tissue is then transported to the mass spectrometer. This is achieved through modification of the surgical electrode to include a 2 m long polytetrafluoroethylene (PTFE) tubing that is interfaced to the orifice of the mass spectrometer. A venturi gas jet pump is then used to clear the surgical site of the charged smoke and drive it toward the entry of the mass spectrometer.

The REIMS system is being developed via a partnership between its inventor, Prof. Zoltan Takats of Imperial College, and Waters Corporation. The recent commercialization of the REIMS system by Waters Corporation (Fig. 8.1) is currently reserved for research use only and not intended for diagnostic or therapeutic purposes [22–24].

The system employs one of two electrosurgical sampling devices, the monopolar electrode (Fig. 8.2) and the bipolar forceps (Fig. 8.3). In the monopolar case, the sample is placed on a return electrode and a high current from a diathermy generator is applied to the sample using the monopolar cutting electrode. At the point of application, the high current density heats the sample to vaporization. In contrast, the large contact area of the return electrode only experiences a low current density and any heating is minimal. The aerosol generated is then carried through the body of the monopolar handheld device and to a quadrupole time-of-flight (Q-ToF) mass spectrometer.

In the case of the bipolar forceps, a separate return electrode is not required and the electrodes are integrated on each tip. Current to vaporize the sample only flows through the tissue between the electrodes on each tip and not the sample itself. Irrigation channels are contained within the arms

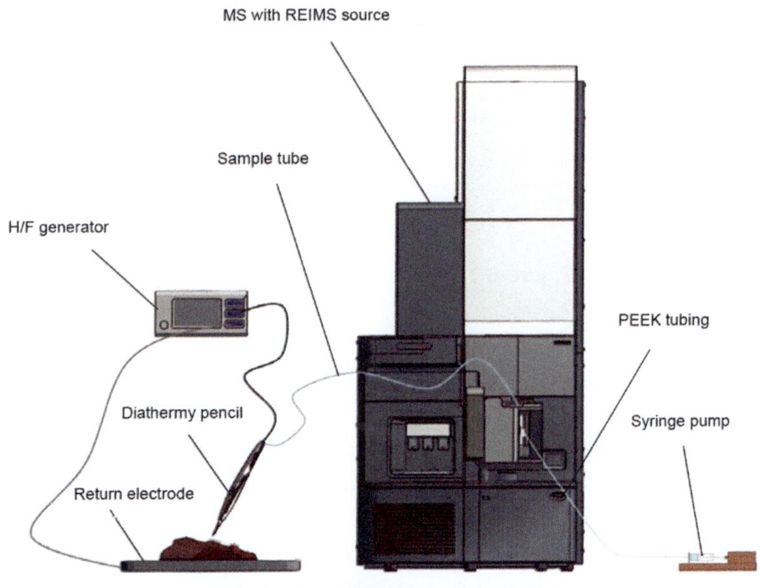

Figure 8.1 REIMS-QToF MS system.

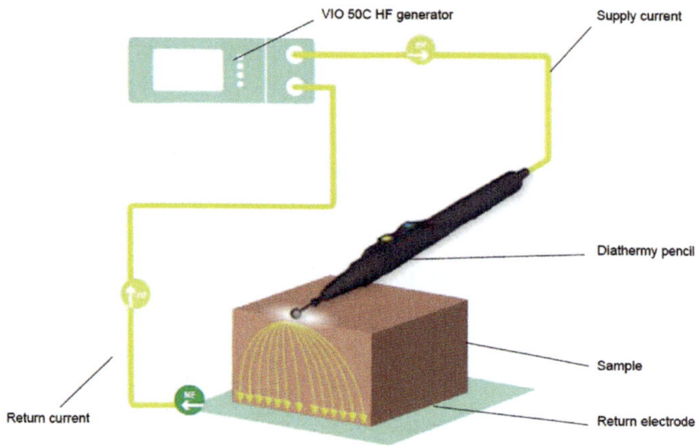

Figure 8.2 Monopolar handheld pencil.

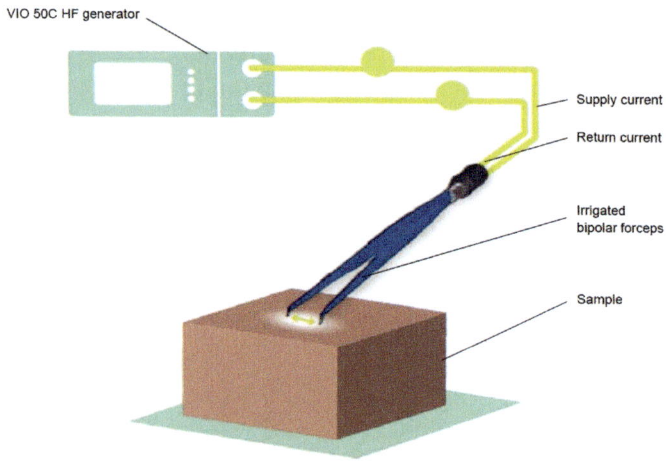

Figure 8.3 Irrigated bipolar forceps.

of the forceps allowing the generated aerosol to be carried to the Q-ToF mass spectrometer.

In both cases, a venturi air jet device actively transports the aerosol from the electrosurgical device via the sample inlet tubing into the transfer capillary of the mass spectrometer (Fig. 8.4).

At this point the aerosol collides with a superheated collision surface (~900°C) disrupting the smoke-ion supramolecular clusters and liberating gaseous phase ions into the mass spectrometer (Fig. 8.5). At the same time,

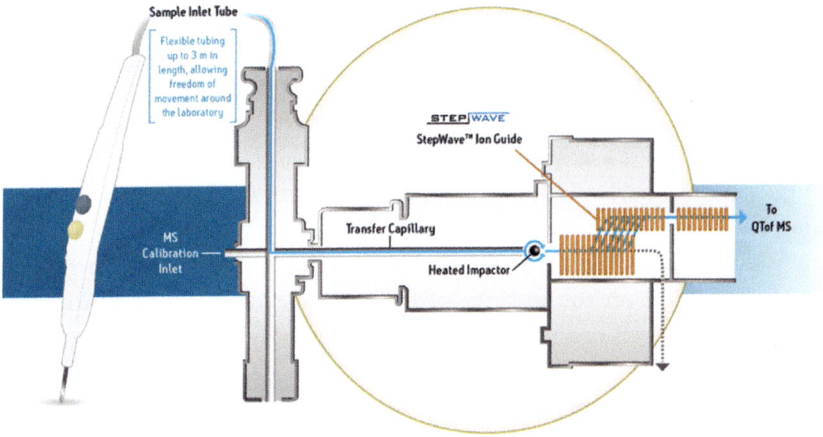

Figure 8.4 REIMS source.

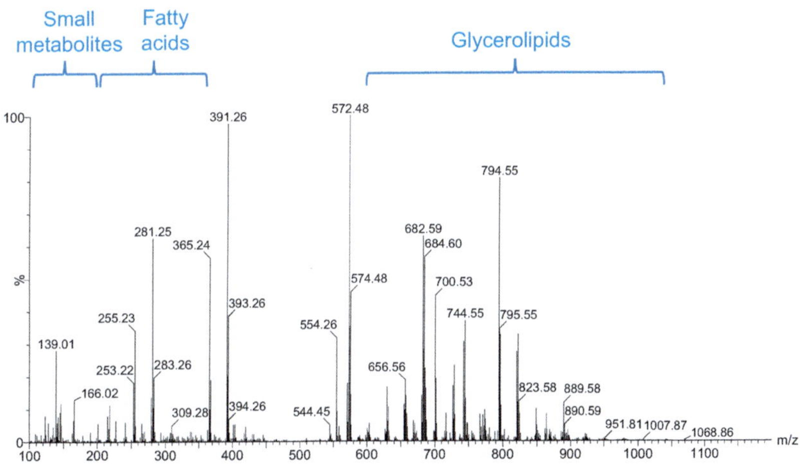

Figure 8.5 Example REIMS cod tissue spectrum.

the heated disruptor prevents any macroscopic particulate matter from entering the instrument and helps prevents contamination of the mass spectrometer.

8.3 REIMS spectra and data handling

Based on experimental results, the ion formation mechanism in REIMS is postulated to be associated with the formation of charged aqueous droplets during the thermal evaporation of the tissue. In this process, the rate of evaporation is matched by the rate of thermal degradation resulting in both the intact molecular species and their primary thermal degradation products being present in the gas phase. However, the amount of thermal degradation products is kept relatively low by the increased efficiency of collisional cooling of ions at atmospheric pressure [1].

Lipids and in particular phospholipids are the most abundant species in tissue readily forming molecular ions. And as in DESI, the REIMS spectrum is similarly dominated by protonated or deprotonated intact lipid species depending on the polarity of the mass spectrometer. The thermal degradation products of these lipids are also present (e.g., lysophospholipids and fatty acids). In addition to lipids, REIMS has also been able to detect amino acids, peptides, and drug molecules from aqueous solution as well as alkali metal and ammonium adducts. However, radical ions have so far not been detected [1].

REIMS spectra derived from different tissue are highly specific. The differences, however, are not immediately distinguishable and differentiation of tissue spectra relies on multivariate statistical tools such as principal component analysis (PCA). In fact, often no tissue-specific marker is observed and identified lipid components are generally detected across tissue types. Therefore, the tissue specificity of these molecular fingerprints is predominantly derived from the distribution of the lipids and not an individual species.

The coupling of electrosurgical REIMS with concomitant multivariate analysis for real-time patient diagnosis is termed the "intelligent knife" (*iKnife*). The use of the REIMS *iKnife* system involves the building a spectral reference database and creation of a multivariate statistical model and identification algorithm for the samples to be differentiated. For example, in the detection of the tumor margins in brain, breast, and liver cancer surgery, data are collected *ex vivo* from benign and malignant brain, breast, and hepatic tissue samples for the database [25]. In food authenticity testing, this may be the differentiation between fish with large differences in price like the more expensive cod and the cheaper whiting [24]. While in microbiology, an example may be the identification of species within the genus *Candida* so that the correct antifungal may be administered [26].

On the commercial Waters system, the process of creating the statistical model, its validation and use for real-time sample recognition is achieved using the LiveID software [27]. MS data from each sample are collected several times and imported into the software, where PCA and linear

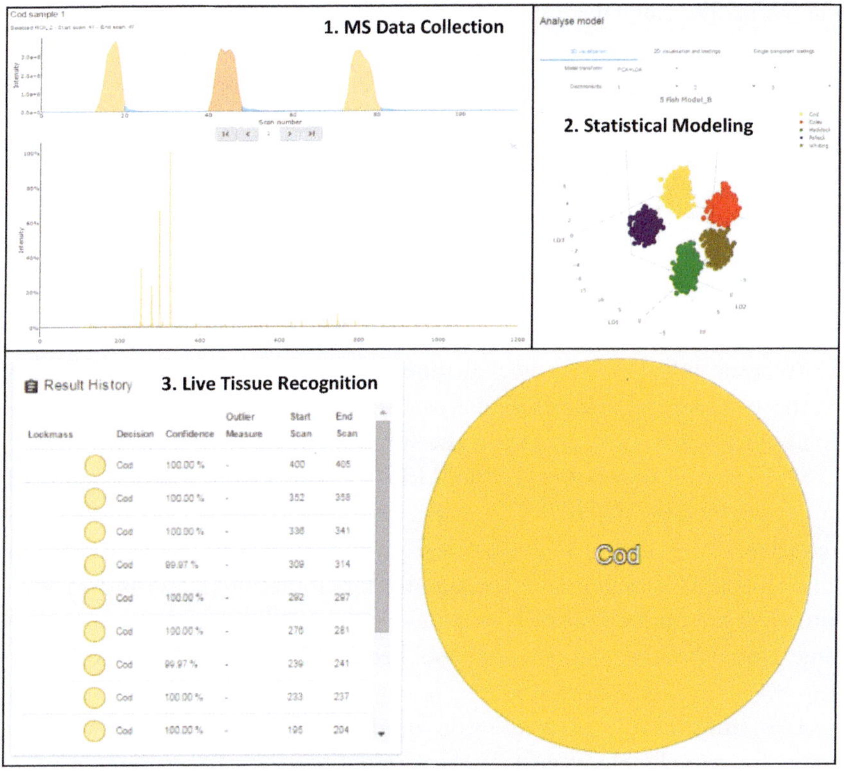

Figure 8.6 LiveID software.

discriminant analysis (LDA) are used to build a statistical model of the samples to be differentiated. The validated model is then used to allow live recognition of sample areas being vaporized by the REIMS sampling electrode (Fig. 8.6).

8.4 Applications

8.4.1 REIMS in surgery

As research continues in the quest for a weapon against cancer, surgery as part of a treatment plan remains one of the best tools available in the treatment arsenal. The ability to precisely excise the cancerous tissue without leaving any diseased tissue behind naturally relies on the skill of the surgeon to identify the exact tumor margins. If unsure, the surgeon often enlists the help of a pathologist who can test if the tissue is malignant or not.

This extra test can take up to 30 min and does not account for retests, occurring while the patient remains under anesthesia. Additional to time constraints, the surgeon is also under pressure to conserve as much as possible the normal tissue but at the same time ensure the surgical margin is clear of cancer cells. Failure to do so increases the risk of a relapse and cancer survival rates suffer. Apart from significantly impacting the patient journey, undesired strain of available resources and extra costs are further placed on the health system necessitating for the introduction of faster and more precise intraoperative margin assessment techniques. By connecting electrosurgical tools directly to MS to allow the collection of in vivo and in situ tissue information and real-time identification, the REIMS system has the potential to overcome many of the barriers with current methodologies and completely transform the surgical paradigm [1,2,25].

Successful demonstration of the REIMS *iKnife* system was first reported on animal models in 2010 [2]. Here, a spectral database and model was created with REIMS data collected from the analysis of ex vivo tissue of canine subjects carrying spontaneous tumors. Subsequent testing in vivo during canine tumor resection surgeries allowed real-time differentiation of tissue from different organs and also between normal and cancer tissue (Fig. 8.7).

This study also illustrated using model rats that nutritional and age factors have negligible impact on tissue identification. This is important in offering real potential for the use of the technology in the real world where diets can be unpredictable. In the case of nutrition, rats were fed with either regular, porcine fat-rich or rapeseed oil-rich feed. The fatty acid composition of each diet would be expected to affect the acyl chain distribution within each phospholipid class and the resultant REIMS spectra However, the in vivo REIMS spectra of various tissues showed little change with diet (with the exception of the myocardial tissue) and did not impact tissue identification in all cases (Fig. 8.8).

On the other hand, although identification is not affected, age was statistically significant and could be explained by the REIMS spectra. Successful testing of the REIMS technique in animals paved the way for validation of the methodology in humans. In 2013 the first study to apply the technology to the human population was conducted with the goal of testing REIMS and its accuracy in the surgical environment [25]. Similar to the animal studies, samples were collected ex vivo in order to build a spectral library and statistical model before subsequent use in the intraoperative margin assessment of patients undergoing resection surgery for

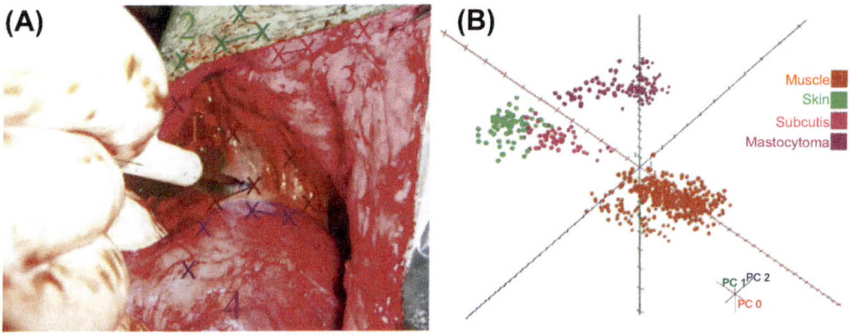

Figure 8.7 (A) REIMS-based canine oncological surgery of a grade III mastocytoma (B) principal component analysis of the spectra from marked places (X). *(Reprinted with permission from J. Balog, T. Szaniszlo, K.-C. Schaefer, J. Denes, A. Lopata, L. Godorhazy, D. Szalay, L. Balogh, L. Sasi-Szabo, M. Toth, Identification of biological tissues by rapid evaporative ionization mass spectrometry, Anal. Chem. 82 (2010) 7343–7350. https://doi.org/10.1021/ac101283x.)*

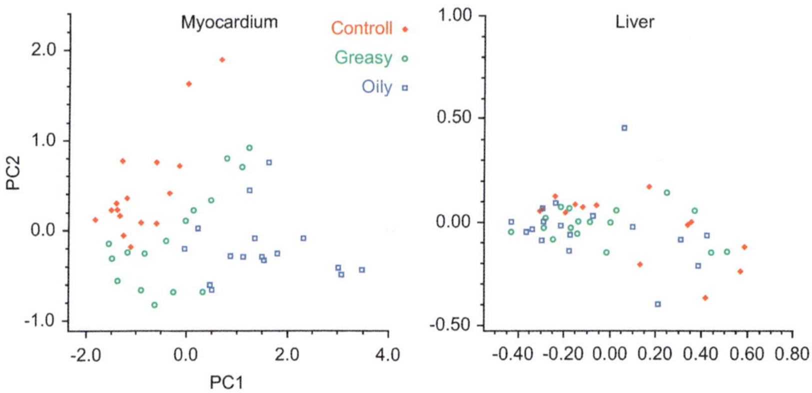

Figure 8.8 Two-dimensional principal component analysis showing the effect of diet on myocardial and liver tissue. *(Reprinted with permission from J. Balog, T. Szaniszlo, K.-C. Schaefer, J. Denes, A. Lopata, L. Godorhazy, D. Szalay, L. Balogh, L. Sasi-Szabo, M. Toth, Identification of biological tissues by rapid evaporative ionization mass spectrometry, Anal. Chem. 82 (2010) 7343–7350. https://doi.org/10.1021/ac101283x.)*

brain, liver, lung, breast or colorectal tumors. A variety of tissue samples from 302 patients were analyzed ex vivo including benign and malignant gastric, colonic, hepatic, breast, lung, and brain tissue samples. In total 2933 database entries were recorded in the spectral database and consisted of 1624 cancerous and 1309 noncancerous entries. In agreement with previous observations, the spectra of each tissue type were dominated by a variety of phospholipids in the *m/z* 600–900 range and was not often characterized by a single unique marker but rather the composite distribution patterns of the detected lipids. It is through the collection of a statistically significant number of tissue samples that allows a PCA-LDA model to be created for REIMS to distinguish between compositional changes and identify with histological specificity different healthy tissue and various malignant tumors. The validated REIMS data from the human study were confirmed to allow unambiguous (100%) tissue identification of the major tissue types (liver, lung, and colon).

From PCA, it was determined that three ionic species contributed the most to tissue specificity. These were identified as phosphatidic acid (PA), phosphatidylethanolamine (PE), and phosphatidylinositol (PI). PA (34:1)-H at m/z 673.48 was found to be most characteristic of alveolar lung tissue, while PE(36:1)-NH_3 at m/z 727.53 was more strongly associated with the colon mucosa. By contrast, PE(34:2)-NH_3 at m/z 697.48 and PI(38:4)-H at m/z 885.55 were represented in all three healthy tissue types.

The REIMS data were also found to be in agreement with histology and able to distinguish between distinct cancer tissue types. The models created from the ex vivo tissue analysis showed clear separation of specific histological regions of different lung tumors (adenocarcinoma, squamous cell carcinoma, mucinous carcinoma, and neuroendocrine carcinoma).

Good separation was similarly observed with data from cholangiocarcinomas allowing tumor regions to be confidently distinguished from surrounding healthy liver tissue as well as be able to identify the gradual metabolic changes as sampling is moved away from visible tumor and between benign and malignant tissue regions (healthy liver parenchyma, hepatocellular carcinoma, and borderline regions within 1 cm of each). Although this metabolic transition was only distinguishable by REIMS in primary and not metastatic tumors, the technique was impressively able to identify the origin of metastatic brain tumors. Bipolar forceps were used to collect small pieces of tissue ($0.1-1.0\ mm^3$) prior to vaporization away from direct contact of the surgical area. This has the benefit of minimizing damage to cortical areas of the brain from leaked electrical currents and unintentional burning. Results were confirmed by histopathology and illustrated the capability to differentiate between brain metastases from colon and from lung as well as the ability to distinguish these from different World Health Organization (WHO) grades of astrocytomas.

The constructed ex vivo spectral database and the derived PCA-LDA model-based algorithm was then tested on 81 patients to assess performance in the intraoperative environment. In total, 864 in vivo spectra were acquired and identified in agreement with postoperative histopathology. The sensitivity and specificity of the technique for differentiating between healthy and cancer samples was high at 97.7% and 96.5%, respectively, corresponding to low false-negative (2.3%) and false-positive (3.5%) rates. The predictive capability of the models was also maintained at the organ level providing >95% sensitivity and specificity rates with the exception of the primary liver and malignant lung tumor models where the specificity rate fell to 94% and 92%, respectively. In fact, since supervised multivariate statistical analysis methods were used, the identification efficiency can be increased by limiting the model to a single organ providing enough tissue is sampled. Overall, only ~0.1 mm^3 of tissue was required for accurate identification and this translated to tissue evaporation rates of 20–250 μg/s or ~1 mm/s. Although the spatial resolution of the REIMS system may not be at the level of MS imaging techniques such as MALDI, DESI, and SIMS, it compares very well with the precision of standard surgical procedures. Furthermore, this tissue identification occurs in real time (0.7–2.5 s) and has the potential to circumvent the 30 min required for intraoperative histopathology altogether. As the first successful human demonstration of the REIMS-*iKnife* technique, the study provided important preliminary evidence and impetus for the further development of the technique into a routine and powerful tool for cancer surgery. More recent studies have focused on more specific cancers like that of the breast while others have focused on the integration of REIMS with other tools like an endoscopic polypectomy snare.

At 25%, breast cancer is the most commonly diagnosed cancer in women. It is the leading cause of cancer death in less developed countries and the second leading cause of death after lung cancer in the United States [28]. Breast conserving surgery (BCS) is often the preferred method of treatment. However, in over 20% of cases, incomplete excision of the cancer means that reoperation is required. Reoperation puts a strain on the physical and psychological health of the patient and can delay the receiving of necessary adjuvant therapies not to mention the additional health-care costs. The REIMS-based model for the identification of breast pathology demonstrated a sensitivity and specificity rate of 93.4% and 94.9%, respectively [29]. The model was constructed from histology validated ex vivo tissue of 113 patients using 932 sampling points from 253 normal breast (B1 and B2) specimens and 226 sampling points from 106 breast tumor (B5a and B5b) specimens (Fig. 8.9A).

Closer examination using tandem MS revealed 24 spectral differences between tissue types identifying 63 glycerophospholipids that were higher

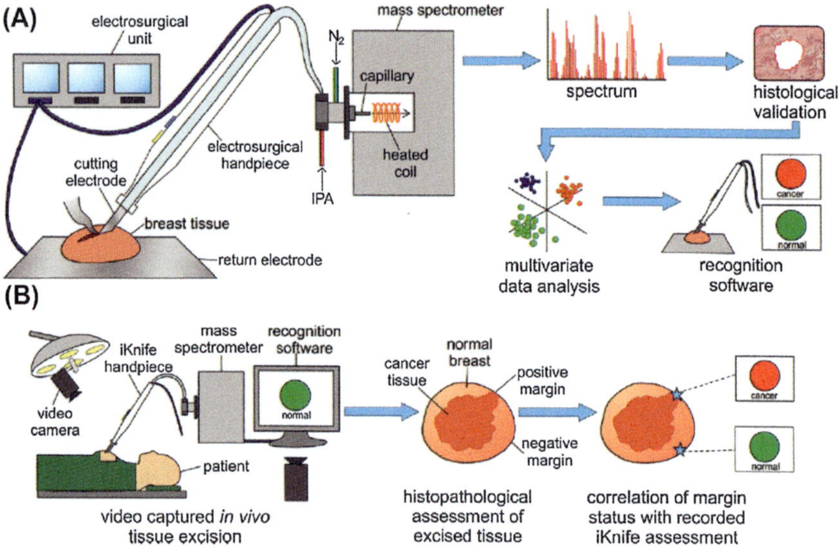

Figure 8.9 REIMS use for breast cancer surgery: (A) Ex vivo workflow for spectrum acquisition and model building, (B) intraoperative workflow.

and 6 triglycerides species that were lower in tumor tissue. The accuracy of the model was additionally confirmed prior to in vivo use against 260 previously untested ex vivo breast specimens providing a recognition sensitivity and specificity of 90.9% and 98.8%, respectively. Finally, as proof of principle, the REIMS *iKnife* was used for the entire operation in six case studies (Fig. 8.9B) allowing surgeons to receive real-time information from the point of sampling in 1.80 s (SD: ±0.40%) with 99.27% or 5422 out of 5462 of the spectra collected being interpretable by the constructed ex vivo model. Further work will still be required to fully determine the accuracy of the *iKnife* for intraoperative margin assessment before it can fully replace existing methodologies but the promise of the technique is obvious. Going beyond margin assessment, the REIMS-*iKnife* also has the potential to allow the customization of treatments based on the chemical information obtained about the tumor biology.

Progress in development of the REIMS technology has also focused on the improvements or modifications to the sampling device. For example, decreasing the size of the monopolar electrode can improve the specificity of sampling so that tumor spectra are less likely to be diluted by chemical information from normal cells [28]. Alternatively, the methods described until now have involved open surgery and there would be obvious benefits to attaching REIMS to noninvasive devices. This is demonstrated in the development of a REIMS method for real-time MS in endoscopic procedures [30]. Endoscopy is commonly used as diagnostic and surgical intervention tool for gastrointestinal (GI) cancers. As a noninvasive tool, endoscopy is commonly combined with electrosurgical diathermy for routine endoscopic mucosal resection (EMR) of early stage GI cancers and premalignant conditions. However, the method is plagued by inaccuracy with up to 41% of patients requiring reoperation. This is anything but ideal when GI cancer accounts for nearly one quarter of all cancer deaths globally. As a diagnostic tool, improvements to the video resolution of standard white light endoscopy and enhancements using spectroscopic characterization have increased polyp detection rates in the GI. Unfortunately, this has had the outcome of placing considerable burden on histopathologists, especially important when one considers that biopsied polyps are mostly found to be benign and that an estimated 7.8% of upper-GI cancers are missed. By coupling REIMS to endoscopic methods to allow accurate real-time tissue identification, it is possible to overcome these shortcomings. In the REIMS endoscopic device, the polypectomy snare acts as the sampling tool and is connected to the mass spectrometer by PTFE tubing. Polyps are captured in the wire loop of the snare and electrocauterization produces a chemical-rich aerosol that is transmitted via the fenestrations in the snare plastic sheath and transmitted via PTFE tubing to the mass spectrometer for analysis (Fig. 8.10).

After optimization of the design and sampling geometry of the device for sensitivity and reproducibility using a food-grade porcine stomach model, the REIMS endoscope was initiated with ex vivo samples from three human patients. This included gastric adenocarcinoma, healthy gastric mucosa, and healthy gastric submucosa tissue samples. Similar to other REIMS studies, phospholipids dominated the m/z 600–1000 region and spectral differences were observed between the three types of tissue analyzed (Fig. 8.11).

Healthy gastric mucosa and gastric adenocarcinoma spectra were characterized by mainly the phospholipid species, while the healthy gastric submucosa–derived spectra showed elevated triglyceride and phosphatidylinositols signals. The ability to differentiate and identify the submucosal layer is potentially useful in minimizing perforation risks during polyp removal or EMR. This is because it would allow the device to alert the operator if there is a breach of the submucosal wall during interventional endoscopic surgery and emergency stop any ongoing electrocauterization.

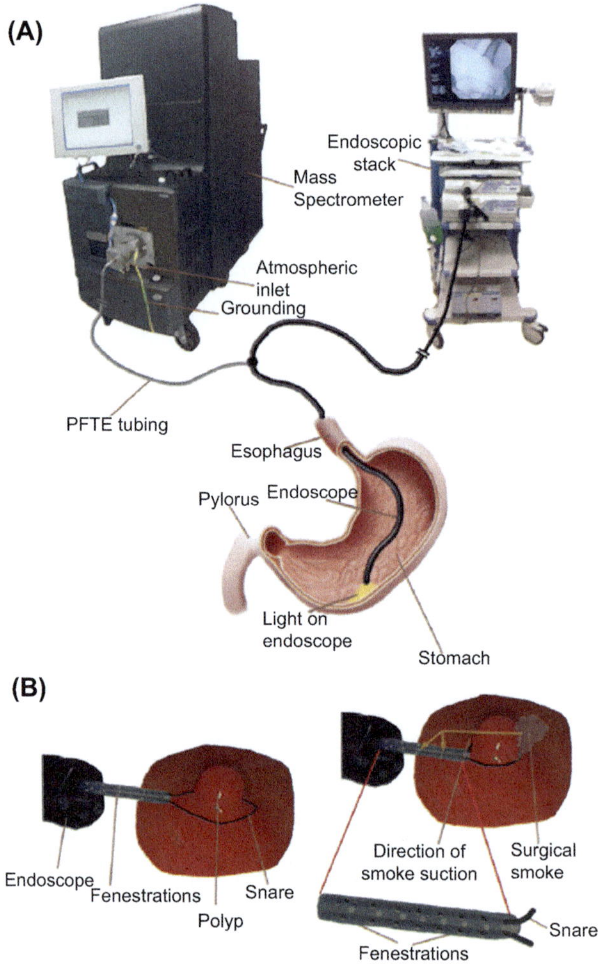

Figure 8.10 (A) REIMS-Endoscopy Setup. (B) Polyp resection using a polypectomy snare. *(Reprinted with permission from J. Balog, S. Kumar, J. Alexander, O. Golf, J. Huang, T. Wiggins, N. Abbassi-Ghadi, A. Enyedi, S. Kacska, J. Kinross, In vivo endoscopic tissue identification by rapid evaporative ionization mass spectrometry (REIMS), Angew. Chem. Int. Ed. Engl. 127 (2015) 11211–11214. https://doi.org/10.1002/ange.201502770.)*

An ex vivo spectral database and model was then constructed using human colonic adenocarcinoma, healthy colonic mucosa, and benign polyp tissue with a total of 62 sampling points from 22 patients providing a specificity rate and sensitivity rate of 95% and 88.5%, respectively. Subsequent in vivo testing on three colonoscopy patients identified that two of the patients had benign polyps while the third had retained a normal healthy mucosa. A

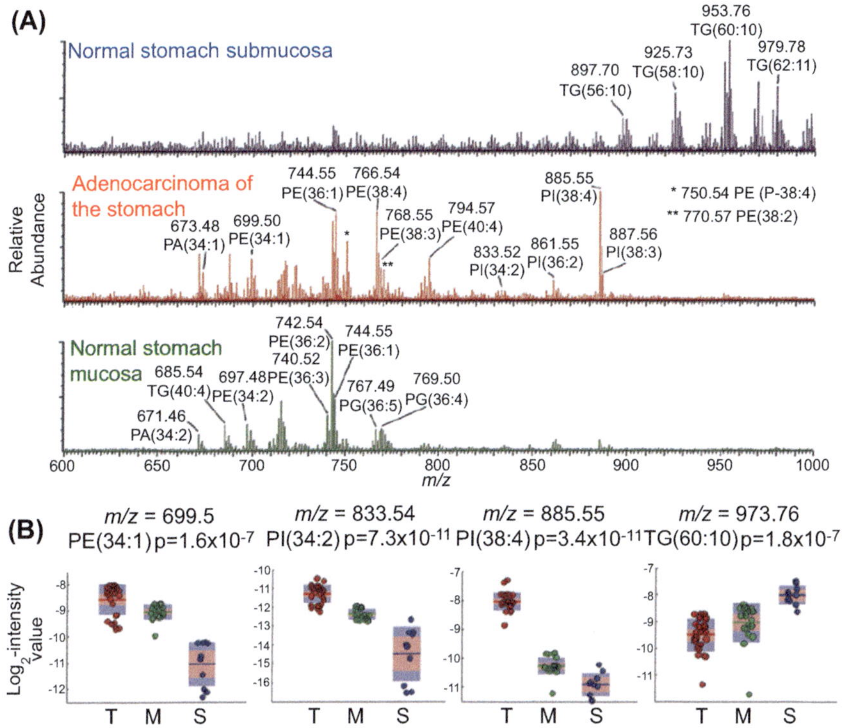

Figure 8.11 Comparison between tumor (T), mucosa (M), and submucosa (S) tissue samples: (A) Spectra (B) four selected lipids. *(Reprinted with permission from J. Balog, S. Kumar, J. Alexander, O. Golf, J. Huang, T. Wiggins, N. Abbassi-Ghadi, A. Enyedi, S. Kacska, J. Kinross, In vivo endoscopic tissue identification by rapid evaporative ionization mass spectrometry (REIMS), Angew. Chem. Int. Ed. Engl. 127 (2015) 11211–11214. https://doi.org/10.1002/ange.201502770.)*

further pilot study to assess the diagnostic accuracy in the real-time analysis of colorectal cancer has also proved successful [31].

The REIMS technology has been modified for use in open and minimally invasive surgery and is readily adapted to a variety of existing electrosurgical tools using RF electric currents including the monopolar *iKnife*, bipolar forceps, and endoscopic snare devices described. This may be further extended to include other sample mobilization methods like lasers [32] which offer their own unique advantages. For example, lasers require no physical contact with the surface and allow accurate amounts of energy to be focused for a better defined sampling area. Finally, the ability to use ex *vivo* tissue samples to create spectral databases and identification models suitable for in vivo deployment will facilitate creation of better defined and validated systems from a larger sampling pool as REIMS moves to fulfill its potential, redefining surgery and changing the clinical decision-making process. As it stands, the system has been commercialized by Waters but since regulatory approval is required for the clinical setting, the system is still reserved for research use only and not intended for diagnostic or therapeutic purposes. At the same time, the REIMS method has been successfully applied to patients during ovarian [33], breast [29], and colorectal [30,31] cancer surgery, providing a platform for its future widespread adoption.

8.4.2 REIMS in microbiology

Besides surgical applications, REIMS has also been examined for use in clinical microbiology. The traditional method of identification relies on the microbiologist being trained to carefully and systematically differentiate between isolates using their phenotypic and biochemical characteristics. For example, this may include colony morphology (e.g., shape, size, elevation, and color), Gram-staining, biochemical reactions (e.g., indole, methyl red, and citrate tests), as well as how it grows on selective media (e.g., Mannitol salt agar, MacConkey agar). However, this is a laborious time-consuming exercise and relies heavily on the experience of the microbiologists [34].

The sequencing of the gene encoding 16S rRNA is the accepted "gold standard" and provides enough phylogenetic information to identify and classify bacteria. The method has the advantage of being culture independent and is especially useful when microorganisms are slow-growing or fastidious. Similar approaches have also been developed for the identification of yeast and molds. However, its widespread use has been hampered by the time and cost associated with sequencing methods [35].

The application of MS to the rapid identification of microorganisms was first demonstrated in the mid-1970s. Direct insertion of the lyophilized bacteria into the ion source of a pyrolysis mass spectrometer was attempted in order to obtain pyrolysis at a lower temperature and detect larger more characteristic molecules for identification. This provided characteristic low molecular weight signal spectra derived from the phospholipids and ubiquinones of gram-negative bacteria [36]. However, the analysis of underivatized intact large biomolecules was not really amenable to MS until the introduction of soft ionization methods such as fast atom bombardment (FAB) [7,8], when bacteria species embedded in a glycerol matrix could be differentiated using their characteristic intact phospholipid signatures [37]. Nevertheless, it was not until the discovery of MALDI [11,12] and its subsequent application to microbial identification that motivated the widespread acceptance of MS as a standardized tool for clinical microbiology [38–41]. These MALDI-TOF methods rely on the measurement of the ribosomal protein signatures which represent over 50% of the proteins observed in the 2–20 kDa range. In comparison to conventional techniques, available commercialized systems are more cost effective and are capable of analyzing samples faster, being able to provide results in a little as 6 min even though the application of a matrix is required to assist ionization and cell lysis.

By contrast, species-specific lipid profiles have also been measured using FAB [37], ESI [42], MALDI [43], and DESI [44]. However, these have not proven specific or robust enough to form the basis of a microbial identification system. Furthermore, the addition of an ionization matrix in MALDI also results in additional signals in the spectra below m/z 1000 that can limit the study of lower molecular weight molecules. The ability to observe these low molecular weight species may prove important as clinical scientists investigate alternative ways to identify microorganisms and even viruses. More recently the development of REIMS in the intraoperative environment has demonstrated that the technique is able to provide specific phospholipid signatures for the identification of different tissue types [1,2,25], and this capability has been developed for the purposes of a lipid-based microbial identification system [45].

Initial tests to optimize the REIMS system for the identification of microbes found that both the monopolar and bipolar diathermy devices were capable of generating characteristic MS profiles from intact bacteria [45]. However, the vaporization of bacteria placed directly on the counter electrode plate produced excessive charring and smoke on contact with the monopolar electrode, resulting in poor MS spectra as well as negatively affecting the robustness of the ion optics. The bipolar forceps were found to provide better distribution of electrical current and facilitated for smaller amounts of bacterial biomass to be controllably vaporized vastly reducing contamination effects. At the same time superior MS spectra were observed rich in features and with reduced fragmentation characteristics, providing better intensity of molecular species above m/z 1000.

The bipolar-REIMS setup (Fig. 8.12) was used to analyze a total of 15 different clinical isolates from each of 28 bacterial species in order to investigate the suitability of the system for bacteria identification [45].

The best quality spectra from most of the bacteria species analyzed were obtained in the negative ion mode. Consequently, 5 separate negative ion MS measurements were obtained for each strain and averaged to form a database entry (Fig. 8.13).

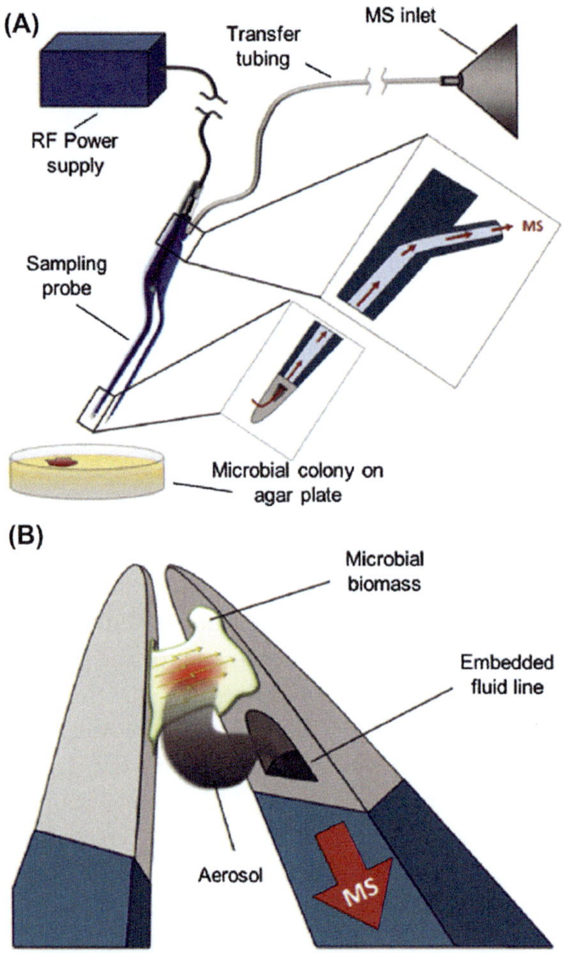

Figure 8.12 (A) REIMS setup for the analysis of microbial biomass. (B) Sampling of microbial biomass. *(Reprinted with permission from N. Strittmatter, M. Rebec, E.A. Jones, O. Golf, A. Abdolrasouli, J. Balog, V. Behrends, K.A. Veselkov, Z. Takats, Characterization and identification of clinically relevant microorganisms using rapid evaporative ionization mass spectrometry, Anal. Chem. 86 (2014) 6555–6562. https://doi.org/10.1021/ac501075f.)*

The entries in the database were then subjected to supervised and unsupervised multivariate statistical analyses. Using only 0.1–1.5 mg of biomass, REIMS was able to detect up to 1600 features within the *m/z* 150–2000 mass range and although the features are dominated by phospholipids (PG, PE, and PA) in the *m/z* 600–900 range, cardiolipins were also detected in all species tested as well as other lipid and lipid related species. For example, fatty acids

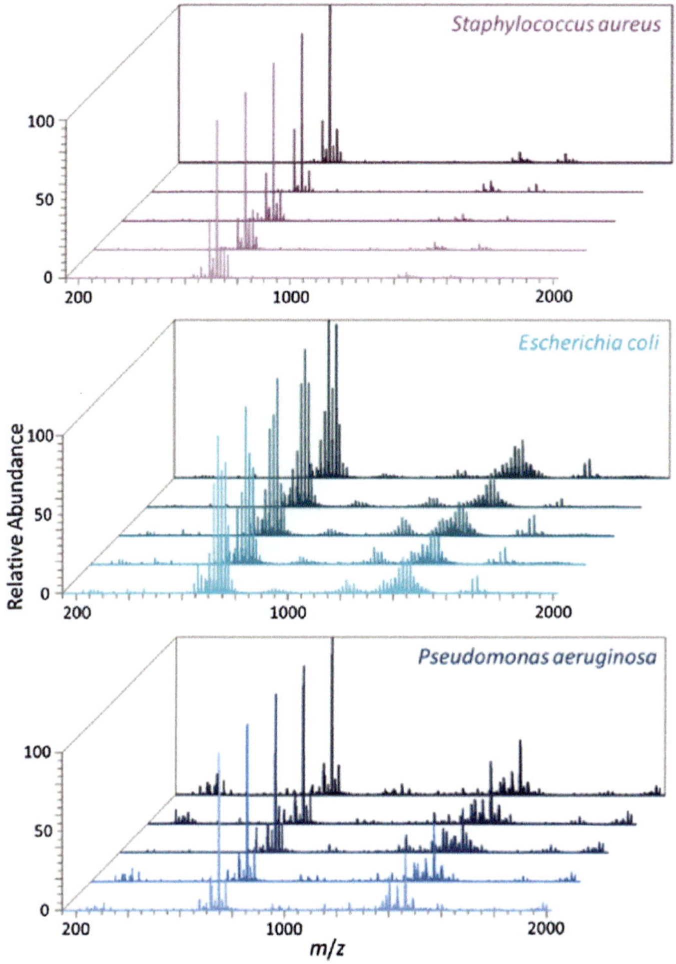

Figure 8.13 Example microbial REIMS spectra. *(Reprinted with permission from N. Strittmatter, M. Rebec, E.A. Jones, O. Golf, A. Abdolrasouli, J. Balog, V. Behrends, K.A. Veselkov, Z. Takats, Characterization and identification of clinically relevant microorganisms using rapid evaporative ionization mass spectrometry, Anal. Chem. 86 (2014) 6555–6562. https://doi.org/10.1021/ac501075f.)*

(C_{12}—C_{20}), mono- and di-rhamnolipids, and hydroxyalkylquinolines-derived quorum sensing molecules were identified in *Pseudomonas aerogenes*, while larger intact lipid A species were observed in *Escherichia coli* (*m/z* 1796) and *Helicobacter pylori* at *m/z* 1796 and *m/z* 1547, respectively. Ceramides (Cer 34:0, Cer 35:0, Cer 36:0) were also found as chloride adducts in *Bacteroides fragilis* as well as a range of mycolic acids (C_{26}—C_3) in *Corynebacterium striatum*. Finally, bacterial polyhydroxybutyrate polymers were detected in *Bacillus cereus* and *Burkholderia cepacia* showing that measurement is not solely characterized by lipid and lipid-related molecules. Hierarchal analysis of the REIMS data also showed that REIMS spectral profiles in most instances followed the bacterial taxonomy as defined by 16S rRNA gene sequences. The supervised models created were able to yield correct cross-validation results on species, genus, and Gram stain level at 95.9%, 97.8%, and 100%, respectively, where misclassification was observed only in closely related bacterial species. The effect of MS resolution on identification accuracy was also investigated by rearranging the original data using various bin sizes (0.01−1 Da) before multivariate statistical analysis. This showed that at even 1 Da, the RIEMS data provided enough identification accuracy to differentiate at the species, genus, and Gram stain level. This is more of an indication of the differential properties of REIMS spectra and does not detract from the use of higher resolution instruments as exact mass data from these instruments are beneficial in the creation of more compact data groupings and the identification of unknowns. Blind identification tests using strains grown on a variety of different culture media provided at least 97.8% accuracy. Strain level specificity was also investigated using seven different *E. coli* strains (NCM3722, MG1655, MC1000, MC4100, DH5a, C600 and OP50) grown in a variety of different culture conditions providing 87.3% accuracy (Fig. 8.14).

The analysis of bacteria using the MALDI-TOF microbial identification method is relatively straightforward and colonies are spotted before the addition of an ionization matrix. However, in the case of fungal isolates, a laborious sample pretreatment step involving a multiple extraction and centrifugation step is necessary to extract the intracellular proteins for profiling and identification [46]. This is undesirable and REIMS was also investigated for the direct identification of pathogenic yeast. A total of 87 different bacterial and 16 different fungal species were analyzed and subjected to multivariate statistical analysis. The REIMS phospholipid profile between bacteria and fungi is noticeably different with bacterial spectra containing mostly PG, PE, and lesser amounts of PA while fungal spectra were dominated by PA, PE, and high amount of PI, which is generally rare in bacteria. REIMS was able to distinguish between five different pathogenic species with a classification accuracy of 98.8%. Further work to evolve the capability of REIMS has meant that it is now possible to directly and rapidly identify 153 clinical *Candida* isolates to the species level with an accuracy of 96% using bipolar diathermy [26]. Furthermore, significant lipid differences have also been observed that could potentially be used as species

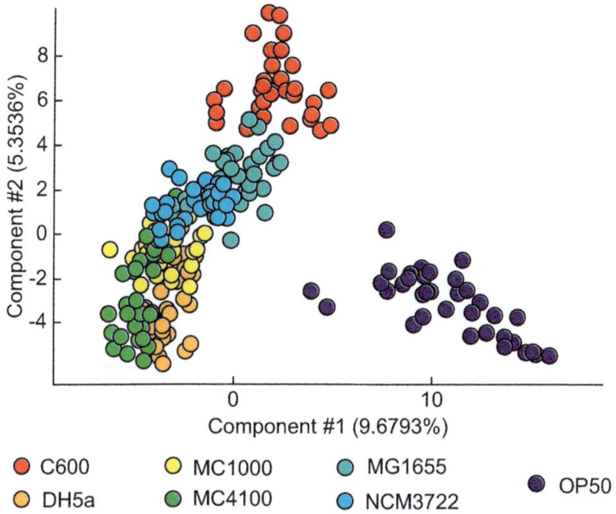

Figure 8.14 Principal component analysis—linear discriminant analysis model of seven different *Escherichia coli* strains. *(Reprinted with permission from N. Strittmatter, M. Rebec, E.A. Jones, O. Golf, A. Abdolrasouli, J. Balog, V. Behrends, K.A. Veselkov, Z. Takats, Characterization and identification of clinically relevant microorganisms using rapid evaporative ionization mass spectrometry, Anal. Chem. 86 (2014) 6555–6562. https://doi.org/10.1021/ac501075f.)*

biomarkers in more complex microbial communities or if the differences are associated with antifungal sensitivities or virulence factors, it could potentially also be useful in the identification of antifungal and virulence phenotypes of *Candida*.

In order to make the REIMS system more suitable for a clinical microbiology laboratory, there has also been focus on the development of an automated high-throughput REIMS system [47]. The system includes an automated colony picker robot equipped with a Pickolo visualization platform for automated analysis. Culture plates are placed in a plate rack and automatically transferred one by one to the Pickolo platform which doubles as a counter electrode for the REIMS diathermy device (Fig. 8.15).

Colonies can then be selected automatically (or manually) and a specialized high-throughput REIMS monopolar stainless-steel electrode probe then moves to touch each of the selected colonies for vaporization. The aerosol is then transferred through a 1.5 m PTFE tubing to the QToF mass spectrometer for analysis. The system was tested on 375 different clinical isolates from 25 microbial species providing an accuracy of 93.9%. The high-throughput system was also used in the *Candida* study where it

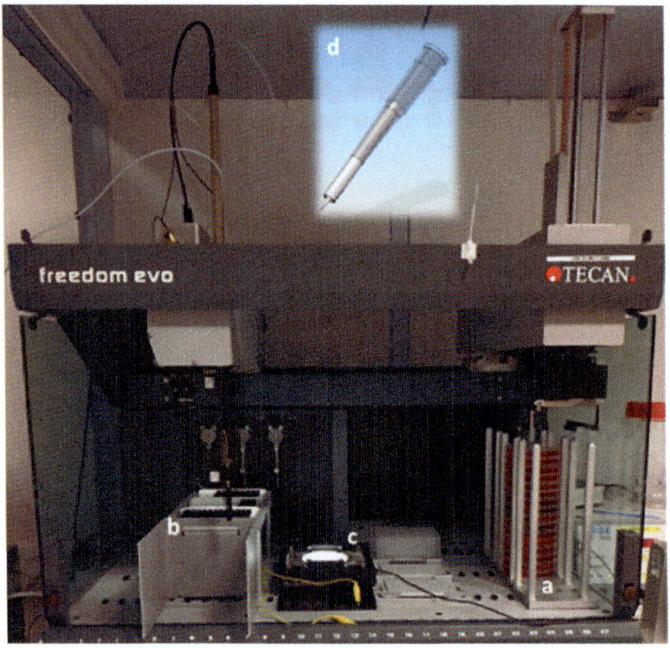

Figure 8.15 Setup for high-throughput REIMS. *(Reprinted with permission from F. Bolt, S.J.S. Cameron, T. Karancsi, D. Simon, R. Schaffer, T. Rickards, K. Hardiman, A. Burke, Z. Bodai, A. Perdones-Montero, Automated high-throughput identification and characterization of clinically important bacteria and fungi using rapid evaporative ionization mass spectrometry, Anal. Chem. 88 (2016) 9419–9426. https://doi.org/10.1021/acs.analchem.6b01016.)*

was able to identify 153 clinical *Candida* isolates with a classification accuracy of 100%. With each analysis and real-time identification taking only 15 s, it can be estimated that the system along with standard routine setup and maintenance procedures would be able to handle around 3000 to 4000 colonies over a 24 h period, representing a considerable improvement over the 6 min it currently takes for a MALDI-TOF identification system.

8.4.3 REIMS in food analysis

Food science has also been investigated as an area of application for the REIMS technology. The security of the modern food supply chain is important to economies globally from both a safety and a cost standpoint as it pertains to the safety and authenticity of the products we trade and consume. Therefore, the ability to be able to accurately and rapidly profile our food and agricultural commodities is potentially useful in a number of areas. These range from food fraud and product authenticity, species identification, and the identification of production methods to that of food quality, where it may be employed to check the consistency of products over time.

An example investigated is related to fish speciation as it relates to food fraud [24]. Fish is one of the most commonly substituted and mislabeled food commodities globally with an average of 30% of fish products being misrepresented. In a seafood industry worth over $400 billion dollars annually, this represents a big issue and is a key motivating factor along with safety factors. A good example of this is the replacement of white tuna with the cheaper substitute escolar. Escolar contains a high amount of wax esters called gempylotoxin which can cause keriorrhea if ingested. The REIMS-QToF system fitted with a monopolar *iKnife* sampling electrode was applied to the analysis of five commercially popular and genetically similar white fish from the Gadidae family (cod, coley, haddock, pollock, and whiting). Cod and haddock are higher value species and can be substituted by cheaper fish like coley, pollock, and whiting. REIMS negative mode ion spectra were acquired in the m/z 200—1200 range from 194 cod, 51 coley, 133 haddock, 50 pollock, and 50 whiting samples and subjected to multivariate analysis. The models created were then validated using 22 cod, 20 coley, 20 haddock, 20 pollock, and 17 whiting samples providing a classification accuracy of 98.99% (Fig. 8.16).

Fortuitously during the investigation, it was also discovered that six samples labeled as haddock were found in the modeling to group with the cod samples. This mislabeling was confirmed by DNA analysis revealing all six samples to actually be cod. This also highlighted the time-saving benefits of REIMS where analysis was seconds, while the polymerase chain reaction (PCR)-based method took 24 h. The model was also validated using six seabass and eight seabream samples providing 100% correct classification as outliers. A preliminary test was also conducted on catch method which is another form of fish fraud. Modeling of the REIMS data from trawl-caught and line-caught fish demonstrated that they could be differentiated into separate groups. However, the reasons for this separation are unclear and may be related to the creation of metabolic stress markers or even explained by diet differences between line-caught fish being from shallower waters versus the deeper environment of trawl-caught fish.

In addition to fish, REIMS has been applied to the differentiation of raw meat products providing accuracies of 100% at the species level and 97% at the breed level (Fig. 8.17).

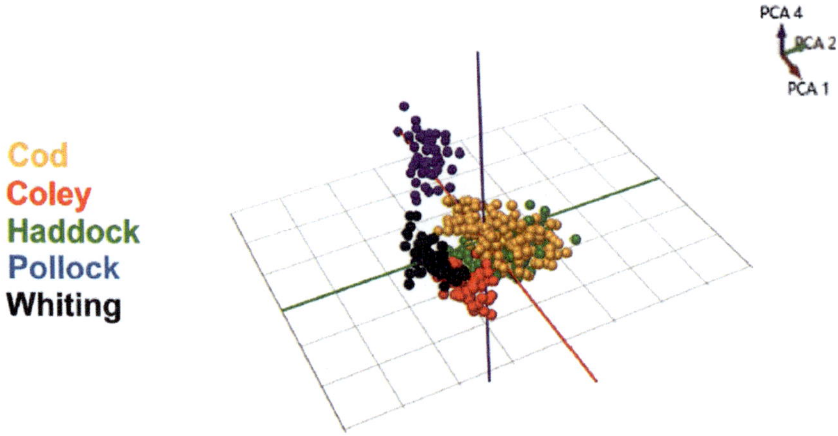

Figure 8.16 Principal component analysis model for the separation of five different species of fish. *(Reprinted with permission from C. Black, O.P. Chevallier, S.A. Haughey, J. Balog, S. Stead, S.D. Pringle, M.V. Riina, F. Martucci, P.L. Acutis, M. Morris, A real time metabolomic profiling approach to detecting fish fraud using rapid evaporative ionisation mass spectrometry, Metabolomics 13 (2017). https://doi.org/10.1007/s11306-017-1291-y.)*

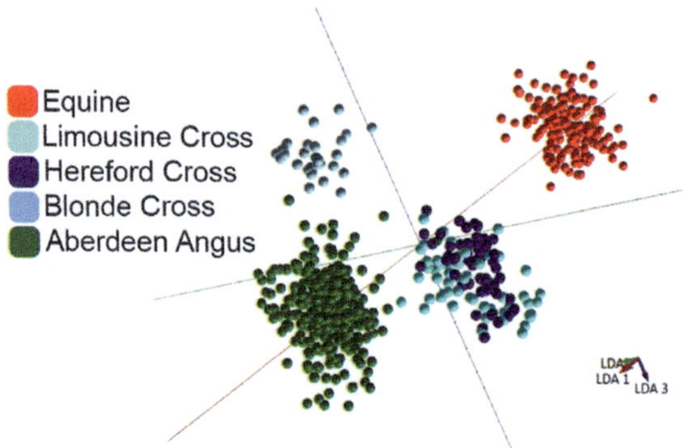

Figure 8.17 Comparison of different bovine and equine breeds (linear discriminant analysis model). *(Reprinted with permission from J. Balog, D. Perenyi, C. Guallar-Hoyas, A. Egri, S.D. Pringle, S. Stead, O.P. Chevallier, C.T. Elliott, Z. Takats, Identification of the species of origin for meat products by rapid evaporative ionization mass spectrometry, J. Agric. Food Chem. 64 (2016) 4793–4800. https://doi.org/10.1021/acs.jafc.6b01041.)*

The effects of cooking were also investigated and found not to interfere with the identification of the species. A driver for this type of application is the example relating to the 2013 horse meat scandal in the United Kingdom where beef labeled products were found to actually contain horse meat. Testing of the meat origin in mixed patties demonstrated that horse,

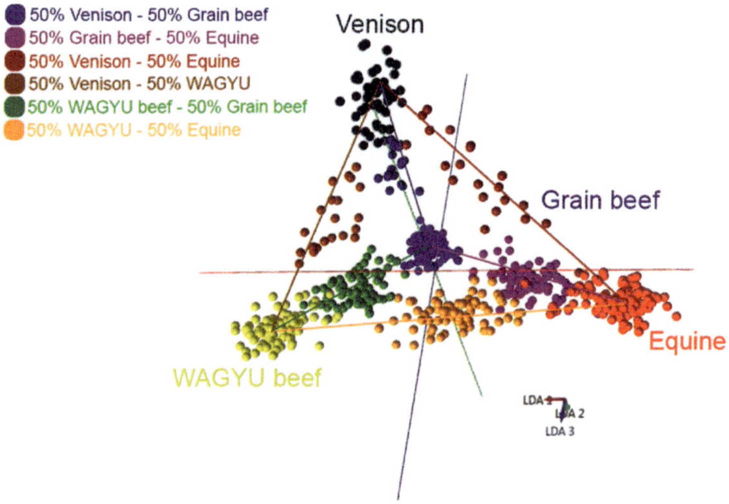

Figure 8.18 Comparing the meat origin of mixed patties (linear discriminant analysis model). *(Reprinted with permission from J. Balog, D. Perenyi, C. Guallar-Hoyas, A. Egri, S.D. Pringle, S. Stead, O.P. Chevallier, C.T. Elliott, Z. Takats, Identification of the species of origin for meat products by rapid evaporative ionization mass spectrometry, J. Agric. Food Chem. 64 (2016) 4793–4800. https://doi.org/10.1021/acs.jafc.6b01041.)*

cattle, and venison meat can be detected when 5% or more is present [48] (Fig. 8.18).

In the case of boar taint, REIMS is used for the in situ detection of food anomalies. Boar taint is an undesirable odor present in pork meat caused by the accumulation of indole, skatole, and androstenone in adipose tissue. Ethical issues, however, have meant that surgical castrations to prevent boar taint are now less common. An alternative method is to detect the boar taint on the slaughter line, but this is challenging in an abattoir where an average of 600 pigs may be killed per hour thus requiring detection in a matter of seconds. REIMS is able to complete each analysis within a 10 s time frame and is used to create a model for the high-throughput identification of boar taint in neck adipose tissue. The orthogonal projections to latent structures—discriminant analysis (OPLS-DA) model derived from the data were able to detect and classify 50 sows, 50 untainted boars, and 50 tainted boars correctly with an accuracy of 99% [49].

The use of ractopamine and other β-agonists as growth promoters in livestock is banned in many countries as they can cause acute toxicity in consumers. Various highly sensitive and selective liquid chromatography/mass spectrometry (LCMS)-based metabolomics methods have been created to meet the regulatory requirements and typically use noninvasive samples like urine and serum. However, high-throughput screening still remains a challenge and the ability to directly analyze tissue rapidly would allow full control of the food commodity at any port of entry. REIMS is well positioned to be able to fulfill these requirements. β-agonists are known to promote the formation of leaner muscle tissue and so lipid profiles are expected to be different. Supervised and unsupervised models were created from the REIMS MS spectra of shoulder, loin, and thigh samples harvested from five control and five treated female pigs. The model was then tested using 11 control and 10 treated samples providing an accuracy of >95% [50].

The application of REIMS to food and agricultural samples will continue to grow and includes crop science where, for example, it can be used to analyze plant tissue in situ. This has been useful in better understanding how to deal with the herbicide resistance of different weed grass population, an important challenge in arable farming. It is capable of not only weed grass speciation but also the differentiation of wild-type and different herbicide-resistant populations from the same species [51]. REIMS has been tested on other food matrices including the analysis of cacao beans and chocolate to determine their origin. This is related to both food fraud and the issues of child slavery within the cocoa industry [52]. Finally, novel sampling methods are being investigated to allow REIMS to better analyze liquid food products. For instance, in the ultrasonic REIMS method, liquid is released in front of an ultrasonic horn and droplets are created using ultrasonic vibration before introduction into the mass spectrometer [53].

8.5 Summary

The ability to rapidly produce characteristic MS spectra in situ from a variety of samples and its amenability to automation provides REIMS with the potential to have a huge impact across a range of application areas in surgery, microbiology, and also food analysis. In doing so, it will be important to not only consider the optimal combination of sampling methods and type of mass spectrometer but also the informatics and database tools required to manage the data and desire for accurate and rapid results. Close collaboration between the instrument manufacturer and industry experts will be key as existing and novel applications are developed based on these innovative technologies opening the door to next-generation solutions within each field.

References

[1] K.-C. Schäfer, J. Dénes, K. Albrecht, T. Szaniszló, J. Balog, R. Skoumal, M. Katona, M. Tóth, L. Balogh, Z. Takáts, In vivo, in situ tissue analysis using rapid evaporative ionization mass spectrometry, Angew. Chem. Int. Ed. Engl. 48 (2009) 8240–8242, https://doi.org/10.1002/anie.200902546.

[2] J. Balog, T. Szaniszlo, K.-C. Schaefer, J. Denes, A. Lopata, L. Godorhazy, D. Szalay, L. Balogh, L. Sasi-Szabo, M. Toth, Identification of biological tissues by rapid evaporative ionization mass spectrometry, Anal. Chem. 82 (2010) 7343–7350, https://doi.org/10.1021/ac101283x.

[3] W. Snedden, R.B. Parker, Determination of volatile constituents of human blood and tissue specimens by quantitative high-resolution mass spectrometry, Anal. Chem. 43 (1971) 1651–1656, https://doi.org/10.1021/ac60306a016.

[4] E.J. Want, B.F. Cravatt, G. Siuzdak, The expanding role of mass spectrometry in metabolite profiling and characterization, ChemBioChem 6 (2005) 1941–1951, https://doi.org/10.1002/cbic.200500151.

[5] H. Beckey, Field desorption mass spectrometry: a technique for the study of thermally unstable substances of low volatility, Int. J. Mass Spectrom. Ion Phys. 2 (1969) 500–502, https://doi.org/10.1016/0020-7381(69)80047-1.

[6] H. Beckey, A. Heindrichs, H. Winkler, Some new field desorption techniques, Int. J. Mass Spectrom. Ion Phys. 3 (1970) A9–A11, https://doi.org/10.1016/0020-7381(70)80009-2.

[7] H.R. Morris, M. Panico, M. Barber, R.S. Bordoli, R.D. Sedgwick, A. Tyler, Fast atom bombardment: a new mass spectrometric method for peptide sequence analysis, Biochem. Biophys. Res. Commun. 101 (1981) 623–631, https://doi.org/10.1016/0006-291x(81)91304-8.

[8] M. Barber, R.S. Bordoli, R.D. Sedgwick, A.N. Tyler, Fast atom bombardment of solids as an ion source in mass spectrometry, Nature 293 (1981) 270–275, https://doi.org/10.1038/293270a0.

[9] R. Macfarlane, D. Torgerson, Californium-252 plasma desorption mass spectroscopy, Science 191 (1976) 920–925, https://doi.org/10.1126/science.1251202.

[10] P. Roepstorff, Plasma desorption mass spectrometry of peptides and proteins, Acc. Chem. Res. 22 (1989) 421–427, https://doi.org/10.1021/ar00168a003.

[11] M. Karas, D. Bachmann, U. Bahr, F. Hillenkamp, Matrix-assisted ultraviolet laser desorption of non-volatile compounds, Int. J. Mass Spectrom. Ion Process. 78 (1987) 53–68, https://doi.org/10.1016/0168-1176(87)87041-6.

[12] K. Tanaka, H. Waki, Y. Ido, S. Akita, Y. Yoshida, T. Yoshida, T. Matsuo, Protein and polymer analyses up to *m/z* 100 000 by laser ionization time-of-flight mass spectrometry, Rapid Commun. Mass Spectrom. 2 (1988) 151–153, https://doi.org/10.1002/rcm.1290020802.

[13] R.M. Caprioli, T.B. Farmer, J. Gile, Molecular imaging of biological samples: localization of peptides and proteins using MALDI-TOF MS, Anal. Chem. 69 (1997) 4751–4760, https://doi.org/10.1021/ac970888i.

[14] Y. Magnusson, P. Friberg, P. Sjövall, F. Dangardt, P. Malmberg, Y. Chen, Lipid imaging of human skeletal muscle using TOF-SIMS with bismuth cluster ion as a primary ion source, Clin. Physiol. Funct. Imaging 28 (2008) 202–209, https://doi.org/10.1111/j.1475-097x.2008.00796.x.

[15] M.-C. Djidja, E. Claude, M.F. Snel, P. Scriven, S. Francese, V. Carolan, M.R. Clench, MALDI-ion mobility separation-mass spectrometry imaging of glucose-regulated protein 78 kDa (Grp78) in human formalin-fixed, paraffin-embedded pancreatic adenocarcinoma tissue sections, J. Proteome Res. 8 (2009) 4876–4884, https://doi.org/10.1021/pr900522m.

[16] Z. Takats, J. Wiseman, B. Gologan, R. Cooks, Mass spectrometry sampling under ambient conditions with desorption electrospray ionization, Science 306 (2004) 471–473, https://doi.org/10.1126/science.1104404.
[17] P. Sjövall, J. Lausmaa, B. Johansson, Mass spectrometric imaging of lipids in brain tissue, Anal. Chem. 76 (2004) 4271–4278, https://doi.org/10.1021/ac049389p.
[18] R. Mirnezami, K. Spagou, P. Vorkas, M. Lewis, J. Kinross, E. Want, H. Shion, R. Goldin, A. Darzi, Z. Takats, Chemical mapping of the colorectal cancer microenvironment via MALDI imaging mass spectrometry (MALDI-MSI) reveals novel cancer-associated field effects, Mol. Oncol. 8 (1) (2014) 39–49, https://doi.org/10.1016/j.molonc.2013.08.010.
[19] D.E. Bauermeister, The role and limitations of frozen section and needle aspiration biopsy in breast cancer diagnosis, Cancer 46 (1980) 947–949, https://doi.org/10.1002/1097-0142(19800815)46:4+<947::AID-CNCR2820461316>3.0.CO;2-N.
[20] S. Santagata, L.S. Eberlin, I. Norton, D. Calligaris, D.R. Feldman, J.L. Ide, X. Liu, J.S. Wiley, M.L. Vestal, S.H. Ramkissoon, Intraoperative mass spectrometry mapping of an onco-metabolite to guide brain tumor surgery, Proc. Natl. Acad. Sci. U.S.A. 111 (2014) 11121–11126, https://doi.org/10.1073/pnas.1404724111.
[21] D. Calligaris, D. Caragacianu, X. Liu, I. Norton, C.J. Thompson, A.L. Richardson, M. Golshan, M.L. Easterling, S. Santagata, D.A. Dillon, Application of desorption electrospray ionization mass spectrometry imaging in breast cancer margin analysis, Proc. Natl. Acad. Sci. U.S.A. 111 (2014) 15184–15189, https://doi.org/10.1073/pnas.1408129111.
[22] Waters Corporation, REIMS Research System With *iKnife* Sampling Brochure, 2015. https://www.waters.com/waters/library.htm?lid=134846772.
[23] Waters Corporation, REIMS System Overview and Maintenance Guide, 2016. http://www.waters.com/waters/support.htm?lid=134909497.
[24] C. Black, O.P. Chevallier, S.A. Haughey, J. Balog, S. Stead, S.D. Pringle, M.V. Riina, F. Martucci, P.L. Acutis, M. Morris, A real time metabolomic profiling approach to detecting fish fraud using rapid evaporative ionisation mass spectrometry, Metabolomics 13 (2017), https://doi.org/10.1007/s11306-017-1291-y.
[25] J. Balog, L. Sasi-Szabo, J. Kinross, M.R. Lewis, L.J. Muirhead, K. Veselkov, R. Mirnezami, B. Dezso, L. Damjanovich, A. Darzi, Intraoperative tissue identification using rapid evaporative ionization mass spectrometry, Sci. Transl. Med. 5 (2013) 194ra93, https://doi.org/10.1126/scitranslmed.3005623.
[26] S.J.S. Cameron, F. Bolt, A. Perdones-Montero, T. Rickards, K. Hardiman, A. Abdolrasouli, A. Burke, Z. Bodai, T. Karancsi, D. Simon, Rapid evaporative ionisation mass spectrometry (REIMS) provides accurate direct from culture species identification within the genus *Candida*, Sci. Rep. 6 (2016), https://doi.org/10.1038/srep36788.
[27] Waters Corporation, Real-time Sample Recognition With LiveID Software, 2018. https://www.waters.com/waters/library.htm?lid=134939556&cid=134528734.
[28] L.A. Torre, F. Islami, R.L. Siegel, E.M. Ward, A. Jemal, Global cancer in women: burden and trends, Cancer Epidemiol. Biomark. Prev. 26 (2017) 444–457, https://doi.org/10.1158/1055-9965.epi-16-0858.
[29] E.R. St John, J. Balog, J.S. McKenzie, M. Rossi, A. Covington, L. Muirhead, Z. Bodai, F. Rosini, A.V.M. Speller, S. Shousha, Rapid evaporative ionisation mass spectrometry of electrosurgical vapours for the identification of breast pathology: towards an intelligent knife for breast cancer surgery, Breast Cancer Res. 19 (2017), https://doi.org/10.1186/s13058-017-0845-2.
[30] J. Balog, S. Kumar, J. Alexander, O. Golf, J. Huang, T. Wiggins, N. Abbassi-Ghadi, A. Enyedi, S. Kacska, J. Kinross, In vivo endoscopic tissue identification by rapid evaporative ionization mass spectrometry (REIMS), Angew. Chem. Int. Ed. Engl. 127 (2015) 11211–11214, https://doi.org/10.1002/ange.201502770.

[31] J. Alexander, L. Gildea, J. Balog, A. Speller, J. McKenzie, L. Muirhead, A. Scott, C. Kontovounisios, S. Rasheed, J. Teare, A novel methodology for in vivo endoscopic phenotyping of colorectal cancer based on real-time analysis of the mucosal lipidome: a prospective observational study of the *iKnife*, Surg. Endosc. 31 (2017) 1361–1370, https://doi.org/10.1007/s00464-016-5121-5.
[32] K.-C. Schäfer, T. Szaniszlo, S. Günther, J. Balog, J. Denes, M. Keseru, B. Dezso, M. Toth, B. Spengler, Z. Takats, In situ, real-time identification of biological tissues by ultraviolet and infrared laser desorption ionization mass spectrometry, Anal. Chem. 83 (2011) 1632–1640, https://doi.org/10.1021/ac102613m.
[33] D.L. Phelps, J. Balog, L.F. Gildea, Z. Bodai, A. Savage, M.A. El-Bahrawy, A.V. Speller, F. Rosini, H. Kudo, J.S. McKenzie, The surgical intelligent knife distinguishes normal, borderline and malignant gynaecological tissues using rapid evaporative ionisation mass spectrometry (REIMS), Br. J. Cancer 118 (2018) 1349–1358, https://doi.org/10.1038/s41416-018-0048-3.
[34] A. van Belkum, G. Durand, M. Peyret, S. Chatellier, G. Zambardi, J. Schrenzel, D. Shortridge, A. Engelhardt, W.M. Dunne, Rapid clinical bacteriology and its future impact, Ann. Lab. Med. 33 (2013) 14, https://doi.org/10.3343/alm.2013.33.1.14.
[35] P. Woo, S. Lau, J. Teng, H. Tse, K.-Y. Yuen, Then and now: use of 16S rDNA gene sequencing for bacterial identification and discovery of novel bacteria in clinical microbiology laboratories, Clin. Microbiol. Infect. 14 (2008) 908–934, https://doi.org/10.1111/j.1469-0691.2008.02070.x.
[36] J.P. Anhalt, C. Fenselau, Identification of bacteria using mass spectrometry, Anal. Chem. 47 (1975) 219–225, https://doi.org/10.1021/ac60352a007.
[37] D.N. Heller, R.J. Cotter, C. Fenselau, O.M. Uy, Profiling of bacteria by fast atom bombardment mass spectrometry, Anal. Chem. 59 (1987) 2806–2809, https://doi.org/10.1021/ac00150a018.
[38] V. Ryzhov, C. Fenselau, Characterization of the protein subset desorbed by MALDI from whole bacterial cells, Anal. Chem. 73 (2001) 746–750, https://doi.org/10.1021/ac0008791.
[39] P. Seng, M. Drancourt, F. Gouriet, B. La Scola, P. Fournier, J. Rolain, D. Raoult, Ongoing revolution in bacteriology: routine identification of bacteria by matrix-assisted laser desorption ionization time-of-flight mass spectrometry, Clin. Infect. Dis. 49 (2009) 543–551, https://doi.org/10.1086/600885.
[40] J.H. Chen, W.-C. Yam, A.H. Ngan, A.M. Fung, W.-L. Woo, M.-K. Yan, G.K. Choi, P.-L. Ho, V.C. Cheng, K.-Y. Yuen, Advantages of using matrix-assisted laser desorption ionization-time of flight mass spectrometry as a rapid diagnostic tool for identification of yeasts and mycobacteria in the clinical microbiological laboratory, J. Clin. Microbiol. 51 (2013) 3981–3987, https://doi.org/10.1128/jcm.01437-13.
[41] K.E. Tan, B.C. Ellis, R. Lee, P.D. Stamper, S.X. Zhang, K.C. Carroll, Prospective evaluation of a matrix-assisted laser desorption ionization-time of flight mass spectrometry system in a hospital clinical microbiology laboratory for identification of bacteria and yeasts: a bench-by-bench study for assessing the impact on time to identification and cost-effectiveness, J. Clin. Microbiol. 50 (2012) 3301–3308, https://doi.org/10.1128/jcm.01405-12.
[42] R. Goodacre, J.K. Heald, D.B. Kell, Characterisation of intact microorganisms using electrospray ionisation mass spectrometry, FEMS Microbiol. Lett. 176 (1) (1999) 17–24, https://doi.org/10.1111/j.1574-6968.1999.tb13636.x.
[43] Y. Ishida, A.J. Madonna, J.C. Rees, M.A. Meetani, K.J. Voorhees, Rapid analysis of intact phospholipids from whole bacterial cells by matrix-assisted laser desorption/ionization mass spectrometry combined with on-probe sample pretreatment, Rapid Commun. Mass Spectrom. 16 (2002) 1877–1882, https://doi.org/10.1002/rcm.802.

[44] Y. Song, N. Talaty, W.A. Tao, Z. Pan, R.G. Cooks, Rapid ambient mass spectrometric profiling of intact, untreated bacteria using desorption electrospray ionization, Chem. Commun. (2007) 61–63, https://doi.org/10.1039/b615724f.
[45] N. Strittmatter, M. Rebec, E.A. Jones, O. Golf, A. Abdolrasouli, J. Balog, V. Behrends, K.A. Veselkov, Z. Takats, Characterization and identification of clinically relevant microorganisms using rapid evaporative ionization mass spectrometry, Anal. Chem. 86 (2014) 6555–6562, https://doi.org/10.1021/ac501075f.
[46] C. Cassagne, A.-L. Cella, P. Suchon, A.-C. Normand, S. Ranque, R. Piarroux, Evaluation of four pretreatment procedures for MALDI-TOF MS yeast identification in the routine clinical laboratory, Med. Mycol. 51 (2013) 371–377, https://doi.org/10.3109/13693786.2012.720720.
[47] F. Bolt, S.J. Cameron, T. Karancsi, D. Simon, R. Schaffer, T. Rickards, K. Hardiman, A. Burke, Z. Bodai, A. Perdones-Montero, Automated high-throughput identification and characterization of clinically important bacteria and fungi using rapid evaporative ionization mass spectrometry, Anal. Chem. 88 (2016) 9419–9426, https://doi.org/10.1021/acs.analchem.6b01016.
[48] J. Balog, D. Perenyi, C. Guallar-Hoyas, A. Egri, S.D. Pringle, S. Stead, O.P. Chevallier, C.T. Elliott, Z. Takats, Identification of the species of origin for meat products by rapid evaporative ionization mass spectrometry, J. Agric. Food Chem. 64 (2016) 4793–4800, https://doi.org/10.1021/acs.jafc.6b01041.
[49] K. Verplanken, S. Stead, R. Jandova, C.V. Poucke, J. Claereboudt, J.V. Bussche, S.D. Saeger, Z. Takats, J. Wauters, L. Vanhaecke, Rapid evaporative ionization mass spectrometry for high-throughput screening in food analysis: the case of boar taint, Talanta 169 (2017) 30–36, https://doi.org/10.1016/j.talanta.2017.03.056.
[50] Y. Guitton, G. Dervilly-Pinel, R. Jandova, S. Stead, Z. Takats, B. Le Bizec, Rapid evaporative ionisation mass spectrometry and chemometrics for high-throughput screening of growth promoters in meat producing animals, Food Addit. Contam. A 35 (2018) 900–910, https://doi.org/10.1080/19440049.2017.1421778.
[51] S. Stead, J. Pope, R. Edwards, M. Brazier-Hicks, C. Tetard-Jones, Z. Takats, REIMS (rapid evaporative ionization mass spectrometry) and multivariant statistics, two tools in support of weed grass speciation and phenotype characterization, Poster Session presented at: 64th Conference on Mass Spectrometry and Allied Topics (ASMS) (2016) June 5–9; San Antonio, TX, United States.
[52] R. Schäffer, T, Karancsi, S.D. Pringle, Z.Takats, V. Varga, Z. Ath-Horvath, J. Balog, Rapid determination of the origin of cocoa bean and chocolate with laser assisted REIMS – The bean to bar project, Poster Session presented at: 66th Conference on Mass Spectrometry and Allied Topics (ASMS) (2018) June 3–7; San Diego, CA, United States.
[53] D. Simon, T. Karancsi, S.D. Pringle, Z. Takats, Rapid analysis of liquid food products and beverages using ultrasonic rapid evaporative ionization mass spectrometry, Poster Session presented at: 65th Conference on Mass Spectrometry and Allied Topics (ASMS) (2017) June 4–8; Indianapolis, IN, United States.

Index

'*Note:* Page numbers followed by "f" indicate figures and "t" indicates tables.'

B

C

D

A7/1021293/3/10/22

CPI Antony Rowe
Eastbourne, UK
November 20, 2019